THE FRANCHISE ERA

Traditions in American Cinema
Series Editors Linda Badley and R. Barton Palmer

Titles in the series include:

The 'War on Terror' and American Film: 9/11 Frames Per Second
Terence McSweeney

American Postfeminist Cinema: Women, Romance and Contemporary Culture
Michele Schreiber

Film Noir
Homer B. Pettey and R. Barton Palmer (eds)

In Secrecy's Shadow: The OSS and CIA in Hollywood Cinema 1941–1979
Simon Willmetts

Indie Reframed: Women's Filmmaking and Contemporary American Independent Cinema
Linda Badley, Claire Perkins and Michele Schreiber (eds)

Vampires, Race and Transnational Hollywoods
Dale Hudson

Who's in the Money? The Great Depression Musicals and Hollywood's New Deal
Harvey G. Cohen

The Style of Sleaze: The American Exploitation Film, 1959–1977
Colin Waddell

Engaging Dialogue: Cinematic Verbalism in American Independent Cinema
Jennifer O'Meara

Cold War Film Genres
Homer B. Pettey (ed.)

The Franchise Era: Managing Media in the Digital Economy
James Fleury, Bryan Hikari Hartzheim, and Stephen Mamber (eds)

edinburghuniversitypress.com/series/tiac

THE FRANCHISE ERA
Managing Media in the Digital Economy

Edited by James Fleury, Bryan Hikari Hartzheim,
and Stephen Mamber

EDINBURGH
University Press

Edinburgh University Press is one of the leading university presses in the UK. We publish academic books and journals in our selected subject areas across the humanities and social sciences, combining cutting-edge scholarship with high editorial and production values to produce academic works of lasting importance. For more information visit our website: edinburghuniversitypress.com

Edinburgh University Press Ltd
The Tun—Holyrood Road
12 (2f) Jackson's Entry
Edinburgh EH8 8PJ

Typeset in 10/12.5pt Sabon by
Servis Filmsetting Ltd, Stockport, Cheshire,
and printed and bound in Great Britain
by CPI Group (UK) Ltd, Croydon, CR0 4YY

ISBN 978 1 4744 1922 2(hardback)
ISBN 978 1 4744 1923 9 (webready PDF)
ISBN 978 1 4744 1924 6 (epub)

CONTENTS

FIGURES

ACKNOWLEDGMENTS

This volume grew out of conversations that took place in several classrooms and at several conferences; we are grateful to the many individuals who helped inspire it in the first place.

We would like to extend our thanks to our commissioning editor Gillian Leslie, series editors Linda Badley and R. Barton Palmer, and the production team at Edinburgh University Press for their continued support throughout this project's development. Also, thank you to Julie Grossman for connecting us to Edinburgh University Press and the Traditions in American Cinema series editors. We wish to acknowledge the institutional assistance provided by the University of California, Los Angeles and Waseda University in Tokyo, Japan. Most of all, we would like to thank our contributors for their insightful and inspirational work, enthusiasm, and support.

NOTES ON THE CONTRIBUTORS

Heather Lea Birdsall is a PhD candidate in UCLA's Cinema and Media Studies program. Her dissertation research considers modern-day theme parks as physically and virtually immersive cinematic, televisual, and game spaces, a phenomenon she calls the "media park."

Alexander Champlin studies spectator video gaming as an emergent media phenomenon. His work focuses on the conjunction of televisual broadcast production and video game play, particularly the way play and game cultures shift as game play is adapted for broadcast. He focuses on esports, video-game livestreaming, and Let's Play media. Alex is a PhD candidate in the Department of Film and Media Studies at the University of California, Santa Barbara.

Rayna Denison is a Senior Lecturer in the School of Art, Media and American Studies at the University of East Anglia, whose research and teaching interests center on Japanese film and animation. She is the author of *Anime: A Critical Introduction* (2015), and the editor of *Princess Mononoke: Understanding Studio Ghibli's Monster Princess* (2018). She is the coeditor of the Eisner Award-nominated *Superheroes on World Screens* (with Rachel Mizsei-Ward, 2015). Her scholarly articles can be found in many leading journals, including *Cinema Journal*, *Velvet Light Trap*, *Japan Forum* and the *International Journal of Cultural Studies*.

James Fleury is a Cinema and Media Studies PhD Candidate at UCLA. His work has appeared in *Mediascape: UCLA's Journal of Cinema and Media Studies* (2012, 2015), *James Bond and Popular Culture: Essays on the Influence of the Fictional Superspy* (2014), the *South Atlantic Review* (2015), and *Content Wars: Tech Empires vs. Media Empires* (forthcoming). His dissertation analyzes the history of video-game development and licensing at Warner Bros.

Jennifer Gillan is Professor of English and Media Studies at Bentley University in the Boston area. She is the author of *Television and New Media: Must-Click TV* (2010) and *Television Brandcasting* (2015). The latter investigates the current and historical role that television content, promotion, and hybrids of the two have played at Disney–ABC Television Group in disseminating brand messaging and influencing consumer decisionmaking. Gillan's articles have appeared in *Cinema Journal*, *Series*, *American Literature*, *Arizona Quarterly*, *Columbia Journal of Gender & the Law*, *American Drama*, and *Mosaic*. She has chapters in *Understanding Reality TV*, *Great American Makeover*, *Reading Makeover Television*, *Teen Television*, and *Analyzing Mad Men*. With Maria Mazziotti Gillan, she coedited four award-winning multicultural literature collections: *Unsettling America*, *Identity Lessons*, *Growing Up Ethnic in America*, and *Italian American Writers on New Jersey*.

Bryan Hikari Hartzheim is an assistant professor in the School of International Liberal Studies at Waseda University where he teaches courses on digital media, games, and animation, with a focus on the Japanese media industries. His work has appeared in *The Journal of Popular Culture*, *Mediascape*, and the edited anthology *Video Games in East Asia* (2017).

Daniel Herbert is an associate professor in Film, Television, and Media at the University of Michigan. He is the author of *Videoland: Movie Culture at the American Video Store* (2014) and *Film Remakes and Franchises* (2017), and is the coeditor (with Derek Johnson) of *Point of Sale: Analyzing Media Retail* (forthcoming). His essays appear in *Canadian Journal of Film Studies*, *Creative Industries Journal*, *Film Quarterly*, *Journal of Film and Video*, *Media Industries*, *Millennium Film Journal*, and *Quarterly Review of Film and Video*, as well as in numerous edited collections.

Derek Johnson is Associate Professor of Media and Cultural Studies in the Department of Communication Arts at the University of Wisconsin—Madison. He is the author of *Media Franchising: Creative License and Collaboration in the Culture Industries* (2013), the editor of *From Networks to Netflix: A Guide to Changing Channels* (2018), and the coeditor of *A Companion to Media Authorship* (2013) as well as *Making Media Work: Cultures of Management in the Entertainment Industries* (2014).

Stephen Mamber is a professor in the UCLA Department of Film, Television, and Digital Media. He is the author of *Cinéma-Vérité in America: Studies in Uncontrolled Documentary* (MIT Press, 1974) and numerous articles about film and new media. He is also the author of four iPad apps: ClipNotes, an annotation and presentation tool, "Who Shot Liberty Valance?," "7 Thursdays: Looking at Brief Encounter," and "The Seventh Race: Kubrick at the Starting Gate." All are available for free on the iOS App Store.

Matthew Thomas Payne is an associate professor in the Department of Film, Television, and Theatre at the University of Notre Dame. He is author of *Playing War: Video Games after 9/11* (2016), and is coeditor of the anthologies *How to Play Video Games* (2019), *Flow TV: Television in the Age of Media Convergence* (2011), and *Joystick Soldiers: The Politics of Play in Military Video Games* (2010).

Andreas Rauscher is an associate professor in the Department for Media Studies at the University of Siegen with a special focus on film studies and video games. He has served as Visiting Professor of Media Studies at the Christian Albrechts–Universität Kiel and the Albert–Ludwigs–University Freiburg. He was the scientific curator for the German Film Museum at Frankfurt am Main (Exhibition *Film & Games—Interactions*. International catalogue available from Bertz Verlag, Berlin 2015). His research deals with film, game, comic, and cultural studies, transmedia aesthetics, and genre theory. Among his book publications are volumes on *The Simpsons*, superhero movies, *Star Wars*, the James Bond series, and the Czechoslovakian Nová Vlna as well as an introduction to game studies. In 2002 he received his PhD for a dissertation on the cultural and cinematic manifestations of the *Star Trek* phenomenon, and in 2011 he got his postdoctoral lecturing qualification (Habilitation) from the Johannes–Gutenberg University Mainz for a study on *Ludic Fictions—Genre Concepts in Video Games* (2012).

Brian Ruh is an independent scholar and author of *Stray Dog of Anime: The Films of Mamoru Oshii* (2004). He holds a PhD in Communication and Culture from Indiana University.

Monica Sandler is a doctoral candidate in Cinema and Media Studies at the University of California, Los Angeles. Her research focuses on historical and contemporary issues in American film industry policy, labor, and financing. Her dissertation examines the cultural and economic development of the Academy Awards, exploring the event's role in public relations and the cultural production of the Hollywood "prestige" film.

FOREWORD

Derek Johnson

At some point in 2007, during the course of my work on the typescript that would ultimately become *Media Franchising: Creative License and Collaboration in the Culture Industries* (2013), a senior scholar suggested to me that I reconsider my embrace of the coarsely commercial terms by which industries have sought to sustain and reuse successful cultural products over time. To acknowledge these media-industry practices as the reproduction of "franchises," this colleague worried, would be to put undue weight on the business side of the equation, emphasizing the unseemly corporate pursuit of profit through risk-averse branded repetition at the expense of attention to the new storytelling possibilities and participatory potentials that might be enabled by the continuing, collaborative extension of media narrative, images, and content. At that moment, in fact, scholars and industry practitioners alike had already invested considerable intellectual energy into theorizing new models and principles of collaborative storytelling that could offer more enriching forms of creativity and engagement than previous, highly managed and hierarchically driven media-entertainment brands allowed. It seemed a reasonable concern, therefore, that as media producers embraced theories of "transmedia storytelling," developed by Henry Jenkins and others, talk of "franchising" risked sounding at best old-fashioned and at worst uninterested in these dynamics of culture and creativity outside the managerial machinations of industry. In its clear association with McDonaldization and the business of managed familiarity, rationalized production, and standardized sameness, franchising risked sounding (to some, at least) like an unappetizing means of studying media as culture.

Fast-forward a decade, and it seems much harder to debate the necessity of a franchise framework when discussing media industries. On the one hand, it is tempting to suggest that, while some innovative creators continue to flirt with the possibilities of transmedia storytelling, the business-minded purveyors of franchising have only cemented their dominance. Sitting through the end credits of a major Hollywood blockbuster, such as Marvel Studios's *Black Panther* (Ryan Coogler, 2018), for example, one may still see the names of director, screenwriters, and executive producers most prominently, but the parade of names ultimately reveals a behind-the-scenes team that includes Vice President of Franchise Creative & Marketing, Franchise Creative & Marketing Executive, Development & Franchise Executive, Franchise Creative Content Executive, Franchise Partnership Manager, and Franchise Marketing Manager. From development to creative content to marketing to partnerships, every phase of the production of the film operates as part of the work of managing a franchise. The significance of franchising as managerial work is also crystallized by the public visibility of figures such as studio head Kevin Feige who does not direct individual Marvel films but is often credited with the consistent vision operating across them. Given the $16 billion (and counting) success of Marvel Studios, it is no surprise that other studios have sought to emulate its franchise-focused strategy (Williams, 2018). In 2016, Warner Bros. established its own Harry Potter Global Franchise Development team to manage the ongoing continuation and exploitation of its Wizarding World brand across film, theme parks, and digital venues (Tartaglione, 2016). Meanwhile, toy companies, such as Hasbro, Mattel, and LEGO, increasingly refigure themselves as media studios to develop content in support of consumer product brands. Franchising has become standard operating procedure—the logic by which the media business is increasingly and explicitly organized, despite the alternatives others have envisioned. Indeed, this reliance on franchising raises important questions about the potential for other kinds of content to thrive.

On the other hand, the dominance of franchising need not be assumed to be a story about the victory of bean counters in business management over the creative processes of cultural production and our engagement with it as consumers. To look at *Black Panther* as just one more product off the franchise assembly line would at the very least miss the political significance of efforts made by director Ryan Coogler and his collaborators to inject into the long-running Marvel Cinematic Universe an Afrofuturist aesthetic in contrast to the white normativity the franchise has previously produced. Of course, we could still critique the limitations of such reclamation—but any such debate demands that we acknowledge within media franchising a potential to support divergence, differentiation, and even diversity. This potential is no cause to celebrate franchising; indeed, we might soberly and cynically envision this incorporation of difference as part of the managerial logic of franchising. Franchising is not only industry machination but also the system through which meaning, identities, and political struggles have increasingly come to be managed in media culture. At the same time, these industry efforts to squeeze every last bit of value out of previous successes can produce new cultural practices. Audiences are not, of course, sticking through the credits of *Black Panther* to give its Franchise Management team proper due; instead, end credits have become a place for franchise managers to tantalize viewers with brief "stinger" scenes that hint at the next franchise offering to come. Beyond the implications for narrative structure and experimentations with serialization, these franchise practices have changed how consumers engage with media, encouraging an orientation as much to the future as the viewer experience of the present. Franchising is a business practice—but its managerial logics and orientation to perpetual reproduction must be accounted for to explain how creativity, consumption, and culture have come to operate in the current moment. In this way, study of franchising compels us to situate culture and politics within, as, and of industry rather than viewing them as opposed.

The chapters that follow similarly insist upon understanding the process of cultural reproduction in the industrial terms of franchise management. Reflecting on *The Franchise Era* that has resulted from the dominance of these media logics, the book helps us see how management is not just the overrationalized work of executives but a complex and vital process through which media culture is continually produced and reproduced across a wide array of industry actors and participants. Rather than sidestep these questions of industry, this book joins a host of exciting works in critical media studies that reveal the managerial efforts of creators, executives, and other collaborative partners as a vital and contested site of meaning, identity, and struggle in their own right. As media franchising continues to serve as standard operating procedure in the corporate world, these chapters will prove vital to making sense of the industries that govern the reproduction of our popular cultures into the future. In addition to helping us understand the continued dominance of media franchising, however, my hope is that these chapters might also model a particular way of conceptualizing and studying media culture in direct relation to industry—where "the franchise era" describes not just what media industries are doing, but also how media scholars are increasingly thinking about their objects of study and the contexts in which they circulate. This is not to say that I hope more scholars will start examining media franchises specifically; the chapters in this book represent an excellent effort on that count. Instead, I mean only that the impulse to consider franchising might encourage media scholars of all kinds to grapple more effectively with the industrial conditions and managerial logics in which media culture is always inseparably embedded.

As it was more than a decade previously, the idea of a "franchise"—in its unseemly invocation of the business of reproduction—refuses to allow us to study media culture without confronting the institutional forces and lines of industry power that bring it into being and manage its future evolution. This book embraces, rather than shies away from, that perspective, producing a rich portrait of contemporary media cultures but also reaffirming a model for situating them in their industrial contexts.

References

Tartaglione, Nancy (2016), "Warner Bros Sets Exec Structure for 'Harry Potter' Franchise Management," *Deadline Hollywood*, 8 June, <https://deadline.com/2016/06/harry-potter-franchise-warner-bros-polly-cochrane-will-harrison-nicole-blake-management-1201769111/> (last accessed 25 July 2018).

Williams, Trey (2018), "How Marvel Became a $16 Billion Franchise," *The Wrap*, 6 May, <https://www.thewrap.com/feige-fandom-and-comic-book-lore-marvels-secret-sauce-to-success/> (last accessed 25 July 2018).

1. INTRODUCTION: THE FRANCHISE ERA

James Fleury, Bryan Hikari Hartzheim, and Stephen Mamber

> It is undeniable that the dawn of the franchise film era is the most meaningful revolution in the movie businesses since the studio system ended, in the 1950s. (Fritz 2018: xv)

Though the above quotation from journalist Ben Fritz refers specifically to the seismic impact of the media franchise on the American film industry, its epicenter extends far beyond the 30-mile zone of Hollywood. As the chapters in this book make clear, multiple media across multiple regions have adopted the franchise—or the expansion of a single intellectual property (IP) into different texts (Johnson, 2013: 6; Herbert, 2017: 86). For decades, franchises have followed a multimedia management strategy in which texts adapt shared material (for example, a novelization or tie-in video game that retells a film's story). More recently, digital technologies—as Henry Jenkins has noted—have encouraged "transmedia storytelling," or the use of various media platforms to tell a single, intertextual narrative (2006: 97–8). Under transmedia, each text contributes different material so that the totality of a franchise presents an overarching story. While both management styles involve collaboration between media industries (for example, film, television, and video games), franchises continue to move from multimedia to transmedia strategies in order to create heightened interest across their brands. In other words, what originally harbingered an innovative shift in constructing multimedia entertainment has become co-opted and incorporated into existing franchise management strategies in response to shifting audience and technological trends.

The chapters in this book examine the role of the franchise in the media industries' responses to ongoing digital technology developments. The contributors collectively posit the franchise as an integral element of American—and, increasingly, global—media culture. As such, the book explores the production, distribution, and marketing of franchises as a historical form of media-making. In particular, the chapters analyze the complex industrial practice of managing franchises across interconnected online platforms with an international scope, presenting a network of scholarly texts that look critically at the collision of new and old industrial logics against an ever more fragmented and consolidated mediascape. The authors address: how traditional incumbents, such as film studios and television networks, have responded to the rise of big data companies such as Facebook, Apple, Amazon, Netflix, and Google (FAANG); the ways in which legacy franchises are adapting to new media platforms and technologies; the significant historical continuities and deviations in franchise-making and how they shape the representation of on-screen texts across digital displays; and, finally, how emerging media formats are expanding the possibility for transmedia experiences. In this regard, *The Franchise Era: Managing Media in the Digital Economy* offers an in-depth analysis of the tectonic shifts that have disrupted entertainment companies in the twenty-first century, demonstrating that the media franchise stands front and center in this high-stakes environment. In fact, the book argues that the franchise remains a tradition in the cultural production of mainstream American and international media.

The growing influence of digital technologies has radically redefined the business of entertainment. With new entrants like FAANG continuing to seize the reigns of production, distribution, and exhibition, content is no longer the exclusive purview of traditional media companies. The tech–media hybrids have embraced a data-driven approach to creative labor that tailors films and television shows to online viewer profiles. These companies also leverage their expertise to reengineer the audience–content formula. At the same time, a new generation of talent has built an empire on social media, translating their popularity into film and television deals and lucrative advertising revenue. In this fragmented and competitive marketplace, film and television industry incumbents struggle to maintain their long-standing monopoly. Not only do they grapple with minimized revenue streams but they also are at risk of losing cultural legitimacy. To reassert and maintain control, their core strategy has involved continued investment in their most valuable asset: the media franchise. While mainstream franchising has been an established practice in Hollywood for decades, the digital age's focus on convergent connectivity has rendered the process much more complex. As the role of transmedia environments in managing interest, buzz, and development has become an integral part of the process, film and television tentpoles—while still centrally important—have seen their roles diminished or recalibrated as but one segment of a larger brand experience.

With the growing ubiquity of franchises, industry pundits anxiously weigh in on how the media industries can angle for continued relevance while voicing cautious skepticism of the new status quo. In the contemporary "Golden Age of Television," think-pieces debate the trend of reviving dormant series (for example, *Twin Peaks* [CBS, 1990–91; Showtime, 2017]) and adapting films (for example, *Fargo* [FX, 2014-present]) (Adalian, 2015; Stubbs, 2018). Similarly, several journalists have written about the reliance of the video-game business on long-running franchises at the expense of innovation (Graser, 2014a; Makuch, 2018). In the context of cinema, media-industry trades continue to publish stories tonally vacillating between celebratory and panicked, depending on the box-office fate of the latest sequel, prequel, remake, reboot, or spin-off. In summer 2017, trade reports diagnosed the American film industry as having succumbed to "sequelitis," whose symptoms included disappointing performances for the newest entries in the *Pirates of the Caribbean* (2003–present) and live-action *Transformers* (2007–present) franchises (Kelly, 2017). Fears persist that film franchises, like those in the gaming sector, represent a threat to original storytelling (Lang, 2016). Hollywood, it seems, has come to regard the franchise as both savior and curse.

To get a sense of the American film industry's fears for its future, look no further than the following excerpt from *Variety*'s breakdown of the 2017 Christmas box-office weekend:

> After a bruising summer, when revenues plummeted more than 6% in the wake of costly flops such as "The Mummy" and "Transformers: The Last Knight," the gap did narrow. Fall and winter hits such as "It," "Thor: Ragnarok," "Coco," and "Murder on the Orient Express," helped make up the difference. The industry was also aided by record ticket prices. Empirically, fewer people made it to the multiplexes. Attendance is expected to hit a 27-year low when official numbers are tallied. (Lang, 2018a)

This reporting underscores the volatility of the franchise model, one that can result in "costly flops" as well as in "hits" that "make up the difference." At the same time that the trades ambivalently embrace media franchises, they also tend to celebrate whenever an original film reigns supreme at the box office, as seen in stories about *Get Out* (Jordan Peele, 2017) and *A Quiet Place* (John Krasinski, 2018)—even if the former had the brand recognition of producer Blumhouse and a sequel to the latter has already been announced (Lang, 2017; Rubin, 2018). Reports such as these suggest that the film industry is still uncomfortable with, but realistically accepting of, turning over the future to franchise production. As we shall see, however, media franchises have been tied to industrial survival for years.

The Franchise, Located

Despite anxieties over the American film industry's reliance on franchises, studio film production has long featured serial entertainments. In the Classical Hollywood era, few films outside of examples such as *Tarzan of the Apes* (Scott Sidney, 1918) and *The Thin Man* (W. S. Van Dyke, 1934) spawned sequels. As Avi Santo reminds us, the early twentieth century nonetheless introduced several "cross-media entertainment franchises, in which stories told across a range of media forms are linked together by a shared affiliation with a brand" (2015: 6). During this time, however, such brands (for example, the Lone Ranger, Buck Rogers, and Superman) tended to appear less often in feature films than in formats with lower production budgets, such as film and radio serials—and, later, television.

Radio and film serials of the 1930s and 1940s, in fact, influenced the development of franchises such as James Bond (1962–present). As Colin Burnett has shown through a concept he terms "threaded seriality," the image and understanding of the cinematic Bond character—introduced in Ian Fleming's novels and short stories (1953–66)—was formed not through any one film or performer but from the collective imagery and narratives from the Bond series. While primarily standalone in nature, the first twenty Bond films from EON Productions—those released prior to the reboot *Casino Royale* (Martin Campbell, 2006)—established franchise practices derived from serials, including recurring characters and settings, formulaic plots, and opening *in media res* (Burnett, 2017).

There are other reasons to trace Hollywood's contemporary reliance on franchise tentpoles back to the studio era. In many ways, the franchise represents a response to the breakup of the industry's old business model. Following the United States Supreme Court's 1948 Paramount Decision, the Big Five studios eventually all consented to the divestiture of their theater chains. With the end of vertical integration, studios no longer could block-book their films to beholden exhibitors, removing the central institution that preserved the majors' oligopoly. The rise of television in subsequent decades further drove audiences away from theaters, leading to profit losses, studio lot selloffs, and the decline of the traditional studio system. To compete with television, Hollywood introduced a number of innovations in content and technology in the 1950s and 1960s, from films with more violence and sex to 3-D cinema and widescreen formats. Formulaic blockbusters represented another strategy by which the studios sought to avert risk; for every hit like Twentieth Century-Fox's *The Sound of Music* (Robert Wise, 1965), though, there was a flop such as Warner Bros.' *Camelot* (Joshua Logan, 1967). Such precariousness sank studio stock prices and allowed for the conglomerate takeover of Hollywood, with films becoming engines responsible for powering other subsidiaries within the same corporate structure (Monaco, 2001: 9–10).

Though the Bond films of the 1960s had demonstrated the profit potential of sequels and tie-in merchandise, film franchises until the late 1970s exhibited little discernible long-term planning. Each Bond film stood on its own, with scant attention paid to continuity of actors or events; for instance, *Diamonds Are Forever* (Guy Hamilton, 1971) not only brings back Sean Connery as Bond after a one-film absence but also appears to ignore the death of Bond's bride at the end of the previous installment. A similar lack of franchise management is evident in Fox's first wave of *Planet of the Apes* films (1968–73) whose pessimistic endings would seem to render any follow-up impossible—such as the eponymous planet literally exploding in the final frames of *Beneath the Planet of the Apes* (Ted Post, 1970).

As Thomas Schatz has argued, the closed industrial system of the vertically integrated studio era had resulted in films whose self-contained narratives did not extend to other texts. The postwar "New Hollywood," however, replaced this closed system with horizontally integrated conglomerates that favored films open to "multimedia reiteration" across corporate holdings (Schatz, 1993: 34). By the 1970s, the increasingly risk-averse, conglomerate-owned studios had found a winning formula of franchising that combined sequels and merchandise with blockbuster budgets, saturation distribution, and television advertising. Helmed by a new generation of young directors, films such as *Jaws* (Steven Spielberg, 1975) and *Star Wars: Episode IV—A New Hope* (George Lucas, 1977) heralded a new age of strategically open, "high-concept" entertainment that could generate long-term audience appeal through franchising.[1]

This cycle accelerated in Hollywood's conglomerate era of the late 1980s and 1990s when the studios rebounded from the corporate takeovers of the 1960s and 1970s to become relevant parts of more tightly diversified multimedia conglomerates. The Warner Bros. film *Batman* (Tim Burton, 1989) exemplifies the model of "synergy" that began to spread during this period, with a comic-book character (published by Warner subsidiary DC Comics) birthing a media franchise whose sequels, animated spin-offs, and ancillary merchandise arrived through the various sectors of the Warner conglomerate.[2] Even at this point, industry insiders remained wary of the newfound mine of franchisable material. In the 1991 "Katzenberg Memo," Walt Disney Studios chairman Jeffrey Katzenberg warned against a post-*Batman* reliance on blockbusters, after Disney had found greater success with the modestly budgeted *Pretty Woman* (Garry Marshall, 1990) than *Dick Tracy* (Warren Beatty, 1990)—an expensive comic-strip adaptation modeled after the Burton film (Schatz, 1993:

28). Disney continued to pursue bigger and bigger films throughout the 1990s regardless, particularly after Katzenberg left the studio in 1994. His manifesto concludes, "If we decide to make a tent pole [*sic*] movie, then let's make it as intelligently as possible, with intelligent risk" (1991)—or, in other words, hedge against colossal failure through cofinancing and profiting in a long tail of consumer products, cable rights, and theme parks. The entry of private equity into Hollywood contributed to the rise of tentpole movies in the late 1990s and 2000s.

The conglomeration of the American film and television industries more and more has privileged franchises over original content. Today's studio executives—unlike those of the Classical Hollywood era—tend to come from a traditional business background, with little emotional attachment to (or, arguably, knowledge about) entertainment products (Galloway, 2006; Fritz, 2018: 82). As part of diversified conglomerates, studios must deliver predictable income to shareholders; formulaic franchises help to create a better sense of predictability. Instead of one-off projects like the Disney historical drama *The Finest Hours* (Craig Gillespie, 2016) that carry high risk and limited value beyond the theatrical and home video markets, conglomerates want franchisable IP capable of generating long-term value in the form of future installments, merchandise, theme-park attractions, and apps (Rainey, 2016). Franchise films and their array of ancillaries also provide a safety net in the case of box-office failure. As *Variety* noted in the wake of the disappointing opening weekend for *Solo: A Star Wars Story* (Ron Howard, 2018):

> [C]alling the film a bomb, as some have labeled it, may be hyperbolic. After all, "Solo" is much more than a film. It's a merchandising apparatus; one designed to move t-shirts, action figures, bedding, and even something as quotidian as plastic cups and paper plates. Disney is unlikely to be taking a write-down, even though the company will lose tens of millions on the theatrical part of "Solo's" life-cycle. (Lang and Rubin, 2018)

In this conception, the film represents just one aspect—and mostly just a promotional one—of the larger franchise's brand.

Digital-age franchises such as *The Matrix* (The Wachowskis, 1999–2003) and the Marvel Cinematic Universe (MCU) have showcased the ability of technology to enable not only cutting-edge visual effects but also transmedia storytelling. As Henry Jenkins has explained, transmedia franchises feature narratives that audiences must seek out across multiple media (for example, the MCU weaves together a single story featured in films, television series, and comic books) (2006: 97–8). Despite digital technology helping to proliferate multi- and transmedia franchises, Katzenberg's anxiety over tentpoles resonates. Decades later, studio executives continue to warn their colleagues about investing solely in blockbusters, particularly superhero movies (Barraclough, 2018). Still, Hollywood shows little indication of turning away from the franchise model.

Though film series appeared before the New Hollywood era both in the United States and elsewhere (for example, the *Apu* trilogy [Satyajit Ray, 1955–59] and the long-running *Gojira* franchise [1954–present]), today, the franchise model has come to dominate the entertainment landscape across film and other media. King Kong, for example, had been subject to legal disputes that allowed the character, starting in 1933, to appear in multiple unrelated film series and television adaptations that cannot be said to constitute a cohesive franchise (Vaz, 2005: 389).[3] As Daniel Herbert has argued, the character's most recent films have adopted a more carefully managed, franchise-friendly approach that has "updated the character for current commercial and cultural expectations" (2017: 84); notably, *Kong: Skull Island* (Jordan Vogt-Roberts, 2017) takes place in the same storyworld as *Godzilla* (Gareth Edwards, 2014) and will receive a crossover sequel in which the two "titans" meet (*Godzilla vs. Kong* [Adam Wingard, 2020]). The unrelated animated television series *Kong: King of the Apes* (Netflix, 2016–present) nonetheless demonstrates that rights issues have continued to limit the character's transmedia potential.

In the shift from multi- to transmedia models of management, the franchise concept has expanded across media industries and regions. Since acquiring Hollywood studios in the 1960s, conglomerates have sought to exploit the IP of their film and television subsidiaries. The media franchise now encompasses a variety of texts, with sequels, often formulaic, still the most common. Even though prequels would become more common in the wake of *Butch and Sundance: The Early Days* (Richard Lester, 1979) and *Indiana Jones and the Temple of Doom* (Steven Spielberg, 1984), the concept of sequels remained novel enough in the 1970s and 1980s to warrant posters for *Jaws 2* (Jeannot Szwarc, 1978) and *Halloween II* (Rick Rosenthal, 1981) to exclaim "ALL NEW" (despite both films retreading their predecessors and offering very little "new"). By the end of the 1980s, sequels had become so prevalent as to become jokes in films such as *Spaceballs* (Mel Brooks, 1987)—which mentions the (currently) fake sequels *Spaceballs 2: The Search for More Money* and *Rocky 5000*—and *Back to the Future Part II* (Robert Zemeckis, 1989), whose "2015" sequence features an advertisement for *Jaws 19*.

Emerging technologies further encouraged media conglomerates to expand their IP through sequels and prequels in entirely new formats. American and international franchises have demonstrated the potential for integrating multiple platforms—such as the British *Carry On* franchise (1958–92) encompassing thirty-one films, as well as television appearances, stage plays, and a comedy music album. In the 1970s, the conglomerates Warner Communications Inc. and Gulf+Western began to leverage their video game subsidiaries as part of the promotional machine for films (for example, *Superman* [Atari, 1978]) and television series (for example, the *Happy Days* [ABC, 1974–84] tie-in *Fonz* [Sega/Gremlin, 1976)]). Subsequent video games would serve not only as direct adaptations of existing texts (for example, *Ghostbusters* [Activision, 1984]) but also as original sequels (for example, *The Goonies II* [Konami, 1987]), prequels (for example, *Star Wars: Knights of the Old Republic* [LucasArts, 2003]), and spin-offs (for example, *Mad Max* [Warner Bros. Interactive Entertainment, 2015]). Furthermore, the growth of home video in the 1980s introduced older films to a new generation. In response, conglomerates pursued belated sequels such as *Psycho II* (Richard Franklin, 1983) and *The Color of Money* (Martin Scorsese, 1986). Together with the rise of the made-for-television movie in the 1970s, home video—and later subscription video on demand (SVOD)—presented a non-theatrical distribution outlet for franchise extensions.

More recently, media conglomerates have experimented with new franchising strategies beyond extending or digging into narrative chronology. The reboot, for instance, has helped to bring back ossifying franchises. Whereas "soft" reboots have existed for decades (for example, whenever a new actor takes over as James Bond), the "hard" reboot gained popularity after *Batman Begins* (Christopher Nolan, 2005) reset the cinematic Batman series. Because hard reboots have developed a negative reputation for abandoning all that came before (for example, beyond sexist criticism against its female cast, part of the fan vitriol that greeted Paul Feig's 2016 *Ghostbusters* reboot centered on its lack of franchise continuity) (Framke, 2016), soft reboots have become more common.

In particular, a type of soft reboot known as the "legacyquel"—a neologism that Matt Singer of *ScreenCrush* coined to describe a sequel featuring both new and "legacy" characters (2015)—serves to support a project's cross-generational appeal. Based on the assumption that viewers will identify with characters like them, legacyquels have sought to lure older, infrequent moviegoers and more dedicated fans simultaneously with narratives in which an older generation of actors, reprising their "legacy" characters, "pass the torch" to a younger set (for example, *Tron: Legacy* [Joseph Kosinski, 2010] and *Creed* [Ryan Coogler, 2015]). Legacyquels have appeared in other media beyond film, including television (for example, *Star Trek: The Next Generation* [1987–94]) and video games (for example, *Gears of War 4* [Microsoft Studios, 2016]).

On the surface, the legacyquel seems like the ideal type of franchise extension for contemporary media conglomerates. Because legacyquels take place, and often are made, many years after a franchise's previous installment, they lend themselves to a transmedia storytelling format capable of generating products that a conglomerate can release through various internal holdings and external partnerships. For example, Warner Bros. and Columbia Pictures digitally distributed anime and live-action short films that contextualized the otherwise unseen events between the original *Blade Runner* (Ridley Scott, 1982) and its legacyquel *Blade Runner 2049* (Denis Villeneuve, 2017), and a series of comics and books and an anime series—all announced in 2018—will continue the story from these films and further explore the interstitial period (Collis, 2018; Otterson, 2018). Still, the concept's viability remains to be seen. The *Mission: Impossible* franchise (1996–present) seemingly has moved beyond Jeremy Renner's legacyquel-intended character in favor of long-time franchise headliner Tom Cruise (Travis and Hewitt, 2018), the upcoming fifth *Indiana Jones* film reportedly won't feature the hero's son introduced in the preceding entry (Nolfi, 2017), and numerous cinematic and televisual efforts have performed below commercial expectations (for example, *Vacation* [Jonathan Goldstein and John Francis Daley, 2015] and *24: Legacy* [Fox, 2017]). Even if the legacyquel trend proves short-lived, however, the franchise model will remain dominant owing to a number of factors that reflect media industry changes in the United States and around the world.

Big Changes in Hollywood and Abroad

Big-budget over Mid-budget

Studio film production has undergone several major changes in the twenty-first century that have contributed to the entrenchment of the blockbuster, franchise model. Consider, for instance, that there are fewer mid-budget films today from the studios, in Hollywood and even in similar industries like Japan's, owing in large part to the fall of the home video market (Schilling, 2018). DVD sales, which helped assure that "all but the biggest flops were profitable," peaked in the mid-2000s and began to fall in 2009 (Fritz, 2018: 17). A variety of factors contributed to the collapse of DVD and the lack of adoption for its successor format, Blu-ray; these have included internet piracy, the rise of cheap rental services such as Redbox and (initially) Netflix, and the proliferation of SVOD services, spearheaded by Netflix.

The fall in physical home video sales hit studios hard. Although they have sought to compensate through electronic sell-through (EST) of "digital copies," consumers have demonstrated a preference for SVOD subscriptions over home video purchases (Garrett, 2017). Struggling physical and digital home video sales have compelled studios to focus on franchises. In fact, blockbusters (particularly those carrying budgets of $75 million or more) continue to sell on home video (Fritz, 2018: 21).

Combined with the fall of the home video market, studios granting theatrical releases to fewer films has fueled the drive toward franchises and away from lower-budgeted projects—including independent films—that carry a greater risk (Bart, 2013; Gottlieb, 2018). The studios have so emphatically embraced blockbuster franchises over indies and modestly budgeted projects that the success of original films tends to generate industry trade buzz (McClintock, 2017). The production company Blumhouse, in fact, has carved out a niche for itself with star-driven, low-cost films, several of which—such as *The Purge* (James DeMonaco, 2013), starring Ethan Hawke—have generated franchises (Fleming Jr, 2017). While Blumhouse has taken up "micro-budget" films for theatrical distribution, mid-budget fare has found a home on SVODs. Netflix, for example, handled distribution for Paramount's sci-fi horror film *Annihilation* (Alex Garland, 2018) outside of North America and globally released *Extinction* (Ben Young, 2018) after Universal dropped the sci-fi thriller from its release schedule (McNary, 2018). Already, however, Netflix has signaled a desire to enter the blockbuster franchise space, with the announcement of a sequel to its $90-million-budgeted fantasy cop film *Bright* (David Ayer, 2017) (Galuppo and Kit 2018) as well as a television prequel to the Indian media franchise *Baahubali* (2015–present) (Ramachandran 2018).

Characters over Stars

The displacement of mid-budget films has reconfigured the global star-centric economy. Stars had dominated Hollywood product since the studio system under which studios signed talent to long-term contracts. The 1950s, however, brought an end to the star system, as actors and filmmakers became free agents hired for individual film projects. Under the new structure, talent would form production companies and become more involved in business-making decisions (McDonald, 2000: 71). For decades, stars commanded the highest salaries and helped decide if a film would be made. Though major actors in prior decades could earn $20 million for nearly any role, such high salaries today tend to be reserved only for franchise participation (Lang and Kroll, 2018). In 2002, Harrison Ford received $25 million, and 20 percent of the box office, for Kathryn Bigelow's submarine drama *K-19: The Widowmaker* (Harris and Dunkley, 2001); more recently, a *Variety* story titled "Harrison Ford Was Paid Over 50x More than 'Star Wars' Co-Stars"—a headline underscoring the increasing rarity of such a high payday—noted that he earned between "upwards of $10 million and as high as $20 million to reprise his role of Han Solo" for *Star Wars: Episode VII—The Force Awakens* (J. J. Abrams, 2015) (Kroll, 2015).

As the example of Ford suggests, characters, rather than stars, now dominate the box office. Following the disappointing theatrical performance of recent star vehicles such as the George Clooney–Julia Roberts thriller *Money Monster* (Jodie Foster, 2016) and the Russell Crowe–Ryan Gosling buddy cop film *The Nice Guys* (Shane Black, 2016), Marc Bernardin in the *Los Angeles Times* opined: "Hollywood has shifted from a star-based economy to a character-based economy. The actor at the top of the credit block has come to matter less than the name of the character or the franchise in the title" (2016). Similarly, Fritz argues that the contemporary franchise era represents "a return to the studio system" and specifies that "[o]nly now the major entertainment companies don't own the most important talent—they own the most important cinematic brands" (2018: 94). Much of Disney's current financial success derives from its acquisitions of Pixar, Marvel, and Lucasfilm, as each production company provides a plethora of franchisable IP (for example, *The Incredibles*, *Avengers*, and *Star Wars*, respectively). Likewise, Warner Bros. has adopted a franchise film strategy supported by the "three pillars" of DC, LEGO, and Wizarding World (that is, *Harry Potter* and its *Fantastic Beasts* spin-offs) (Bond, 2015).

Though the franchise era has left little room for mid-budget, star vehicles, this is not to say that stars no longer matter in Hollywood. Some actors can become so identified with particular characters as to be considered indispensable to their franchises. This can, as in Ford's case, lead to historically high salaries. Robert Downey Jr, whose casting as the titular hero in *Iron Man* (Jon Favreau, 2008) proved integral to starting the MCU, earned "around $40 million plus backend participation" for replaying the character in *Captain America: Civil War* (Joe and Anthony Russo, 2016) (Graser, 2014). His identification with the role stands in contrast to how Marvel Studios recast the Hulk and War Machine owing to contract disputes with Edward Norton and Terrence Howard, respectively. The *Mission: Impossible* franchise presents a similar actor–character dynamic. For reprising the role of elite spy Ethan Hunt in the sixth installment, *Fallout* (Christopher McQuarrie, 2018), Tom Cruise's salary demands proved high enough to delay production. Still, because of Cruise's longtime connection to the role, *Deadline Hollywood* noted in its reporting of the contract negotiations, "In the long term, this amounts to a hiccup, because there is no way Paramount doesn't make this film and no way they would do it without Cruise, who receives an eight-figure upfront sum against break even back end" (Fleming Jr and Busch, 2016).

Playing such franchise marquees, however, can result in other consequences, as both actors' careers have come to reflect their most successful franchise personas. By 2015, with the release of *Mission: Impossible: Rogue Nation* (Christopher McQuarrie, 2015), *Forbes* claimed that the franchise was "the only thing maintaining Cruise's status as one of the top movie stars in Hollywood" (Pomerantz, 2015). Notably, in most of his recent roles, Cruise has played variations of Hunt in action-adventure scenarios (for example, *Jack Reacher* [Christopher McQuarrie, 2012], *Edge of Tomorrow* [Doug Liman, 2014], and *The Mummy* [Alex Kurtzman, 2017]). Downey Jr, meanwhile, has had little time to star in anything other than the MCU and the *Sherlock Holmes* franchise (2009–present) while upholding the persona of Tony Stark.

As Wesley Morris in *The New York Times* puts it, “The brand comes first. Marvel is the star” (2016). The same also increasingly can be said of other actors and franchises, as Cruise and *Mission: Impossible* demonstrate,

For non-tentpole content, many stars simply have taken shelter in other media. Once-marquee talent such as Catherine Zeta-Jones and Kevin Costner have expanded into television while other actors have embraced the creative, and financial, possibilities of video games (for example, Gary Oldman), podcasts (for example, Anna Faris), and virtual reality (for example, John Cusack). Some stars (such as Dwayne Johnson, Kevin Hart, and Melissa McCarthy) still do make star film vehicles but these projects have become more inconsistent than franchise installments in terms of attracting audiences. Whereas Johnson found a hit in the soft reboot *Jumanji: Welcome to the Jungle* (Jake Kasdan, 2017), his subsequent, non-franchise release *Skyscraper* (Rawson Marshall Thurber, 2018) stumbled at the North American box office (McClintock, 2018). Stars can no longer rely on the body of their film work to deliver audiences. A 2014 survey, indicating that social influencers had higher awareness than traditional film stars among the youth audience, prompted panic in Hollywood (Ault, 2014). In response, studios have acquired web platforms (for example, Warner Bros. buying Machinima to support its media franchises) and also have begun casting artists from YouTube and other digital platforms in films and television projects (Spangler, 2016; Jarvey, 2016). Film stars, for their part, have worked to cultivate social media followings. Though Johnson’s contract for the upcoming *Red Notice* (Rawson Marshall Thurber, 2020) includes “a $1 million social-media fee” (Lang and Kroll, 2018), the film’s basis on an original concept rather than a preexisting franchise may spell further trouble for not just Johnson as a dependable star but for Hollywood’s star system more broadly.

Cinematic Universes over Tentpole Films

Despite screenwriter William Goldman’s maxim that “nobody knows anything” (1983), studios have learned that franchises work. While tentpole, “event” films have existed at least since *Jaws*, they represented just one part of a studio’s production strategy. In fact, as Fritz reminds us, “Tentpoles got that name because they were supposed to hold up a structure that also contained dramas, romantic comedies, adult thrillers, and even totally original ideas” (2018: 150). Today, however, studios have come to consider standalone franchises such as *Jaws* as less valuable than cinematic universes—extended film narratives that connect multiple movies, characters, and storylines. From the 1930s to the 1950s, Universal presented a prototype shared universe, as its “Universal Monsters” franchise included series focused on individual characters as well as crossovers that brought these characters together in a single film (for example, *House of Frankenstein* [Erle C. Kenton, 1944] whose print advertisements declared, "All together! Frankenstein's monster! Wolf Man! [*sic*] Dracula! Hunchback! Mad Doctor!"). Though cinematic universes would appear sporadically in subsequent decades (for example, the "View Askewniverse" that began with Kevin Smith's 1994 film *Clerks*), the concept exploded in popularity after Marvel Studios launched the MCU with *Iron Man*. Modeled after Marvel's characters that inhabit the same fictional storyworld in the comic books, the MCU has shown studios the predictable profits that cinematic universes can deliver. One reason for this shift from standalone franchises to cinematic universes, therefore, lies in this predictability. Like sequels, cinematic universes provide reliable revenue and predictable costs but reduce the waiting time between theatrical releases and the potential for diminishing returns.

For multimedia conglomerates, cinematic universes represent, in many ways, a natural extension of something they already do: television production. First, cinematic universes often employ "writers rooms," a television production practice in which a team of writers collectively collaborates on episode scripts and season arcs. Toy company Hasbro, for example, has created a writers room for its *Transformers* franchise and larger cinematic universe (Kroll, 2016). Second, under cinematic universe production, the role of the director is often less central than that of the producer—who, like a television series "showrunner," retains creative oversight. In practical terms, directors are less instrumental to a cinematic universe because of audience expectation of films releasing in relatively rapid succession; with each production taking approximately two years, any director would find it impossible to manage the total enterprise (Fritz, 2018: 175). Owing to the diminished importance of directors, cinematic universes have inspired a trend of hiring directors for studio franchise productions with only indie film or television projects to their name—such as Patty Jenkins (2017's *Wonder Woman*) and Taika Waititi (2017's *Thor: Ragnarok*). This trend even has extended beyond cinematic universes to franchise films in general (for example, Colin Trevorrow helming 2015's *Jurassic World*) and to non-Hollywood contexts. In Japan, for instance, the decision was made to hire directors whose résumés were made up more from work in television and even video games for the live-action adaptations of best-selling *Shōnen Jump* manga *Rurouni Kenshin* (Keishi Ōtomo, 2012, 2014), *Death Note: Light up the World* (Shinsuke Sato, 2016), and *Gintama* (Yuichi Fukuda, 2017).

Decisions such as these have raised concerns about the effect of the cinematic universe concept on filmmaking in the future. The "director problem" on *Star Wars* and MCU films, whereby numerous filmmakers have been replaced owing to "creative differences," gives the cynical impression that, under the cinematic universe model, directors have become largely interchangeable

(Dibdin, 2015; Gonzalez, 2017). If movie stars already were being squeezed out of larger paydays for standard franchise fare, then they face even greater challenges with cinematic universes. Tom Cruise, for one, may have found continued commercial and creative success with the *Mission: Impossible* franchise, on which he serves as a producer, but his management skills did not translate to *The Mummy*. Despite Universal developing the film as part of a planned "Dark Universe" reboot of its earlier "Universal Monsters" cinematic universe, Cruise "exerted nearly complete creative oversight" throughout the production (Setoodeh and Lang, 2017). Even though Downey Jr has received top billing (and a salary to match) for his appearances in Marvel's *Avengers* films (2012–present), the case of *The Mummy* suggests the seeming incompatibility of movie stars and cinematic universes. Critics also lament the effect that cinematic universes will have on storytelling, arguing that crossovers such as *Avengers: Infinity War* (Joe and Anthony Russo, 2018) feature too many characters to develop properly (Rozsa, 2018). Finally, while cinematic universes might seem like an ideal component of transmedia storytelling, their very name underscores the reality that they often support a narrative and production hierarchy in which films reign supreme over any non-cinematic text; for example, the films of the MCU exert a one-way exchange of influence over the creative decision-making on the universe's television series (Whitbrook, 2016).

Television over Film Content

The rise of transmedia and cinematic universe franchise strategies, as well as the move of mid-budget films and star vehicles to television, demonstrate that the boundaries between film and other media have become more permeable. Just as cinematic universes have adopted production and narrative aspects from television, television series (such as *Big Little Lies* [HBO, 2017–present]) feel like extended films. Although Fritz notes that television has replaced film as the home for "original, risk-taking" content (2018: 150), the overwhelming selection of programming—a concept FX head John Landgraf has referred as "peak TV"—has encouraged networks, cable and premium channels, and SVODs to embrace the franchise concept (Rose and Guthrie, 2015; O'Connell, 2018).

Because peak television provides space not just for franchise fare but also the type of adult, character-driven storytelling that film studios prioritized in the pre-*Star Wars* era, the film industry has sought to differentiate itself. Similar to how film studios had chased blockbusters in the 1950s and early 1960s to compete with the then nascent medium, the contemporary television industry has encouraged Hollywood to present blockbuster franchises as a form of product differentiation. That is, if television offers the complex narrative of *Westworld* (HBO, 2016–present) and the rich characterization of

Mad Men (AMC, 2007–15), then the film industry presents the spectacular set pieces and special effects of franchises such as the MCU—which, along with exhibition innovations such as IMAX screens and Dolby Atmos sound systems, are designed to deliver audiences to theaters and away from their home theater setups (Hughes, 2015). Ironically, therefore, even as film studios have turned to franchises to distinguish themselves from original television content, networks, cable channels, and SVODs have also turned to franchises to distinguish their own content.

Global over Domestic (or The Rise of China)

It has become increasingly apparent that worldwide audiences want franchise films. The cumulative global box office now vastly outpaces declining US box-office receipts, which have consistently comprised less than 30 percent of the global box office since 2014 (Lang, 2018b). Nowhere is this change more evident than in the rising box office of China. The growth of the Chinese market starkly contrasts with the decline of box offices in movie industries with longer histories. Between 2012 and 2015, the North American box office rose from $10.8 billion to $11.1 billion whereas China's grew from $2.7 billion to $6.8 billion during that same period (Kokas, 2017: 9). And, while Hollywood has long exported blockbuster cinema throughout the world, the rise of the Chinese exhibition market has further fueled demand for big-budget product.

According to Fritz, international audiences, particularly those in developing theatrical markets, prefer "big-budget films full of visual effects" and proven "brands" over "most American dramas, comedies, or other mid-budget films" (2018: 210). Behind this is the implicit assumption that worldwide audiences want the same predictability of branded entertainment as American audiences. In response, Hollywood studios have engineered blockbuster franchise films with international non-US markets—particularly China—explicitly in mind. Despite the *kaiju*-mech beat-em-up *Pacific Rim* (Guillermo del Toro, 2013) underperforming in the United States and even in Japan, where its mecha aesthetics were expected to find a sympathetic reception, the sequel *Pacific Rim Uprising* (Steven S. DeKnight, 2018) went into production because the original had grossed nearly as much in China as in the United States and Japan combined, and because China's Dalian Wanda Group had purchased production company Legendary Pictures in 2016 (Wojnar, 2018).

Previously, China had exerted indirect control over Hollywood filmmaking. Studios reedited films for the Chinese market (such as *Skyfall* [Sam Mendes, 2012]), added scenes (for example, *Iron Man 3* [Shane Black, 2013]), or changed content during the scripting phase (for example, *Doctor Strange* [Scott Derrickson, 2016] depicts the Ancient One as Celtic instead of Tibetan as in the Marvel comics) as executives considered how a film would best play in China. When Chinese companies began investing in American films, the indirect became direct. As an example, Hollywood films increasingly feature Chinese stars in supporting roles, such as Angelababy in *Independence Day: Resurgence* (Roland Emmerich, 2016) and Li Bingbing in *The Meg* (Jon Turteltaub, 2018). Since the mid-2010s, Chinese companies have sought even more control, investing in Hollywood productions in exchange for a percentage of the film's profits and to be credited as a producer (Fritz, 2018: 201).

Because domestic and non-US productions have begun to compete with American imports in China, studios worry that Chinese audiences may be growing weary of Hollywood blockbusters—especially as *Star Wars: The Last Jedi* (Rian Johnson, 2017) lost the Chinese box-office battle to local franchise installment *The Ex-File 3: The Return of the Exes* (Tian Yusheng, 2017) (Chow, 2017; Brzeski, 2018). To compensate, Hollywood studios have partnered Chinese companies. In 2016, Dalian Wanda Group began investing in and promoting individual films from Sony Pictures; the deal proved short-lived, however, after Wanda came under government scrutiny at home, signaling another sign of the continued struggle to unite the Chinese and American film markets (Frater, 2017). Wanda's partnership with Sony seemed fitting, given that the Japanese consumer technology giant was one of the first foreign companies to purchase Hollywood holdings such as Columbia Pictures and the leftovers of the MGM lot. In some ways, the partnership with Sony and acquisitions of Legendary and exhibition chain AMC all suggest Wanda's attempts to follow in Sony's footsteps. Furthermore, in 2016, Wanda almost bought a 49 percent stake in Paramount Pictures but infighting within corporate parent Viacom ended the deal (Fritz, 2018: 203). Despite this setback, Paramount had several Chinese partnerships for *Transformers: The Last Knight* (Michael Bay, 2017)—a franchise entry that failed to match the box-office heights of its 2014 predecessor (Rahman, 2017). For Paramount and other foreign media companies, Chinese coproductions such as *The Last Knight* qualify as "local" productions in the eyes of regulators; as such, these films tend to bypass import quotas while also providing foreign producers with a higher than average box-office intake (typically 25 percent in China, but 55 or 60 percent in the United States) (Kokas, 2017: 67).

In addition to censorship, import restrictions, and comparatively low cuts of distribution revenue, further challenges remain for developing Hollywood films for the Chinese market. The Chinese groups in charge of importing American films (such as SARFT, its subsidiary China Film Bureau, and the state-owned China Film Group) are mercurial (Fritz, 2018: 212). China's protectionist economic policies also limit imports in order to support Chinese films—such as opening *The Amazing Spider-Man* (Marc Webb, 2012) and *The Dark Knight Rises* (Christopher Nolan, 2012) on the same day (Tartaglione,

2012). In February 2018, three Chinese franchise films—*Monster Hunt 2* (Raman Hui, 2018), *Detective Chinatown 2* (Chen Sicheng, 2018), and *The Monkey King 3* (Cheang Pou-soi, 2018)—also opened on the same day; they still all managed to achieve commercial success, however, and thus demonstrated the massive market for homegrown franchises in China (Mendelson, 2018). Even if Chinese blockbusters, such as *The Great Wall* (Zhang Yimou, 2016), have struggled at the North American box office, as long as China and other global markets remain viable and fast growing, Hollywood will continue to target those audiences with franchise films (Yimou, 2017).

Big Tech over the Big Six (or The Rise of Silicon Valley)

Silicon Valley tech–media hybrids (such as the FAANG group) have entered Hollywood, poised to take over existing studios. At first, Amazon and Netflix, in particular, appeared to be saviors for the struggling independent film market. By the time the former's Amazon Studios film and television division opened in 2010, most major American studios had abandoned indie films in favor of franchises (Fritz, 2018: 188). Recently, however, both Amazon Studios and Netflix themselves have indicated a growing interest in franchisable blockbuster content. After Netflix and Amazon had picked ten and twelve titles, respectively, at the 2017 Sundance Film Festival, both companies opted to forgo all acquisitions at the 2018 festival, suggesting a shift to mainstream features (Guerrasio, 2018). Whereas Netflix had adopted a franchise approach early on (for example, its television series set within the MCU), Amazon has only more recently embraced a blockbuster content strategy.

As such, even though Amazon Studios and Netflix have not yet acquired a traditional Hollywood studio, they have come to resemble traditional studios themselves. Amazon Studios has shifted from ordering television series based on audience testing of pilots in favor of straight-to-series orders, ended its film and television script solicitation program, and hired several veteran industry executives (Ramos, 2018). Amazon's moves into traditional decision-making and announcement of a megabudget television series based on the *Lord of the Rings* suggest a future strategy that, like established Hollywood studios, privileges tentpoles and franchises over riskier, mid- and low-budget projects (Siegel, 2018). Similarly at Netflix, along with its own continued focus on franchise content, such as *Bright* and the MCU television series, its distribution deal for *Mowgli: Legend of the Jungle* (Andy Serkis, 2019) from Warner Bros. sends a clear signal of how tough it has become not just for modest productions but also for non-franchise blockbusters to find an audience (Lang, 2018c).

If Amazon, Netflix, and other tech–media hybrids move more aggressively into the traditional blockbuster space, they will pose a significant threat to traditional Hollywood studios. Each has the financial resources not only to invest in film and television in the long term but also to withstand short-term losses. In particular, Amazon's ecosystem can converge its Amazon Video SVOD service with its other products; we already see this with the IMDb-powered X-ray feature, and the potential exists to integrate the chat function from its livestreaming service, Twitch. By leveraging the totality of its ecosystem, Amazon can support an expansive media franchise strategy in the future. The company's "massive database of hundreds of millions of customers' interests and tastes" also means that it can "market movies with precision, saving money and reducing financial risk" (Fritz, 2018: 194), while the ability to rewatch films on Amazon Video helps to cultivate a passionate fan base for particular films—and franchises.

In other words, Amazon makes movies to drive user engagement. The ultimate goal is for consumers to stay engaged with the Amazon ecosystem and to buy products from its online retail service (Ad Age, 2016). Just as Amazon has begun to pursue a franchise strategy, its use of films to encourage consumer product sales also resembles the established conglomerate practice of positioning films as synergy engines. The Walt Disney Company uses films to sell their own consumer products and theme-park tickets, while Comcast NBCUniversal uses Universal films to provide content for its cable and VOD systems. These movements underscore the fact that films have become a less central product at the conglomerate-owned film studios and more a driver of content consumption in all shapes and sizes.

Franchises in the Digital Economy: The Essays

Transmedia storytelling, the collapse of the theatrical mid-budget film market, the shift of stars to television, the cinematic universe strategy, the permeability of film and television, the growing significance of the international market, and the convergence of Silicon Valley and Hollywood all indicate that the media industries are entrenched in the franchise era. As such, today's approach to franchise management must consider multiple consumer touchpoints that are constantly in flux:

- **Story/Mythology:** The story constitutes the central driver of the franchise economy. While some franchises display a consistency of vision, others undergo changes in the form of reboots, remakes, and spin-offs.
- **Brand:** The franchise brand refers to the overall value and meaning in the public eye. While some franchises manage to sustain a positive image, others navigate bad press and publicity.
- **Audience:** Franchises are built on audience growth, engaging old fans and gaining new ones along the way.
- **Format:** Formats comprise an area of constant transformation as fran-

chises enter new stages of technological evolution. To remain relevant, franchises are open to embracing change (for example, virtual, and augmented reality).

- **Platform:** As content moves away from an analog infrastructure and toward a digital ecosystem, franchises expand their content to online platforms (for example, social, mobile, and SVOD).
- **Software/Data:** In the evolving network of software-driven automation and information, franchises are exposed to unprecedented forms of data that fundamentally inform all other touchpoints. The beleaguered movie-ticketing service MoviePass, for instance, turned to selling the data of its subscribers, on the basis that studios would find value in this consumer information (such as optimizing when to release or how to market a particular film) (Andrews, 2018).

The chapters in this book cumulatively examine these consumer touchpoints across five sections. Each section centers on a particular aspect of franchise management, its role in the domestic and global media and entertainment industry, as well as its development over time. The respective chapters in each section highlight historical continuities and shifts that have shaped the management of media franchises. Overall, the sections offer insights into franchise films, television shows, video games, consumer products, and surrounding fan communities.

The Franchise: Defining and Historicizing

The first chapter, by James Fleury and Stephen Mamber, works toward a definition of the media franchise and uses the *Alien* franchise (1979–present) to consider how digital technologies have influenced a shift in franchise management from multimedia to transmedia. Next, Daniel Herbert presents a historical examination of New Line Cinema's *Critters* (1986–present) franchise. The chapter looks at the role that home video played early on in the development of media franchises and also explains the centrality of *Critters* and other franchises to the development of New Line itself. Both chapters consider how traditional film franchises have struggled to adjust to shifting economic conditions, either by piggybacking off existing trends or fighting for studio survival by innovating from the industry's margins.

Video Games: Software, Hardware, and Space

In the second section, the authors present video game industry case studies for a broader perspective of media franchising. Heather Lea Birdsall explores the Disney theme parks as a branded franchise space through a number of its video-game appearances. Her chapter makes it clear that franchise management unites multiple physical and digital spaces in its strategic global expansion. Matthew Thomas Payne's chapter continues the thread of video games and franchise management, using the case study of Nintendo's NES and SNES microconsoles. His chapter posits that franchises can refer to both software *and* hardware, as the built-in games on Nintendo's miniconsoles function as a form of franchise management and corporate canonizing. Closing this section, Andreas Rauscher analyzes the complex convergence of film, video games, and other media within the *Star Wars* franchise. He argues that stylistic influence within the franchise extends in multiple directions as, while the games bear cinematic elements, the recent film entries remediate ludic concepts.

Animation: Adaptation across Nations, Industries, and Platforms

In the next section, the authors turn to animation. Brian Ruh's chapter analyzes the representational politics in the Japanese-originated *Ghost in the Shell* franchise (1989–present). Media franchises continue to struggle with representation both in front of the camera (such as the marginalization of LGBTQ characters in franchise films) and behind it (for example, a lack of female directors on franchise projects, which, of course, speaks to a broader gender imbalance throughout Hollywood) (Couch, 2018; Lincoln, 2015). As Ruh explains, Rupert Sanders's 2017 American, live-action *Ghost in the Shell* adaptation sparked a controversy in representation after casting Scarlett Johansson in the lead role of Motoko Kusanagi. Then, Rayna Denison analyzes how and in what ways the *How To Train Your Dragon* franchise (2010–present) has been developed in relation to the success or struggles of DreamWorks Animation. She argues that the franchise has developed primarily in relation to three specific factors: the evolution of its studio; attempts by its production studio to capitalize on new distributive technologies; and tensions between producers that can be read at the level of the franchise's narrative flows. Both chapters attempt to correct an imbalance in the discussion of franchises, which often exclude animation in favor of live-action content. The two chapters' examination of technological and ideological issues exclusive to animation franchises present an argument for scholars to consider more strongly animated franchises for future study.

Television: New Directions for a Legacy Medium

The penultimate section focuses on the presence of television in managing media franchises. Jennifer Gillan's chapter looks at how television sitcoms have become another promotional arm by which parent companies prop up their franchises. She considers how sitcoms can play a part in what she calls "transmedia marketing circuits," looking specifically at the content-as-promotion strategies connected to *Black-ish* (ABC, 2014–present) and *Parks and Recreation* (NBC, 2009–15). Monica Sandler focuses on SVOD services, using the case study of Comcast NBCUniversal's failed SeeSo platform. Her chapter examines how television networks can cultivate a franchise identity for themselves by expanding their brand into emerging formats. As both chapters make clear, what television means—and where it begins and ends—is becoming difficult to define but also increasingly irrelevant. With the rapid spread of SVOD services that are either directly owned by or are supplied with content from the studios, television content is but one group of repurposable and reusable assets in massive, integrated platforms.

Emergent Platforms: Possible Futures of the Franchise

In the final section, the authors continue to explore new media formats and their support of franchises. Bryan Hikari Hartzheim posits the licensed mobile game as a flexible, promotional paratext within Japanese video game and anime transmedia franchises such as *Dragon Ball* and *Final Fantasy*. As such, his chapter considers the increasingly important role of ancillary, low-budget texts and how they can quickly adapt to changing market conditions to monetize effectively audience interests (Gray, 2010). Next, Alexander Champlin focuses on esports, one of the fastest-growing entertainment markets today. His chapter specifically analyzes the relationship between esports franchises (that is, teams), branded esports studio spaces, and video game franchises (for example, *Overwatch* [Activision, 2016]). Lastly, James Fleury examines virtual reality (VR) in relation to Hollywood promotional strategies. As his chapter explains, the VR format may be struggling to gain mainstream adoption but media companies continue to use it in service of their existing film, television, and video game franchises.

Conclusion

In conclusion, each chapter in this collection speaks to the history and continued presence of the media franchise in the face of cultural, industrial, and technological changes. The broad range of scholarship featured in these chapters displays both the heightened interest in studying media franchises and the continued effects of the digital economy on multiple creative industries in Hollywood and abroad. Franchises represent global media industries' response to dynamic changes, and understanding how such franchises are managed, sustained, and disseminated will be centrally important to understanding how and what entertainment is produced in the decades to come. Welcome to the franchise era.

Notes

1. See Wyatt, 1994: 81. As Justin Wyatt explains, "high concept" refers to films whose inherently marketable style can be replicated across other media (for example, posters and television advertisements) and ancillary merchandise (for example, soundtracks and video games).
2. In January 1990, Warner Communications, Inc. completed its merger with Time Inc. to become Time Warner.
3. Specifically, these series have included a pair of films from RKO (both 1933); another pair from Toho (1962; 1967); yet another pair from Paramount Pictures and De Laurentiis Entertainment Group (1976; 1986); a remake from Universal (2005); and the most recent series from Legendary Pictures and Warner Bros.

References

Ad Age (2016), "Amazon Using Twitch Audience as Focus Group for New TV Shows," *Ad Age*, 2 September, <http://adage.com/article/media/amazon-twitch-audience-focus-group-tv-shows/305696/> (last accessed 3 August 2018).

Adalian, Josef (2015), "Full House, Coach, The X-Files: Why TV Remakes Are Everywhere Right Now," *Vulture*, 8 April, <http://www.vulture.com/2015/04/tv-remake-full-house-coach-x-files.html> (last accessed 2 August 2018).

Andrews, Travis M. (2018), "The Rise and Decline of MoviePass, the Subscription Service that Flew too Close to the Sun," *The Washington Post*, 1 August, <https://www.washingtonpost.com/news/arts-and-entertainment/wp/2018/08/01/the-rise-and-decline-of-moviepass-the-subscription-service-that-flew-too-close-to-the-sun/?utm_term=.e08a92d2ba45> (last accessed 12 August 2018).

Ault, Susanne (2014), "Survey: YouTube Stars More Popular Than Mainstream Celebs Among U.S. Teens," *Variety*, 5 August, <https://variety.com/2014/digital/news/survey-youtube-stars-more-popular-than-mainstream-celebs-among-u-s-teens-1201275245/> (last accessed 2 August 2018).

Barraclough, Leo (2018), "Fox Film Chief Stacey Snider Warns Hollywood Not to Rely Solely on 'Caped Crusaders'," *Variety*, 11 June, <https://variety.com/2018/film/news/fox-stacey-snider-hollywood-caped-crusader-movies-1202840288/> (last accessed 1 August 2018).

Bart, Peter (2013), "Development Slates Shrinking As Film Remakes, Sequels Dominate," *Variety*, 27 June, <https://variety.com/2013/biz/news/development-slates-shrinking-as-film-remakes-sequels-dominate-1200502662/> (last accessed 2 August 2018).

Bernardin, Marc (2016), "Marvel, 'Star Wars,' 'Harry Potter,' and More: Why the Movie Star No Longer Shines as Bright as the Franchise," *Los Angeles Times*, 17 June, <http://www.latimes.com/entertainment/movies/la-ca-mn-why-movie-stars-dont-matter-25-franchises-20160616-snap-story.html> (last accessed 1 August 2018).

Bond, Paul (2015), "Warner Bros. CEO Dismisses Talk of Superhero Fatigue," *The Hollywood Reporter*, 4 March, <https://www.hollywoodreporter.com/news/warner-bros-ceo-dismisses-talk-779183> (last accessed 1 August 2018).

Brzeski, Patrick (2018), "Why Disney's 'Last Jedi' Mind Trick Isn't Working in China," *The Hollywood Reporter*, 4 March, 16 January, <https://www.hollywoodreporter.com/heat-vision/star-wars-china-box-office-why-disneys-jedi-mind-trick-isnt-working-china-1075177> (last accessed 2 August 2018).

Burnett, Colin (2017), "Richard Maibaum's 007 Screenplays: 'Lost' James Bond Stories and the Intricacies of Threaded Serialism," paper presented at Society of Cinema and Media Studies Conference, Chicago, 25 March.

Chow, Vivienne (2017), "Chinese Audiences Move Away From Hollywood Pictures," 5 November, <https://variety.com/2017/biz/markets-festivals/bad-genius-chinese-audiences-move-away-from-hollywood-films-1202606930/> (last accessed 2 August 2018).

Collis, Clark (2018), "*Blade Runner* Universe to Be Explored in Series of Comics and Graphic Novels," *Entertainment Weekly*, 12 July, <http://ew.com/books/2018/07/12/blade-runner-comics/> (last accessed 5 August 2018).

Couch, Aaron (2018), "'Jurassic World: Fallen Kingdom' Cut Lesbian Reveal," *The Hollywood Reporter*, 19 June, <https://www.hollywoodreporter.com/heat-vision/jurassic-world-2-cut-lesbian-reference-says-daniella-pineda-1121454> (last accessed 3 August 2018).

Dibdin, Anna (2015), "Does Marvel Studios Have a Director Problem?," *Digital Spy*, 23 August, <http://www.digitalspy.com/movies/feature/a664341/does-marvel-studios-have-a-director-problem-why-a-list-filmmakers-are-fleeing-the-mcu/> (last accessed 2 August 2018).

Fleming, Mike, Jr (2017), "How Jason Blum Honed His Micro-Budget Blockbuster Formula," *Deadline Hollywood*, 16 May, <https://deadline.com/2017/05/jason-blum-blumhouse-jordan-peele-disruptors-interview-news-1202094624/> (last accessed 2 August 2018).

Fleming, Mike, Jr and Anita Busch (2016), "'Mission: Impossible 6' Halts Prep Until Tom Cruise and Other Deals Get Worked Out," *Deadline Hollywood*, <https://deadline.com/2016/08/mission-impossible-6-halts-prep-until-tom-cruise-and-others-deals-get-worked-out-1201806220/> (last accessed 17 August 2018).

Framke, Caroline (2016), "How the All-Female Ghostbusters Reboot Became a Lightning Rod of Controversy," *Vox*, 15 July, <https://www.vox.com/2016/6/30/12027882/ghostbusters-reboot-all-female-backlash-sexism-sony> (last accessed 3 August 2018).

Frater, Patrick (2017), "Wanda–Sony Pictures Financing and Marketing Deal Is Over," *Variety*, 21 November, <https://variety.com/2017/film/asia/wanda-sony-pictures-finance-deal-halted-1202620411/> (last accessed 2 August 2018).

Fritz, Ben (2018), *The Big Picture: The Fight for the Future of Movies*, New York: Houghton Mifflin.

Galloway, Stephen (2006), "MBAs in Hollywood," *The Hollywood Reporter*, 7 November, <https://www.hollywoodreporter.com/news/mbas-hollywood-142210> (last accessed 2 August 2018).

Galuppo, Mia and Borys Kit (2018), "'Bright' Sequel Officially a Go at Netflix," *The Hollywood Reporter*, 3 January, <https://www.hollywoodreporter.com/heat-vision/bright-sequel-will-smith-david-ayer-expected-return-netflix-movie-1071188> (last accessed 2 August 2018).

Garrett, Diane (2017), "Studios Pull Out All the Stops to Pump Up Digital Movie Purchases in the Streaming Age," *Variety*, 5 December, <https://variety.com/2017/tv/spotlight/studios-pull-out-all-stops-pump-up-digital-movie-purchases-streaming-age-1202630371/> (last accessed 1 August 2018).

Giardina, Carolyn and Pamela McClintock (2017), "Is the Golden Age of 3D Officially Over?," *The Hollywood Reporter*, 3 August, <https://www.hollywoodreporter.com/behind-screen/is-golden-age-3d-officially-1025843> (last accessed 4 August 2018).

Goldman, William (1983), *Adventures in the Screen Trade: A Personal View of Hollywood and Screenwriting*, New York: Warner Books.

Gonzalez, Umberto (2017), "Inside the 'Star Wars' Director Problem, and How Lucasfilm Might Try to Fix It," *The Wrap*, 6 September, <https://www.thewrap.com/star-wars-colin-trevorrow-lucasfilm-director-rian-johnson/> (last accessed 2 August 2018).

Gottlieb, Akiva (2018), "Indies Fight to Survive as Marketplace Undergoes Radical Changes," *Variety*, 23 April, <https://variety.com/2018/film/features/indies-fight-to-survive-world-of-blockbusters-1202754499/> (last accessed 15 August 2018).

Graser, Marc (2014a), "Franchise Fatigue Gives Way to Fresh Ideas From Videogame Creators," *Variety*, 7 March, <https://variety.com/2014/voices/opinion/franchise-fatigue-gives-way-to-fresh-ideas-from-videogame-creators-1201126098/> (last accessed 2 August 2018).

Graser, Marc (2014b), "Robert Downey Jr to Join 'Captain America 3'," *Variety*, 13 October, <https://variety.com/2014/film/news/robert-downey-jr-to-join-captain-america-3-exclusive-1201312229/> (last accessed 2 August 2018).

Gray, Jonathan (2010), *Show Sold Separately: Promos, Spoilers, and Other Media Paratexts*, New York: New York University Press.

Guerrasio, Jason (2018), "After Dominating Sundance for 2 years, Netflix and Amazon Bought Zero Movies at the Festival this Time Around—and Insiders Told Us Why," *Business Insider*, 29 January, <https://www.businessinsider.sg/sundance-2018-amazon-netflix-buys-2018-2018-1/> (last accessed 2 August 2018).

Harris, Dana and Cathy Dunkley (2001), "Par Eyes Raising 'K-19'," *Variety*, 31 May, <https://variety.com/2001/film/news/par-eyes-raising-k-19-1117800446/> (last accessed 1 August 2018).

Herbert, Daniel (2017), *Film Remakes and Franchises*, New Brunswick, NJ: Rutgers University Press.

Hughes, Mark (2015), "Dolby, IMAX Announce Expansions as Global Box Office Grows," *Forbes*, 24 July, <https://www.forbes.com/sites/markhughes/2015/07/24/dolby-imax-announce-expansions-as-global-box-office-grows/#5ff2c1e42e8b> (last accessed 15 August 2018).

Jarvey, Natalie (2016), "How Studios Are Cashing In on YouTube Stars," *The Hollywood Reporter*, 18 March, <https://www.hollywoodreporter.com/news/youtube-stars-invade-hollywood-how-874812> (last accessed 2 August 2018).

Jenkins, Henry (2006), *Convergence Culture: Where Old and New Media Collide*, New York: New York University Press.

Johnson, Derek (2013), *Media Franchising: Creative License and Collaboration in the Culture Industries*, New York: New York University Press.

Katzenberg, Jeffrey (1991), "The World Is Changing: Some Thoughts on Our Business," *Letters of Note*, 11 January, <http://www.lettersofnote.com/2011/11/some-thoughts-on-our-business.html> (last accessed 1 August 2018).

Kelly, Seth (2017), "Summer Box Office Officially Worst in Over a Decade," *Variety*, 5 September, <https://variety.com/2017/film/news/summer-box-office-2017-worst-decade-1202548272/> (last accessed 31 July 2018).

Kinder, Marsha (1991), *Playing with Power in Movies, Television, and Video Games: From Muppet Babies to Teenage Mutant Ninja Turtles*, Los Angeles: University of California Press.

Kokas, Aynne (2017), *Hollywood Made in China*, Los Angeles: University of California Press.

Kroll, Justin (2015), "Harrison Ford Was Paid Over 50x More Than 'Star Wars' Co-Stars," *Variety*, 21 December, <https://variety.com/2015/film/news/harrison-ford-star-wars-pay-force-awakens-cast-1201666236/> (last accessed 1 August 2018).

Kroll, Justin (2016), "Hasbro Cinematic Universe Assembles Writers Room," *Variety*, 21 April, <https://variety.com/2016/film/news/hasbro-cinematic-universe-paramount-writers-1201758497/> (last accessed 2 August 2018).

Landreth, Jonathan (2013), "Hollywood and China: In Figures," *The Atlantic*, 7 November, <https://www.theatlantic.com/china/archive/2013/11/hollywood-and-china-in-figures/281222/> (last accessed 2 August 2018).

Lang, Brent (2016), "How 'Iron Man' and Superhero Movies Are Causing Headaches at Cannes," *Variety*, 10 May, <https://variety.com/2016/film/markets-festivals/iron-man-cannes-superhero-movies-1201770087/> (last accessed 2 August 2018).

Lang, Brent (2017), "Will 'Get Out' Make Blumhouse the Pixar of Horror Filmmakers?," *Variety*, 26 February, <https://variety.com/2017/film/box-office/get-out-jordan-peele-blumhouse-pixar-horror-1201996703/> (last accessed 15 August 2018).

Lang, Brent (2018a), "Box Office: 'Star Wars: The Last Jedi' Holds Off 'Jumanji' on New Year's Weekend," *Variety*, 1 January, <https://variety.com/2018/film/news/box-office-star-wars-the-last-jedi-holds-off-jumanji-on-new-years-weekend-1202650549/> (last accessed 31 July 2018).

Lang, Brent (2018b), "Global Box Office Hits Record $40.6 Billion in 2017; U.S. Attendance Lowest in 23 Years," *Variety*, 4 April, <https://variety.com/2018/digital/news/global-box-office-hits-record-40-6-billion-in-2017-u-s-attendance-lowest-in-23-years-1202742991/> (last accessed 5 August 2018).

Lang, Brent (2018c), "'Mowgli's' Sale to Netflix Signals Changing Times," *Variety*, 30 July, <https://variety.com/2018/film/news/mowgli-netflix-sale-jungle-book-1202889639/> (last accessed 2 August 2018).

Lang, Brent, and Justin Kroll (2018), "Leonardo DiCaprio, Jennifer Lawrence and Other Star Salaries Revealed," *Variety*, 8 May, <https://variety.com/2018/film/news/celebrity-salaries-daniel-craig-jennifer-lawrence-leonardo-dicaprio-1202801717/> (last accessed 1 August 2018).

Lang, Brent, and Rebecca Rubin (2018), "'Solo': How Big a Box Office Dud Is the 'Star Wars' Spinoff?," *Variety*, 30 May, <https://variety.com/2018/film/news/solo-a-star-wars-story-box-office-losses-1202825432/> (last accessed 2 August 2018).

Lincoln, Ross A. (2015), "Colin Trevorrow Comments Spark Discussion on Women in Hollywood," *Deadline Hollywood*, 22 August, <https://deadline.com/2015/08/colin-trevorrow-star-wars-episode-ix-women-hollywood-1201503834/> (last accessed 3 August 2018).

Loffhagen, Matthew (2016), "15 Movies That Made Drastic Changes For A Chinese Audience," Screen Rant, 13 May, <https://screenrant.com/movies-changed-for-china/> (last accessed 2 August 2018).

McClintock, Pamela (2017), "'Girls Trip" Box Office Breaks R-Rated Comedy Curse, Proves Power of Diversity," *The Hollywood Reporter*, 24 July, <https://www.hollywoodreporter.com/news/girls-trip-box-office-breaks-r-rated-comedy-curse-proves-power-diversity-1023740> (last accessed 2 August 2018).

McClintock, Pamela (2018), "Box Office: Is 'Skyscraper' Dying Hard Due to Dwayne Johnson Overexposure?," *The Hollywood Reporter*, 16 July, <https://www.hollywoodreporter.com/news/skyscraper-box-office-stumble-is-dwayne-johnson-overexposed-1127192> (last accessed 1 August 2018).

McDonald, Paul (2000), *The Star System: Hollywood's Production of Popular Identities*, New York: Wallflower Press.

McNary, Paul (2018), "Netflix Buys Michael Pena-Lizzy Caplan Thriller 'Extinction' From Universal," *Variety*, 8 February, <https://variety.com/2018/film/news/netflix-buys-michael-pena-lizzy-caplan-thriller-extinction-1202692114/> (last accessed 2 August 2018).

Makuch, Eddie (2018), "E3 2018: Elijah Wood Talks Video Game Sequels And The Importance Of Taking Risks," *Gamespot*, 13 June, <https://www.gamespot.com/articles/e3-2018-elijah-wood-talks-video-game-sequels-and-t/1100-6459792/> (last accessed 2 August 2018).

Meehan, Eileen (2015), "'Holy Commodity Fetish, Batman!': The Political Economy of a Commercial Intertext," in Roberta Pearson, William Uricchio, and

Will Brooker (eds), *Many More Lives of the Batman*, New York: Palgrave, pp. 69–87.

Mendelson, Scott (2018), "'Monster Hunt' and 'Monkey King' Pushed China Box Office to a Record-Crushing Week," *Forbes*, <https://www.forbes.com/sites/scottmendelson/2018/02/22/local-franchises-and-sequels-pushed-chinese-box-office-to-a-record-crushing-week/#30794e7f26bc> (last accessed 15 August 2018).

Monaco, Paul (2001), *The Sixties: 1960–1969*, Los Angeles: University of California Press.

Morris, Wesley (2016), "The Superhero Franchise: Where Traditional Movies Go to Die," *The New York Times*, 19 May, <https://www.nytimes.com/2016/05/22/movies/in-x-men-apocalypse-and-captain-america-superheroes-versus-movie-stars.html> (last accessed 17 August 2018).

Nolfi, Joey (2017), "Indiana Jones 5 Won't Feature Shia LaBeouf's Character," *Entertainment Weekly*, 4 September, <http://ew.com/movies/2017/09/04/indiana-jones-5-shia-labeouf-not-returning/> (last accessed 3 August 2018).

O'Connell, Michael (2018), "The New Economics of TV's Reboot Craze," *The Hollywood Reporter*, 21 February, <https://www.hollywoodreporter.com/features/new-economics-tvs-reboot-craze-1086797> (last accessed 2 August 2018).

Otterson, Joe (2018). "Blade Runner' Anime Series Set at Adult Swim, Crunchyroll." *Variety*, 29 November, <https://variety.com/2018/tv/news/blade-runner-anime-series-adult-swim-crunchyroll-1203047083/> (last accessed 30 November 2018).

Pomerantz, Dorothy (2015), "Tom Cruise Needs 'Mission: Impossible' More than 'Mission: Impossible' Needs Him," *Forbes*, 23 March, <https://www.forbes.com/sites/dorothypomerantz/2015/03/23/tom-cruise-needs-mission-impossible-more-than-mission-impossible-needs-him/#70712df76ef5> (last accessed 17 August 2018).

Rahman, Abid (2017), "Has Pandering to Chinese Audiences Hurt 'Transformers 5'?," *The Hollywood Reporter*, 2 July, <https://www.hollywoodreporter.com/heat-vision/transformers-last-knight-has-pandering-chinese-audiences-hurt-transformers-5-1016926> (last accessed 2 August 2018).

Rainey, James (2016), "Disney Expects $75 Million Loss on 'The Finest Hours'," *Variety*, 8 March, <https://variety.com/2016/film/news/disney-expects-to-lose-75-million-on-the-finest-hours-1201725537/> (15 August 2018).

Ramachandran, Naman (2018), "Netflix Adds Prequel Series to 'Baahubali' Indian Hit Movie Franchise," *Variety*, 2 August, <https://variety.com/2018/digital/asia/netflix-prequel-series-indian-hit-movie-bahubaali-1202893181/> (last accessed 2 August 2018).

Ramos, Dino-Ray (2018), "Amazon Studios Ends Open Call For Script Submissions," *Deadline Hollywood*, 14 April, <https://deadline.com/2018/04/amazon-studios-open-call-script-concept-roy-price-jennifer-salke-1202364429/> (last accessed 3 August 2018).

Rehak, Bob (2018), *More than Meets the Eye: Special Effects and the Fantastic Transmedia Franchise*, New York: New York University Press.

Rose, Lacey and Marisa Guthrie (2015), "FX Chief John Landgraf on Content Bubble: 'This Is Simply Too Much Television'," *The Hollywood Reporter*, 7 August, <https://www.hollywoodreporter.com/live-feed/fx-chief-john-landgraf-content-813914> (last accessed 2 August 2018).

Rozsa, Matthew (2018), "'Avengers: Infinity War': Not Just a Mediocre, Muddled Mess—but Possibly a Dangerous Precedent," *Salon*, 27 April, <https://www.salon.com/2018/04/27/the-avengers-infinity-war-not-just-a-mediocre-muddled-mess-but-possibly-a-dangerous-precedent/> (last accessed 2 August 2018).

Rubin, Rebecca (2018), "'A Quiet Place': Why the Horror Film Roared at the Box

Office," *Variety*, 9 April, <https://variety.com/2018/film/box-office/a-quiet-place-box-office-anaylsis-john-krasinski-emily-blunt-1202746988/> (last accessed 31 July 2018).

Santo, Avi (2015), *Selling the Silver Bullet: The Lone Ranger and Transmedia Brand Licensing*, Austin, TX: University of Texas Press.

Schatz, Thomas (1993), "The New Hollywood," in Jim Collins, Hilary Radner, and Ava Collins, *Film Theory Goes to the Movies*, New York: Routledge, pp. 8–36.

Schilling, Mark (2018), "Japan Debates the Ethics of Making Films on the Cheap," *Variety*, 30 May, <https://variety.com/2018/film/asia/japan-ethics-of-making-cheap-films-1202825001/#article-comments> (last accessed 5 August 2018).

Setoodeh, Ramin and Brent Lang (2017), "Inside 'The Mummy's' Troubles: Tom Cruise Had Excessive Control," *Variety*, 14 June, <https://variety.com/2017/film/news/the-mummy-meltdown-tom-cruise-1202465742/> (last accessed 2 August 2018).

Siegel, Tatiana (2018), "Inside Amazon's $250M 'Lord of the Rings' Deal: 'It's Very Much a Creature of the Times'," *The Hollywood Reporter*, 5 April, <https://www.hollywoodreporter.com/live-feed/how-lord-rings-tv-series-landed-at-amazon-not-netflix-1099213> (last accessed 3 August 2018).

Singer, Matthew (2015), "Welcome to the Age of the Legacyquel," *Screencrush*, 23 November, <http://screencrush.com/the-age-of-legacyquels/> (last accessed 3 August 2018).

Spangler, Todd (2016), "Warner Bros. Acquires Full Control of Machinima," *Variety*, 17 November, <https://variety.com/2016/digital/news/warner-bros-acquires-machinima-1201920793/> (last accessed 2 August 2018).

Steinberg, Marc (2012), *Anime's Media Mix: Franchising Toys and Characters in Japan*, Minneapolis: Minnesota University Press.

Stubbs, David (2018), "Reboot, Revive, Repeat: Has Retromania Really Taken Over TV?," *The Guardian*, 24 May, <https://www.theguardian.com/tv-and-radio/2018/may/24/reboot-revive-repeat-has-retromania-really-taken-over-tv> (last accessed 2 August 2018).

Tartaglione, Nancy (2012), "U.S. Movie Influx Has 'Shaken' Chinese Industry, Official Says," *Deadline Hollywood*, 11 November, <https://deadline.com/2012/11/u-s-movie-influx-has-shaken-chinese-industry-official-says-369996/> (last accessed 4 August 2018).

Tempest, Rone (1994), "How Do You Say 'Boffo' in Chinese? 'The Fugitive'," *Los Angeles Times*, 29 November, <http://articles.latimes.com/1994-11-29/entertainment/ca-2665_1_american-movie> (last accessed 4 August 2018).

Thompson, Kristin (2007), *The Frodo Franchise: The Lord of the Rings and Modern Hollywood*, Los Angeles: University of California Press.

Travis, Ben and Chris Hewitt (2018), "Mission Impossible Director Christopher McQuarrie Reveals Why Jeremy Renner Didn't Return in Fallout," *Empire*, 1 August, <https://www.empireonline.com/movies/news/mission-impossible-director-christopher-mcquarrie-reveals-jeremy-renner-return-f/> (last accessed 3 August 2018).

Vaz, Mark Cotta (2005), *Living Dangerously: The Adventures of Merian C. Cooper, Creator of King Kong*, New York: Villard.

Waxman, Sharon (2012), "How Hollywood and Joe Biden Got China to Drop a 20-Year Movie Quota," *Reuters*, 19 February, <https://www.reuters.com/article/idUS22096264620120220> (last accessed 3 August 2018).

Whitbrook, James (2016), "*Agents of SHIELD* Star Says Marvel Doesn't Care Enough About Its Own TV Show," *io9*, 17 May, <https://io9.gizmodo.com/agents-of-shield-star-says-marvel-doesnt-care-enough-ab-1777057136> (last accessed 2 August 2018).

Wojnar, Zak (2018), "Pacific Rim Uprising's China Gamble Didn't Pay Off," *Screen Rant*, 25 April, <https://screenrant.com/pacific-rim-uprising-box-office-success-china/> (last accessed 2 August 2018).

Wyatt, Justin (1994), *High Concept: Movies and Marketing in Hollywood*, Austin, TX: University of Texas Press.

Yimou, Zhang (2017), "What Hollywood Looks Like from China," *The New York Times*, 4 December, <https://www.nytimes.com/2017/12/04/opinion/china-film-industry-hollywood.html> (last accessed 2 August 2018).

PART I

THE FRANCHISE: DEFINING AND HISTORICIZING

2. THE (IM)PERFECT ORGANISM: DISSECTING THE *ALIEN* MEDIA FRANCHISE

James Fleury and Stephen Mamber

In the Twentieth Century-Fox film *Alien* (Ridley Scott, 1979), the android Ash (Ian Holm) refers to the titular creature as "the perfect organism" and admits to "admir[ing] its purity." Although the film gave birth to a long-running media franchise, the development of the franchise cannot be described as perfect or pure. The same can be said of many venerable franchises because the term "media franchise"—as the chapters in this book demonstrate—describes a variety of management strategies applied to a variety of texts and contexts. In this chapter, we break down the term "media franchise" by tracing the complex history of the *Alien* franchise. It, in fact, remains one of the few media franchises to follow the story continuity that its originating text established. While *Shaft* (1971–present), *Rocky* (1976–present), *Star Wars* (1977–present), and *Indiana Jones* (1981–present) also maintain their original continuity, other franchises either have rebooted to a new storyline (for example, *Planet of the Apes*) or established an alternative timeline to serve as a simultaneous reboot and sequel (for example, *Star Trek*). Today, the long-running *Alien* franchise—following a number of cinematic sequels, prequels, and spin-offs as well as video game, literary, and comic book installments—has struggled to unify its disparate texts while adapting to changes in media-franchise management. Hollywood's turn from multimedia replication to transmedia expansion, in particular, has led to conflicts of textual continuity, creative ownership, and public relations within the *Alien* franchise. We dissect these conflicts through the video game *Aliens: Colonial Marines* (Sega, 2013) as a case study of media franchise mismanagement.

Multimedia and Transmedia Franchise Management

The *Alien* franchise illustrates how the transition from multimedia to transmedia has brought new demands, such as stricter continuity among texts, stronger collaboration between licensors and licensees, and regular engagement with fans. Under multimedia, franchises follow a top–down form of management in which ancillary texts (such as a tie-in video game) repeat the narrative of a primary text (for example, a film). The digital era, however, has encouraged a more integrated and less hierarchical form of franchise management. Henry Jenkins has noted that our contemporary convergence culture involves "the flow of content across multiple media platforms, the cooperation between multiple media industries, and the migratory behavior of media audiences who will go almost anywhere in search of the kinds of entertainment experiences they want" (2006: 2). According to Jenkins, transmedia storytelling disperses a single narrative across multiple media platforms and therefore suggests how digital technology has encouraged franchises to create more complex, intertextual stories and more expansive storyworlds (2006: 97–8). Put another way, whereas multimedia franchises involve narrative repetition, transmedia franchises involve narrative expansion. To sustain itself over the past four decades, the *Alien* franchise has had to adapt from the demands of multimedia to those of transmedia.

Though transmedia would appear ideally suited to contemporary franchise management, it remains problematic. In his case study of the *Matrix* franchise, Jenkins argues that "[n]o film franchise has ever made such demands on its audience" (2006: 96). In other words, the *Matrix* transmedia strategy meant that each franchise text withholds certain narrative information that can be found only in other *Matrix* products; for example, the tie-in video game *Enter the Matrix* (Atari, 2003) helped make sense of the film *The Matrix Reloaded* (The Wachowski, 2003). Furthermore, Jonathan Gray has analyzed how transmedia franchises blur distinctions between "mothership," or primary, texts and "paratexts," or ancillary texts (2010: 6). In an ideal transmedia network, each text can serve as an "entryway" into a franchise but paratexts more commonly occupy a lower industrial and cultural rung than their mothership (Gray, 2010: 23). Finally, because each paratext affects a franchise's storyworld, transmedia raises the stakes for each text to add financial and narrative value. In multimedia networks, where the narrative does not extend beyond mothership texts (for example, a series of films), tie-ins could be written off as peripheral, but *Aliens: Colonial Marines* (*A:CM*) demonstrates how, under transmedia, paratexts perceived to be lacking in added value can be discredited through public criticism or even have their narrative contributions retroactively written out of continuity. The troubled production context of *A:CM* suggests that the *Alien* franchise has remained rife with conflict even as transmedia has replaced multimedia as the dominant organizational logic. We argue that this conflict stems from a mismanagement of mothership texts and paratexts with divergent continuities that have led to negative critical and audience reception.

PROBLEMATIZING THE *ALIEN* FRANCHISE

Although the intertextual network that began with the film *Alien* demonstrates characteristics associated with both multimedia and transmedia franchising, the term "media franchise" should not be taken as self-evident. Derek Johnson has noted that a media franchise denotes "multiplied cultural production," and Daniel Herbert has specified that the term refers to a particularly "industrial" logic designed "to expand and spread" an intellectual property (IP) into an array of texts (Johnson, 2013: 6; Herbert, 2017: 86). Building upon Johnson and Herbert's definitions, we analyze the overlapping cultural and industrial aspects of the *Alien* franchise; in essence, we use *Alien* to provide our own guide for defining a media franchise as an intertextual network of media extensions that converge to create a unique storyworld, or fictional universe. We begin by breaking down *Alien* as a franchise and describing its development in response to shifts in management strategies. We then present *A:CM* as a case study of media franchise mismanagement and argue that it bears evidence of struggling to navigate these shifts.

An Intertextual Network of Media Extensions

Alien demonstrates that a media franchise represents a collection of texts from different media based on the same IP. That is, while a series consists of works in a single medium, a franchise migrates a brand across multiple media. The original film, 1979's *Alien*, features Ellen Ripley (Sigourney Weaver) and the other crew members of the commercial space vessel *Nostromo* fighting to survive against an alien creature that appears equal parts biological and mechanical. When Twentieth Century-Fox followed up the film, the franchise circuitously transitioned between multimedia and transmedia. At first, a novelization and a pair of video games, *Alien* (Fox Games, 1982) and *Alien* (Argus Press Software, 1984), took a multimedia approach by adapting the film's plot—directly in the case of the book and, owing to hardware limitations of the time, abstractly with the games.

The cinematic sequel, *Aliens* (James Cameron, 1986), picks up with Ripley awakened fifty-seven years after having destroyed the creature aboard the *Nostromo*. Along with a squad of Colonial Marines aboard the USS *Sulaco*, Ripley faces off against a wave of the same alien species from the first film, here referred to as "xenomorphs." *Aliens* both expanded the franchise as a film series and extended its multimedia presence, with another novelization and multiple tie-in video games that again repeated, rather than added to, the film's narrative. Shortly after the release of *Aliens*, Dark Horse Comics published a number of comic books between 1988 and 1991 that continued the narrative of *Aliens* with the film's surviving characters: Ripley, Newt (Carrie Henn), Corporal Hicks (Michael Biehn), and the android Bishop (Lance Henriksen).

The second film sequel, *Alien 3* (David Fincher, 1992), splintered the franchise. Its production included a constantly changing screenplay and clashes between the studio and director David Fincher, a music video veteran without feature-filmmaking experience. The result was a film that continues to divide fans, partly because of the path it took the franchise down. Specifically, whereas the comic books had been showcasing the survivors from *Aliens*, *Alien 3* retroactively revised the continuity of, or retconned, the comics by killing Newt and Hicks offscreen during the opening credits and by dispatching with Bishop early on, thereby positioning Ripley as the sole remaining character. To maintain the film's revised continuity, Dark Horse comics published after *Alien 3* replaced Newt and Hicks with new, but similar, characters named Billie and Wilks, respectively (Bishop, 2016). This retconning followed the textual hierarchy of multimedia and therefore contradicted the previous transmedia approach in which each *Alien* text occupied the same level of narrative importance. As a result, *Alien 3* created the first of several divergent continuities within the franchise.

Furthermore, the behind-the-scenes conflicts inspired alternative versions of the film. To add value to the 2003 "*Alien* Quadrilogy" home video box set, Ridley Scott provided a "Director's Cut" of *Alien* and Jean-Pierre Jeunet oversaw a "special edition" of *Alien: Resurrection* (Jean-Pierre Jeunet, 1997) to accompany James Cameron's 1991 "special edition" of *Aliens*; because of lingering tensions from the making of *Alien 3*, however, Fincher opted not to participate in any aspect of the box set. This left him as the only director not to provide an audio commentary for his film, to be involved in retrospective interviews, or to have supervised a new cut of his film. Instead, the studio itself used alternative and new footage to create the *Alien 3* "assembly cut," which made several changes to the theatrical version.[1] *Alien 3*, then, not only conflicts with the continuity established in the previous paratexts but its multiple cuts also conflict with one another. On the one hand, the contradictory nature of the *Alien 3* assembly cut becomes more apparent when considering that Scott, in a booklet included with the "*Alien* Quadrilogy" box set, dismisses the idea of his film's "director's cut" as the definitive version:

> The traditional definition of the term "director's cut" suggests the restoration of a director's original vision, free of any creative limitations. It suggests that the filmmaker has finally overcome the interference of

> heavy-handed studio executives, and that the film has been restored to its original, untampered form. Such is not the case with *Alien: The Director's Cut* . . . For marketing purposes, this version is being called "The Director's Cut." To film purists everywhere, rest easy. The original 1979 theatrical version isn't going anywhere. It remains my version of choice.

Through this dismissal, Scott claims ownership over both editions of *Alien*, so that the director's cut does not supersede the original in terms of franchise canonicity. On the other hand, Fincher, by not taking part in the box set's assembly cut of *Alien 3*, has surrendered claim over the film to the studio, rendering the new version as the studio's preferred version. In this way, *Alien 3* has become an explicit example of what Jerome Christensen has called "studio authorship," by which a corporation, rather than a filmmaker, represents the driving force behind a film (2012). In contrast to the *Alien* director's cut, the *Alien 3* assembly cut revises the franchise's continuity.

Following the next film *Alien: Resurrection*, the franchise further split its storylines. First, the Dark Horse comics and the novels from DH Press and Titan Books followed their own narrative events not tied to the events or characters of the films. Second, the subsequent films have all served as prequels. Based on a series of comics launched in 1989, the spin offs *Aliens vs. Predator* (Paul W. S. Anderson, 2004) and *Aliens vs. Predator: Requiem* (The Brothers Strause, 2007) showcase conflicts between xenomorphs and the species from the *Predator* franchise (1987–present). Meanwhile, the more recent *Prometheus* (Ridley Scott, 2012) and *Alien: Covenant* (Ridley Scott, 2017) establish some links to *Alien* while also exploring their own, more existential themes. Third, the video games have followed different narrative paths as well, including film adaptations such as *Alien: Resurrection* (Fox Interactive, 2000), spin-offs such as *Aliens vs. Predator* (Sega, 2010), film continuations like *Alien: Isolation* (Sega, 2014), and virtual reality experiences including "*Alien:* Descent" (FoxNext, 2018).[2] Finally, in 2015, director Neill Blomkamp revealed a proposed film that would continue the storyline of *Aliens* and ignore its sequels. Though Fox ultimately cancelled this project to focus on *Alien: Covenant*, it would have become another instance of the franchise vacillating between multimedia and transmedia approaches over the course of its development and therefore would have created additional, competing continuities (Lincoln, 2015; Shirey, 2018).[3]

As such, the *Alien* franchise has become less a cohesive narrative spread across texts than a series of diverging narratives that demand audiences to choose their preferred story path. Audiences can, for instance, follow the storyline of *Alien* and *Aliens* before concluding with the comic books, or they can limit themselves to the films without turning to the ancillary material. On one level, this narrative divergence, or what might be termed "negative convergence," tends to emerge from a perceived hierarchy of mothership texts retconning the contributions of paratexts; the film *The Amazing Spider-Man 2* (Marc Webb, 2014), for example, revises the characterization of the Rhino (Paul Giamatti), a villain introduced in *The Amazing Spider-Man* (Activision, 2012)—a video game sequel to the 2012 film of the same name. On another level, revisionism also tends to affect franchises whose textual extensions—whether films, television series, comic books, or video games—have produced creative or commercial diminishing returns. We see this, for instance, in the five, at present, competing timelines of *Halloween* (1978–present) whose sequels and reboots have met mixed receptions.[4] In the *Alien* franchise, divergence began with *Alien 3* retconning the *Aliens* comic books, as well as from the film's consequent divisiveness. As we explain later, *A:CM* represents a mismanaged attempt at redirecting the franchise away from *Alien 3*.

Creation of a Unique Storyworld

Throughout its intertextual network, a media franchise presents a unique storyworld. Each media extension contributes to this storyworld by featuring related elements such as locations, architecture, costumes, manners of speech, and ideological values. The *Alien* franchise, for instance, typically takes place in the future and in outer space (although some texts, such as the *Alien vs. Predator* films, are set on contemporary Earth), features a "used universe" aesthetic, and explores themes such as the relationship between humans and technology. It is not enough for the storyworld of a set of films to belong to a genre; some combination of genre elements may help to define it, however. As in the case of the *Alien* franchise, films within the same fictional universe can flexibly combine genre qualities.

The *Alien* films, in other words, have adopted characteristics of distinct genres at different times. Just as the titular creature blends biological and mechanical elements, the first *Alien* installment narratively, iconographically, and formally synthesizes the genres of science fiction (that is, an outer space setting, an oppressive corporation, an android) and horror (that is, an isolated setting, a monster that hunts down the characters one by one, low-contrast lighting). Though the tagline on the poster encapsulates this genre blending ("In space, no one can hear you scream"), the original *Alien* nonetheless left some critics divided at the time of its release. John Brosnan of science-fiction magazine *Starburst*, even declared, "[A]s a science fiction film it's seriously flawed, but as a horror film it works perfectly" (1979: 5). With James Cameron replacing Ridley Scott in the director's chair, the sequel *Aliens* fused science fiction with action-film traits (for example, military protagonists, an enemy army, a climactic confrontation with the enemy leader). Next, *Alien 3* brought the science-fiction tropes into a "prison-picture" genre format, with a setting on a penal colony planet. Though subsequent entries—including the spin-off *Alien vs. Predator* and the prequel *Prometheus*—have returned to the first film's blend of science fiction and horror, the franchise's genre flexibility has helped attract different audiences and consequently sustain long-term viability. Other venerable franchises have taken a similarly genre-blending approach, such as James Bond and the Marvel Cinematic Universe, to foster sustainability and audience appeal.[5] As *Alien* and these other examples indicate, a franchise's identity depends more on adherence to storyworld conventions than to the conventions of any particular genre.

Just as the *Alien* films have shifted genres, the franchise's video games have also alternated between and blended genres. Like many media tie-ins, some have mimicked the gameplay and aesthetics of popular titles, such as 1982's *Alien* resembling *Pac-Man* (Namco/Midway, 1980) and *Frogger* (Sega-Gremlin, 1981).[6] Other *Alien*-branded games have taken genre cues from their associated film as a narrative and stylistic framework and as a strategy for delivering an authentic genre experience to fans of their mothership film. Whereas *A:CM* follows a first-person shooter (FPS) format to align with the action-oriented *Aliens*, the survival horror *Alien: Isolation* remains faithful to the first film's creation of terror through atmosphere and moments of shock. Genre flexibility, then, extends not just to mothership texts within a franchise but also to paratexts. While the expectations of tie-ins derive from their own genre as well as from the genre of their mothership, expectations also extend from a paratext's relationship with its franchise's storyworld. *A:CM*, as we argue below, suggests how this relationship can problematize a paratext.

The Franchise Mismanagement of *Aliens: Colonial Marines*

Aliens: Colonial Marines serves as a case study in franchise mismanagement, involving multiple parties. Announced in 2006 but not released until 2013, *A:CM* received such a poor reception that consumers filed a class-action false advertising lawsuit. Beyond not meeting consumer expectations of game quality, its divergent continuity contributions also failed to meet the expectations of *Alien* franchise fans. In other words, the game shows the extent to which seemingly ancillary merchandise—such as a tie-in game—can reshape understandings of mothership texts.

Since 1982, so many *Alien*-related video games have appeared that a *Wikipedia* entry titled "List of *Alien*, *Predator*, and *Alien vs. Predator* games" has been organized into decades (Wikipedia, 2008). In addition, the franchise has influenced unrelated games as well. The website *Gamesradar* has even listed twelve elements—from sequences to characters to environmental design—that video games have borrowed from *Alien* and its successors (Houghton, 2013). This influence derives from the films' inherently ludic nature: in *Alien*, for instance, a single character must rid an isolated spaceship of an alien invader, and in *Aliens*, a squad of marines must rid an isolated settlement of an alien horde.

Owing to the franchise's longevity in, and influence on, the video game industry, it came as little surprise in December 2006 when Sega announced that it had acquired the *Alien* license and would be publishing two video games from third-party developers: a role-playing game (RPG) from Obsidian and a first-person shooter from Gearbox Software (Perry 2006). In addition to growing out of the *Alien* franchise's enduring popularity, the announcement also reflected a trend at the time of publishers pursuing media licenses due to escalating game budgets in order to stand out among their competitors. For Sega, already suffering from its recent exit out of the video game hardware business and from maligned software such as the 2006 *Sonic the Hedgehog* reboot, licenses would help fight off the domination of original franchises such as Microsoft's *Halo* (2001–present) and Rockstar's *Grand Theft Auto* (1997–present). In November 2006, Sega revealed that it had partnered with Marvel to release tie-ins for its upcoming films, beginning with *Iron Man* (Sega, 2008) (Sinclair and Boyes 2009). Sega's announcement of the *Alien* games one month later demonstrates that, instead of releasing tie-ins only for current films, publishers at this time also turned to games based on older titles, including *The Godfather* (Electronic Arts, 2006) and *Scarface: The World Is Yours* (Vivendi Interactive, 2006). Retro licenses, such as these and for *Alien*, allowed publishers to cash in on audience-tested IP for a significantly reduced price and without having to meet a film's theatrical release date. For film studios, retro tie-ins helped introduce older IP to younger audiences, which would, in theory, encourage home video sales and stimulate new merchandising opportunities (Gaudiosi, 2004: 13; Holson, 2005).

Planned for a 2009 release, *A:CM* would accompany a rising public consciousness of the *Alien* franchise and would become its first home-console video game since 2003. Also, in 2009, the original film would celebrate its thirtieth anniversary, with a commemorative "*Alien* Anthology" Blu-ray box set arriving the next year. Furthermore, after a pair of *Alien vs. Predator* films in 2004 and 2007, Ridley Scott returned to the franchise for the first time since 1979 with the 2012 prequel *Prometheus*. SEGA's announced titles would anchor themselves to the older, more popular (and, therefore, less risky from a licensing perspective) cinematic installments. That is, *A:CM* would offer an alternative *Alien* franchise text to the prequel films; as such, the *Alien* franchise adopted the concept of multiplicity, by which multiple versions of the same storyworld exist simultaneously (Uricchio and Pearson, 2015: 232).[7] Because of the diminishing domestic box office returns of the *Aliens* cinematic successors, it made sense for Sega to turn to such a highly regarded entry in the franchise (*Box Office Mojo*, 2018).[8]

Beyond just adding a subtitle to *Aliens*, *A:CM* leverages its narrative, aesthetic, and personnel associations. Narratively, the game's campaign, or story mode, serves as a direct sequel to the film, taking place seventeen weeks later. In this transmedia narrative, players control Corporal Christopher Winter, who, along with a squad of other Colonial Marines, responds to a distress call from the USS *Sulaco*. Once aboard, the Marines discover a group of xenomorphs and mercenaries from Weyland-Yutani, the same corporation featured in the films that seeks to study and profit from the creatures. Fleeing the destroyed *Sulaco*, Winter and his allies escape to LV-426, the home of the Hadley's Hope colony and the moon on which the xenomorphs had been found in the first two films. Players then must escape from LV-426 while also attempting to survive an onslaught of xenomorphs and mercenaries. Aesthetically, the development team collaborated with *Aliens* concept artist Syd Mead to replicate the film's architectural design and to create new areas of the *Sulaco* (Gaudiosi, 2013). The developers also worked with Kodak to replicate the colors of the film stock, Kodak 5294, used in *Aliens* (Stuart, 2012). In addition to Mead, actors Michael Biehn (Hicks) and Lance Henriksen (Bishop) lent their likeness and voice to the game. The score, from composer Kevin Riepl, drew inspiration from Jerry Goldsmith and James Horner's work on *Alien* and *Aliens*, respectively (Dyer, 2012).

Owing to its association with the fan favorite *Aliens*, expectations for *A:CM* remained high throughout its development—a process that would extend for nearly seven years. By 2009, Sega's financial troubles led to the cancellation of the Obsidian *Alien* RPG (Schreier, 2013). That same year saw the release not of *A:CM* but of *Borderlands* (2K Games, 2009), a Gearbox-owned IP. Gearbox's decision to have prioritized its internal project reflects a trend that began in the mid-2000s, as escalating development costs for high-definition consoles like the Sony PlayStation 3 and Microsoft Xbox 360 encouraged game companies to begin shifting away from media properties carrying licensing fees in order to focus on original IP from which they stood to make greater profits (Kohler, 2013). In addition to Gearbox attending first to *Borderlands*, Sega also delayed *A:CM*'s development to concentrate its marketing efforts on Rebellion Developments' *Aliens vs. Predator*, a reboot of the *Aliens vs. Predator* video game series that had launched in 1993 and a product whose release date could synergistically capitalize from the prerelease hype of the Twentieth Century-Fox film *Predators* (Nimród Antal, 2010) (Fahey, 2009). Though *A:CM* had been rescheduled for a fall 2012 debut, in May 2012, Gearbox President Randy Pitchford announced a further delay to February 2013. In a press statement, Pitchford claimed the game's development period was no longer than those of the studio's other projects but that it only seemed longer because Sega and 20th Century-Fox had announced the project so early, back in December 2006 (Yin-Poole, 2012).

More likely, Gearbox's mismanagement of *A:CM* had caused the additional delay. After finding critical and commercial success with *Borderlands*, the developer immediately pursued a follow-up that took precedence over the already years-in-development *Alien* title. To focus on both projects at the same time, Gearbox outsourced much of *A:CM*'s development to TimeGate Studios. When TimeGate received the assets from Gearbox in late 2010, however, the new team discovered the game to be in a surprisingly rough state. As an anonymous source involved with *A:CM* disclosed to *Kotaku*, "There was obviously not four years of work done on the game" (Schreier 2013). Between 2007 and 2010, Gearbox allegedly had prioritized its internal IP—including *Borderlands* and its 2012 sequel as well as the long-in-development *Duke Nukem Forever* (2K Games, 2011) (Schreier, 2013). According to exposés published after *A:CM*'s eventual release, TimeGate represented just one of several developers that Gearbox had subcontracted to contribute to the game. On Reddit, an individual allegedly involved with the project's development revealed:

> GBX [Gearbox] outsourced a good portion of the game to outside companies. Initially, the plan was for TimeGate to take the majority of campaign, GBX would take [multiplayer], [developers] Demiurge and Nerve would handle DLC [downloadable content] and various other focused tasks. This decision was made mostly so that most of the developers at GBX could continue working on *Borderlands 2*, while a small group of [lead developers], coders and designers dealt with [*Aliens: Colonial Marines*]. (Throwawayacm, 2013)

Despite these revelations, Pitchford has asserted that Gearbox still contributed "roughly 80 percent" to the final product (Morris, 2012). In mid-2012, as *Borderlands 2* neared the end of development, Gearbox reclaimed *A:CM* from TimeGate and "changed everything" out of dissatisfaction with the project's quality. Unwilling to ask Sega for an extension on a game already six years in the making, Gearbox chose to push on with a project in which it lacked confidence. As a source told *Kotaku*, "The game feels like it was made in nine months, and that's because it was" (Schreier, 2013). Moreover, according to the individual on Reddit, Gearbox decided to rush *A:CM* because "SEGA was pretty close to taking legal action against GBX" (Throwawayacm, 2013). While Sega and Gearbox took no legal action against one another, consumers did file a class action lawsuit against both of them for misleading advertising.

This consumer deception underscored the companies' mismanagement of the game, including their treatment of the *Alien* franchise and their promotional strategy. Leading up to its release, *A:CM* had been marketed as a significant expansion of the franchise. At the 2012 New York Comic Con, Pitchford described *A:CM* as "pure *Alien* canon" (Corriea, 2012); furthermore, in a video "developer diary," Gearbox's Chief Creative Officer Brian Martel claimed, "I think *Alien* fans, people who love this franchise, are going to be really happy with the way in which we brought this forward" (SegaAmerica, 2012). In contrast to these public relations statements, the average critic score on review aggregator *Metacritic* ranged between 43 and 48 percent across platforms; users of the site evaluated the game even more negatively, with average scores ranging from 3.5 to 3.9 (Metacritic, 2018). Reviews opined that the game's quality reflected poorly on the franchise, highlighting *A:CM*'s deviations from continuity. Kevin VanOrd of *Gamespot*, for instance, commented, "The first two *Alien* films are steeped in mystery and anxiety, qualities all but absent in developer Gearbox's lackluster interpretation" (2013), and YouTube personality "Angry Joe" (Joe Vargas) admitted, "Nobody was looking forward to this game more than I. I really, really wanted it to be good . . . So, after many years, the game is finally here, and this is going to break my heart: *Aliens: Colonial Marines* is a colossal fucking disappointment!" (AngryJoeShow, 2013). In his 33-minute video review, which counts more than five million views at the time of writing, Angry Joe goes on to call *A:CM* "borderline insulting to any fan of the franchise" (AngryJoeShow, 2013). Such comments indicate the franchise's passionate and opinionated audience.

Critics and fans have voiced their displeasure with the game's departure from the established *Alien* franchise canon. As Jonathan Gray has argued, paratexts can enrich but also reshape or even detract from our understanding

Figure 2.1 Angry Joe's review of *Aliens: Colonial Marines*.

of mothership texts within a franchise (2010). Instead of fans finding value in *A:CM*'s transmedia enrichment, many took issue with its reshaping of the franchise through deviating from its genre elements and through instances of retconning. Despite the development team in promotional material referring to the game as "scary" and "frightening," critics found *A:CM* to be decidedly lacking in horror owing to its absence of difficulty and uninspired settings (IGN, 2012). As Angry Joe noted, "The major issue with this game is that it's far too fucking easy! . . . You can completely run through entire levels in less than five minutes, ignoring aliens along the way and never firing a shot" (AngryJoeShow, 2013). A *Polygon* op-ed similarly noted, "*Aliens: Colonial Marines* is the diametric opposite of both Ridley Scott's original *Alien* and horror films in general" (Plante, 2013). *Polygon*'s piece points to the game's dark corridors that have replaced the films' seemingly safe environments and to the game having presented the xenomorphs as "harmless tin ducks in a shooting gallery" (Plante, 2013). In addition to lacking the horror aspect of the first film, the game also lacks the action of the second; according to *Polygon*, numerous glitches, imprecise aiming, and "joyless battles" break any sense of immersion (Plante, 2013). Thematically, *A:CM* departs from the films in favor of superficiality, with *Kotaku*'s review noting,

> The sophistication of the writing in the game can be boiled down to quick one-liners and occasional quote [*sic*] from the movies; there is no meaningful exploration of ideas like motherhood in the game. It comes down to you, your gun, and aliens that need to die. (Hernandez, 2013)

As these reviews suggest, an intertextual network of media extensions within a long-running franchise breeds expectations, and paratexts—already considered less central to a franchise's textual hierarchy—bear a great responsibility for meeting these expectations.

Aside from not meeting the franchise's genre and thematic expectations, *A:CM* also failed to faithfully match the films' typical level of characterization. The human characters, critics noted, appear far less developed than those of the films, despite the game's five- to six-hour story. As Angry Joe commented, "Try to describe the personality of any of the marines in the game. You can't!" (AngryJoeShow, 2013). The game also misrepresents the xenomorphs, with *A:CM* tasking players to combat not only the aliens but also Weyland–Yutani mercenaries. In fact, only three out of the campaign's eleven levels focus solely on fighting against xenomorphs. Conflicts in the films, by comparison, tend to take place between human characters and xenomorphs, along with the occasional android, rather than against other humans.[9]

These generic, thematic, and characterization discrepancies emphasize the game's mismanaged production. According to the anonymous Reddit user, problems in development negatively affected the game's story: "[T]he script was approved by 20th Century FOX [*sic*], and . . . the rush to throw a playable product together came at the cost of the story" (Throwawayacm, 2013). Several sources have disclosed that, as late as 2011, the game's script continued to undergo rewrites, leading to the discarding of entire scenes and levels (Schreier, 2013). While Gearbox's outsourcing to, and consequent lack of oversight over, numerous companies did little to ensure consistency, Sega's interference in the script and gameplay only worsened the situation. The publisher had wanted *A:CM* to emulate Activision's *Call of Duty*, a franchise whose games recently had broken sales records (Schreier, 2011; Crecente, 2011).[10] To meet Sega's mandate, the human mercenaries joined the xenomorphs as the game's adversaries; this decision prompted criticism from fans, with Angry Joe asking, "What is this? *Call of Duty: Weyland-Yutani*?" (AngryJoeShow, 2013). These discrepancies reflect Trevor Elkington's observation that licensed video games tend to involve conflicts between multiple stakeholders; in *A:CM*'s case, this included not just Fox the licensor, Sega the publisher, and Gearbox the developer but also the three companies to which Gearbox had outsourced (Elkington, 2009: 224).

Although, or perhaps because, the script went through so many changes, the retail version of *A:CM* features a story that retcons *Aliens* and its sequels. Gray has argued that paratextual video games enrich mothership texts by allowing fans to explore diegetic spaces both seen and unseen (2010). Similarly, Gearbox's Pitchford has said that *A:CM* creates "fun" not just in its FPS action but also in "feeling that sense of exploration and discovery" because "[y]ou're in the movie, and you control the camera—you're like a virtual tourist in these sets" (Corriea, 2012). Specifically, players can explore Hadley's Hope from *Aliens* and the "Derelict" spacecraft that the *Nostromo* had discovered on LV-426 in *Alien*. That the game permits exploration of these spaces contradicts the events of *Aliens*, however. At the end of that film, Ripley and crew escape LV-426 just before a thermonuclear explosion with a yield, as Bishop explains, "equal to about forty megatons." Despite this, in *A:CM*, Hadley's Hope and the nearby Derelict remain surprisingly intact. In a developer diary, Gearbox's Brian Martel justifies the decision to feature these settings that presumably would have been destroyed:

> You'll be able to explore places you are familiar with, like Hadley's Hope. It's after the atmospheric processor has blown up, so it's been very torn up. And it . . . had an explosion that explodes up, not out, but nonetheless it still had an impact on what happened to some of those areas you were familiar with. So now you get to come back in there and see them in a destroyed or distressed state that you didn't see before. (SegaAmerica, 2012)

Martel's justification implies that the developers chose to ignore the likely outcome of *Aliens* in order to set the game in "familiar" environments from the favored, first two films in the franchise while simultaneously, as a transmedia product, not merely repeating the events of those films.

In contrast to its close connections with *Alien* and *Aliens*, *A:CM* most notably breaks from the continuity of *Alien 3*. First, the plenitude of xenomorphs on LV-426 contradicts the events of *Alien 3* and *Alien: Resurrection*. Specifically, at the conclusion of *Alien 3*, Bishop II (Lance Henriksen) from Weyland–Yutani wants to extract the xenomorph queen embryo from inside Ripley; in *Alien: Resurrection*, military scientists seeking to harness the alien's power have cloned Ripley to retrieve the embryo. In *A:CM*, however, LV-426 remains infested with xenomorphs and several alien queens. Second, in an even more explicit example of retconning, *A:CM* shows that Hicks actually had not died in the opening scene of *Alien 3*. In the game's revised version of events, Hicks awakes from hypersleep after Weyland–Yutani mercenaries have overtaken the *Sulaco*. A conflict between Hicks and the mercenaries causes an electrical fire that jettisons an escape pod. The game reveals that, while the mercenaries apprehend Hicks and return him to LV-426 as prisoner, the pod that crashes on the penal colony planet from *Alien 3* carries Ripley, Newt, and—in contrast to the film—one of the mercenaries. According to Pitchford, one of the Gearbox developers had suggested this change, given that "the only evidence we have that Hicks is dead is they showed [in *Alien 3*] a charred body in an escape pod." Pitchford also admits, ironically, "Maybe it's a bad kind of retconning. At the time I thought this was fixing a mistake and doing it in a way that's absolutely plausible with the storyline" (Yin-Poole, 2015). From Pitchford's perspective, *A:CM* works to correct other "mistakes" from *Alien 3*, a film he has alternately referred to as "great" and "terrible" (Peele, 2012; Yin-Poole, 2015). The game, he has argued, provides the "bigger, wider experience" and "an escalation" missing from *Alien 3* (Correia, 2012; Yin-Poole, 2015). Believing that the threequel had created "[i]nconsistencies with the canon," Pitchford has asserted that *A:CM*'s retconning enriches, rather than detracts from, the third film: "After this game, *Alien 3* is a better film" (Peele, 2012).

Such claims to canonical superiority proved problematic because critics called out not just the game's continuity departures but also Sega's and Gearbox's deceptive marketing campaign. At trade, press, and fan events, such as the 2011 Electronic Entertainment Expo, Sega and Gearbox had featured video and timed-playable previews of *A:CM*. As some critics pointed out, these promotional demos did not match the final product; *Polygon*'s review even featured a gallery highlighting differences between the retail version of *A:CM* and a 2012 walkthrough that Gearbox had created (Gies, 2013). In April 2013, two months after the game's release, consumers filed a class-action

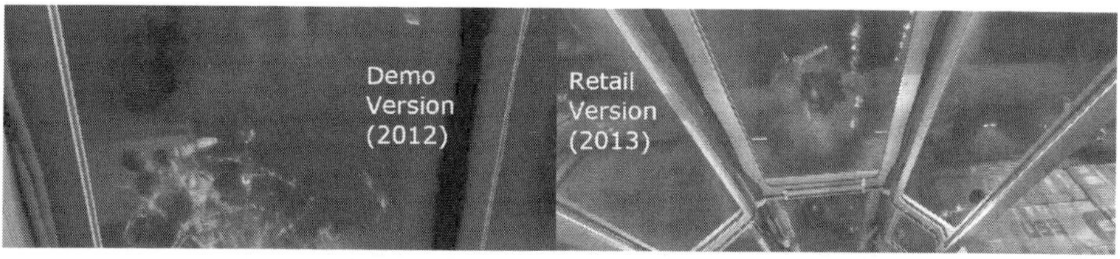

Figure 2.2 The *Aliens: Colonial Marines* demo vs. the final product.

lawsuit against Sega of America and Gearbox for having falsely advertised *A:CM* with unrepresentative demos. As reviewers and the lawsuit contended, the final product differed in visual fidelity, character behavior, and level design (Crecente, 2014b).

Significantly, the lawsuit drew attention to manipulative practices that have remained relatively commonplace in the video-game industry. An embargo from Sega stipulated that no reviews could be posted until the morning of the game's February 12, 2013 release. Because of the review embargo, consumers who had preordered the game or purchased it on launch day lacked knowledge of how drastically the game differed from its previews (McElroy, 2013). In response to the game's unsatisfactory launch state, individuals involved in *A:CM*'s development and journalists contextualized the inherent deception that game companies perform in the lead up to a game's release. A source at TimeGate revealed that the *A:CM* demos ran "on a high-end computer with specs that would be unfeasible for a normal console game" (Schreier, 2013). In other words, because the demos surpassed the visual capabilities of available PCs and home consoles, the developers argued that they needed to downgrade the performance of the consumer version. In a comment posted to *Kotaku*'s review of *A:CM*, Jason Schreier noted that the promotional imperatives and restrictive context of preview events tend to encourage misrepresentation:

> [A] preview is based on a heavily controlled slice of a game that is designed to make it look as good as possible. It's hard to really judge what a game will be like when you're sitting (or standing) with a 20-minute demo while a PR person stands over your shoulder and talks you through everything. (Hernandez, 2013)

The misleading nature of video-game previews adversely affects consumers, especially those already loyal to the franchise associated with the previewed game.[11]

As a result of having deceived consumers, Sega, in August 2014, agreed to a $1.25 million settlement. According to reports, Sega had settled not out of an admission of guilt but rather out of a desire to avoid further legal fees. While the settlement cleared Sega, Gearbox had, a month earlier, filed a motion to be dropped from the case, describing itself as an independent contractor and therefore as not liable (Crecente, 2014b). Though Sega formally charged Gearbox as equally responsible for the game's marketing and promotional strategy, the judge presiding over the case, in May 2015, dismissed Gearbox from the suit and changed the suit from a class action to representing only the two consumers who had originally filed the claim (Crecente, 2014a; Crecente, 2015). In an interview published two months after Gearbox's dismissal, Pitchford admitted to having lost "between $10–$15m" on *A:CM* but that his fandom for the franchise had made the experience worthwhile:

> I wouldn't trade it for the world, because you know what I got to do? I got to work with Syd Mead. We got to recreate the *Sulaco*. I got to play with Stan Winston's original prop of the Alien queen. I got to fucking touch it and interact with it and use it. We got to rebuild Hadley's Hope with Syd Mead's direction, for all the rooms that were never built in the sets, all the rooms that were never shot on film. We got to do that. (Yin-Poole, 2015)

While Pitchford may have felt enriched by his involvement with the *Alien* franchise, the franchise did not fare as well.

Conclusions

In the years following the release of *Aliens: Colonial Marines*, the franchise has made several course corrections. While Twentieth Century-Fox halted development on Neill Blomkamp's film that would have ignored the events of *Alien 3* and *Alien: Resurrection*, the studio did move forward with Ridley Scott's *Alien: Covenant*. Through its title and off-screen death of *Prometheus*'s protagonist (Elizabeth Shaw, played by Noomi Rapace), the film attempts to break from its predecessor and to meet more fully the expectations of *Alien* franchise fans. Furthermore, in 2014, Sega and developer Creative Assembly debuted *Alien: Isolation*, a video game follow-up to the first film, with players taking on the role of the Ripley's daughter. Unlike those of *A:CM*, *Isolation*'s reviews praised the accurate adaptation of the franchise's defining elements; *Games™*, for example, rated the game a "nine" out of ten and noted, "The highest compliment that you can pay Creative Assembly's uncompromising, nerve-plucking and lengthy entry in the *Alien* canon is that even without any prior knowledge of its source material, *Alien: Isolation* is superb" (Games™, 2014). This statement suggests that, because it stands on its own while also enriching the franchise, *Alien: Isolation* represents a model for transmedia storytelling and therefore contrasts with *A:CM*'s franchise mismanagement.

The *Alien* franchise in general and *A:CM* in particular suggest that media franchises place immense pressure on individual texts in having to meet expectations; this becomes especially relevant in transmedia configurations. Daniel Herbert argues that *Star Wars: Episode VII—The Force Awakens* (J. J. Abrams, 2015) "may be a single film, but it held the responsibility of representing and supporting an entire media franchise" (2017: 13). *A:CM* shows that paratexts can hold the same responsibility. To contribute valuably to their respective franchise, mothership texts and paratexts tend to situate themselves in relation to fan- and critic-favored works. Doing so suggests a franchise's development over time and its enduring popularity. This becomes especially true with belated extensions of celebrated franchise texts, such as *Alien* and *Aliens*. Failure to meet expectations, as in the case of *A:CM*, tends to intensify distinctions between mothership texts and paratexts within franchises, while successes, such as *Alien: Isolation*, tend to blur such hierarchies.

Just as it had introduced transmedia storytelling as a more cohesive form of franchise management, the digital era has also increased the level of pressure placed on franchise texts. As the situation of *A:CM* demonstrates, the process of scrutinization began with Sega's December 2006 announcement, continued throughout its protracted development and previews, and finally culminated in the vitriol that greeted its February 2013 release. Every step in this process had been examined, raising issues of how *A:CM* would compare with other *Alien* games on one level and to the wider *Alien* franchise on another. While *A:CM* met derision from critics and from fans, subsequent texts have helped maintain the *Alien* franchise. As it enters into its fortieth anniversary and as its ownership changes hands from Twentieth Century-Fox to Disney, it remains to be seen how the *Alien* franchise will reconcile the deviations of its multimedia and transmedia past (Warner and Fletcher, 2018).

Notes

1. Among its changes to the theatrical version, the *Alien 3* assembly cut replaces a dog with an ox as the xenomorph's host.
2. For more on *Alien: Descent*, see James Fleury's chapter in this book.
3. According to speculation in media industry trades and movie blogs, Blomkamp's film had been cancelled in favor of *Alien: Covenant* because of Ridley Scott's comparatively higher industrial status compared with Blomkamp, who had most recently directed the critically and commercially disappointing *Chappie* (Neill Blomkamp, 2015).
4. Other examples of media franchises with branching continuities include *Superman* (1978–2006), *Highlander* (1986–2007), and *X-Men* (2000–present).
5. Certain Bond films have combined spy thriller elements with those of other genres, including martial arts in *The Man with the Golden Gun* (Guy Hamilton, 1974) and science fiction in *Moonraker* (Lewis Gilbert, 1979). Likewise, the Marvel Cinematic Universe has expanded its superhero conventions into other genres, such as the high school comedy in *Spider-Man: Homecoming* (Jon Watts, 2017) and the spy thriller in *Black Panther* (Ryan Coogler, 2018).

6. To meet tight development deadlines, tie-in video games tend to follow conventions established in preexisting, popular titles instead of establishing their own. Examples include *The Lawnmower Man* (THQ, 1993) emulating the platformer-shooting mechanics of *Contra* (Konami, 1987), *The Simpsons: Hit and Run* (Vivendi Universal Games, 2003) mimicking the open world of *Grand Theft Auto III* (Rockstar Games, 2001), and *Despicable Me: Minion Rush* (Gameloft, 2013) following in the footsteps of endless runner *Temple Run* (Imangi Studios, 2011).
7. As Uricchio and Pearson explain, contemporary media franchises rely on multiplicity as a management strategy: "The multiplicity strategy perfectly suits a twenty-first century mediascape characterized by marked industrial and technological convergence . . . Producers seek to maximize revenue from their intellectual property by creating and spreading new iterations while at the same time continuing to circulate old iterations as widely as possible." The authors examine the multiplicity of the Batman franchise, one in which the satirical, animated *The LEGO Batman Movie* (Chris McKay, 2017) appeared the same year as more mature versions of the Batman character in the video game *Injustice 2* (Warner Bros. Interactive Entertainment, 2017) and the film *Justice League* (Zack Snyder, 2017).
8. According to *Box Office Mojo*, until *Prometheus*, only 2004's *Alien vs. Predator* generated more money than the film that had preceded it.
9. The films do feature some human antagonists, notably the duplicitous Carter Burke (Paul Reiser) in *Aliens*.
10. Both *Call of Duty: Black Ops* (Activision, 2010) and *Call of Duty: Modern Warfare 3* (Activision, 2011), for instance, set records for launch-day sales figures.
11. Despite consumer manipulation leading to the lawsuit against Sega and Gearbox, game companies have not halted misleading promotional practices. In October 2016, Bethesda Softworks instituted a policy of not providing review copies until a day before one of their games hits retail (Kuchera, 2016). As with *A:CM*, this means that consumers who preorder a particular game have little opportunity to learn of its quality prior to purchase; also as with *A:CM*, this most adversely affects fans whose passion for a franchise or new title would inspire them to preorder in the first place.

References

AngryJoeShow (2013), "*Aliens: Colonial Marines* Angry Review", *YouTube*, 14 February, <https://www.youtube.com/watch?v=UGX2WE4QUw8> (last accessed 6 June 2018).

Bishop, Bart (2016), "Game Over, Man!: The Love/Hate Relationship Between 'Aliens' and its Tie-In Books," *Litreactor*, 14 April, <https://litreactor.com/columns/the-love-hate-relationship-between-aliens-and-its-tie-in-books> (last accessed 6 June 2018).

Box Office Mojo (2018), "Franchises: Alien," *Box Office Mojo*, 1 May, <http://www.boxofficemojo.com/franchises/chart/?id=alien.htm> (last accessed 6 June 2018).

Brosnan, John (1979), "Illustrated Review of Alien," *Starburst*, 14 (October), pp. 4–6.

Christensen, Jerome (2012), *America's Corporate Art: The Studio Authorship of Hollywood Motion Pictures*, Stanford, CA: Stanford University Press.

Corriea, Alexa Ray (2012), "Aliens: Colonial Marines Fills In Story Gaps Left by Alien 3, Says Gearbox," *Polygon*, 14 October, <https://www.polygon.com/2012/10/14/3500562/aliens-colonial-marines-fills-in-story-gaps-left-by-alien-3-says> (last accessed 6 June 2018).

Crecente, Brian (2011), "*Call of Duty: Modern Warfare 3* Shatters All Sales Records, Nears Half a Billion Dollars in Day One Sales," *Kotaku*, 11 November, <https://

kotaku.com/5857400/call-of-duty-modern-warfare-3-shatters-all-sales-records-surprises-no-one> (last accessed 6 June 2018).
Crecente, Brian (2014a), "Sega Blames Gearbox for Mismanaging Aliens: Colonial Marines Marketing," *Polygon*, 3 September, <https://www.polygon.com/2014/9/3/6102035/sega-gearbox-lawsuit-aliens-colonial-marines> (last accessed 6 June 2018).
Crecente, Brian (2014b), "Sega to Tentatively Pay Out $1.25M in Aliens Suit while Gearbox Fights On," *Polygon*, 11 August, <https://www.polygon.com/2014/8/11/5993509/aliens-colonial-marines-class-action-settlement> (last accessed 6 June 2018).
Crecente, Brian (2015), "Aliens: Colonial Marines Lawsuit Loses Class Status, Gearbox Dropped from Suit," *Polygon*, 28 May, <https://www.polygon.com/2015/5/28/8683827/aliens-colonial-marines-lawsuit-loses-class-status-gearbox-dropped> (last accessed 6 June 2018).
Dyar, Amanda (2012), "Exclusive Interview: Composer Kevin Riepl Discusses Aliens Colonial Marines and Resistance: Burning Skies," *Dread Central*, 9 May, <http://www.dreadcentral.com/news/33564/exclusive-interview-composer-kevin-riepl-discusses-aliens-colonial-marines-and-resistance-burning-skies/> (last accessed 6 June 2018).
Elkington, Trevor (2009), "Too Many Cooks: Media Convergence and Self-Defeating Adaptations," in Bernard Perron and Mark J. P. Wolf (eds), *The Video Game Theory Reader 2*, New York: Routledge, pp. 213–35.
Fahey, Mike (2009), "Aliens Fight Predators Again in 2010," *Kotaku*, 11 February, <https://kotaku.com/5151456/aliens-fight-predators-again-in-2010> (last accessed 6 June 2018).
Games™ (2014), "Alien: Isolation Review" (2014), Games™, 3 October, <https://www.gamestm.co.uk/reviews/alien-isolation-review/> (last accessed 6 June 2018).
Gaudiosi, John (2004), "Game Makers Mining H'wood's Past," *The Hollywood Reporter*, 14-20 September, p. 13.
Gaudiosi, John (2013), "Visual Futurist Syd Mead Discusses 'Aliens: Colonial Marines' and Video Games," *The Hollywood Reporter*, 22 January, <https://www.hollywoodreporter.com/news/visual-futurist-syd-mead-discusses-414326> (last accessed 6 June 2018).
Gies, Arthur (2013), "Aliens: Colonial Marines Review: The Worse Species," *Polygon*, 12 February, <https://www.polygon.com/2013/2/12/3978652/aliens-colonial-marines-review-the-worse-species> (last accessed 6 June 2018).
Gray, Jonathan (2010), *Show Sold Separately: Promos, Spoilers, and Other Media Paratexts*, New York: New York University Press.
Herbert, Daniel (2017), *Film Remakes and Franchises*, New Brunswick, NJ: Rutgers University Press.
Hernandez, Patricia (2013), "*Aliens: Colonial Marines*: The *Kotaku* Review," *Kotaku*, 12 February, <https://kotaku.com/5983463/aliens-colonial-marines-the-kotaku-review> (last accessed 6 June 2018).
Holson, Laura M. (2005), "Blockbuster With a Joystick," *The New York Times*, 7 February, <http://query.nytimes.com/gst/fullpage.html?res=9A07E6DC1E3BF934A35751C0A9639C8B63&pagewanted=all> (last accessed 6 June 2018).
Houghton, David (2013), "12 Big Ways that Video Games Have Ripped Off the Alien Movies," *Gamesradar*, 11 February, <https://www.gamesradar.com/12-big-ways-that-video-games-have-ripped-off-the-alien-movies/> (last accessed 6 June 2018).
IGN (2012), "Aliens: Colonial Marines Developer Diary—Authentic Aliens," *YouTube*, 1 August, <https://www.youtube.com/watch?v=VdKsDgQ7Wh0> (last accessed 6 June 2018).

Jenkins, Henry (2006), *Convergence Culture: Where Old and New Media Collide*, New York: New York University Press.
Johnson, Derek (2013), *Media Franchising: Creative License and Collaboration in the Culture Industries*, New York: New York University Press.
Kohler, Chris (2013), "Why Games Based on Movies Disappeared (and Why They're Coming Back)," *Wired*, 15 February, <https://www.wired.com/2013/02/movie-games/> (last accessed 6 June 2018).
Kuchera, Ben (2016), "Bethesda Wants Your Money Before the Reviews Hit," *Polygon*, 25 October, <https://www.polygon.com/2016/10/25/13409774/bethesda-review-policy-dont-preorder> (last accessed 6 June 2018).
Lincoln, Ross A. (2015), "Neill Blomkamp-Directed 'Alien' Film 'Holding' in Favor of 'Prometheus 2'," *Deadline Hollywood*, 29 October, <http://deadline.com/2015/10/neill-blomkamp-aliens-sequel-set-aside-for-prometheus-2-1201598007/> (last accessed 6 June 2018).
McElroy, Griffin (2013), "Gearbox and Sega Falsely Advertised Aliens: Colonial Marines with Press Demos, According to Lawsuit," *Polygon*, 30 April, <https://www.polygon.com/2013/4/30/4287382/aliens-colonial-marines-lawsuit-class-action-sega-gearbox> (last accessed 6 June 2018).
Metacritic (2018), "Aliens: Colonial Marines," *Metacritic*, 28 March, <http://www.metacritic.com/game/xbox-360/aliens-colonial-marines> (last accessed 6 June 2018).
Morris, Chris (2012), "Gearbox's Pitchford on Aliens' Evolution," *Gamasutra*, 4 June, <http://www.gamasutra.com/view/news/171665/Gearboxs_Pitchford_on_Aliens_evolution.php> (last accessed 6 June 2018).
Peele, Britton (2012) "The Acid-Blooded Devil Is In the Details for Aliens: Colonial Marines," *Engadget*, 4 April, <https://www.engadget.com/2012/04/04/the-acid-blooded-devil-is-in-the-details-for-aliens-colonial-ma/> (last accessed 6 June 2018).
Perry, Douglass C. (2006), "Obsidian, Gearbox and Aliens, Oh My!," *IGN*, 12 December, <http://www.ign.com/articles/2006/12/12/obsidian-gearbox-and-aliens-oh-my> (last accessed 6 June 2018).
Plante, Chris (2013), "Opinion: How Aliens: Colonial Marines Is Horror, Inverted," *Polygon*, 14 February, <https://www.polygon.com/2013/2/14/3988656/how-aliens-colonial-marines-is-horror-inverted> (last accessed 6 June 2018).
Schreier, Jason (2010), "Call of Duty: Black Ops Smashes Day-One Sales Record," *Wired*, 11 November, <https://www.wired.com/2010/11/black-ops-sales/> (last accessed 6 June 2018).
Schreier, Jason (2013), "From Dream To Disaster: The Story Of *Aliens: Colonial Marines*," *Kotaku*, 25 February, <https://kotaku.com/5986694/from-dream-to-disaster-the-story-of-aliens-colonial-marines> (last accessed 6 June 2018).
SegaAmerica (2012), "Aliens: Colonial Marines Making Of—Dev Diary 3," *YouTube*, 14 August, <https://www.youtube.com/watch?v=hSPqgDtM0lU> (last accessed 6 June 2018).
Shirey, Paul (2018), "Neill Blomkamp on the Future of Oats Studios & Funding a Firebase Sequel," *JoBlo*, 11 April, <http://www.joblo.com/movie-news/ex-neill-blomkamp-on-the-future-of-oats-studios-funding-a-firebase-sequel-190> (last accessed 6 June 2018).
Sinclair, Brendan and Emma Boyes (2009), "Aliens to Spawn on Next-Gen Systems," *Gamespot*, 31 March, <https://www.gamespot.com/articles/aliens-to-spawn-on-next-gen-systems/1100-6162961/> (last accessed 6 June 2018).
Stuart, Keith (2012), "Aliens Colonial Marines—Preview," *The Guardian*, 25 April, <https://www.theguardian.com/technology/gamesblog/2012/apr/25/aliens-colonial-marines-preview> (last accessed 6 June 2018).
Throwawayacm (2013), "A Lot of You Are (Rightfully) Upset at the Final Product that

is A:CM. Maybe I can Shed Some Light as to How It Got the Way It Is," *Reddit*, 13 February, <https://www.reddit.com/r/LV426/comments/18ewf4/a_lot_of_you_are_rightfully_upset_at_the_final/> (last accessed 6 June 2018).

Uricchio, William and Roberta E. Pearson (2015), "I'm Not Fooled by That Cheap Disguise," in Roberta Pearson, William Uricchio, and Will Brooker (eds), *Many More Lives of the Batman*, New York: Palgrave, pp. 205–36.

VanOrd, Kevin (2013), "Aliens: Colonial Marines Review," *Gamespot*, 12 February, <https://www.gamespot.com/reviews/aliens-colonial-marines-review/1900-6403683/> (last accessed 6 June 2018).

Warner, Sam and Rosie Fletcher (2018), "Ridley Scott Wants to Make More Alien Movies with Disney—and He Already Knows What He Wants to Do with the Next One," *Digital Spy*, 5 January, <http://www.digitalspy.com/movies/alien/news/a846768/ridley-scott-wants-more-alien-movies-with-disney/> (last accessed 6 June 2018).

Wikipedia (2008), "List of *Alien*, *Predator*, and *Alien vs. Predator* Games," *Wikipedia*, 28 February <https://en.wikipedia.org/wiki/List_of_Alien,_Predator_and_Alien_vs._Predator_games> (last accessed 6 June 2018).

Yin-Poole, Wesley (2012), "Aliens: Colonial Marines Delayed to 2013, New Trailer Released," *Eurogamer*, 21 May, <https://www.eurogamer.net/articles/2012-05-21-aliens-colonial-marines-delayed-to-2013-new-trailer-released> (last accessed 6 June 2018).

Yin-Poole, Wesley (2015), "An Hour with Randy Pitchford," *Eurogamer*, 24 July, <https://www.eurogamer.net/articles/2015-07-23-an-hour-with-randy-pitchford> (last accessed 6 June 2018).

3. EVIL SPAWN OR GOOD BUSINESS? NEW LINE CINEMA, *CRITTERS*, AND FILM FRANCHISING AT THE MARGINS

Daniel Herbert

Introduction

The narrative of *Gremlins* (Joe Dante, 1984) could be read as an allegory of the movie's industrial identity as a transmedia franchise built upon a studio-authored film (Christensen, 2011: 14).[1] The film tells the story of a man who gives his son a small furry pet, called a "mogwai." The son has been given strict instructions about how to care for the little guy, whom they name "Gizmo," including not exposing him to daylight, not getting him wet, and not feeding him after midnight. Nevertheless, a friend of the boy spills water on Gizmo, at which point puffs of fur spring from the animal's back, before each grows into another fully formed mogwai. When the son shows the litter of furry animals to his father, the older man says, entrepreneurially, "You know what? I'll bet every kid in America would like to have one of these."

There was, in fact, an attempt to sell merchandise based on *Gremlins*, including cute little mogwai dolls, simultaneous with the film's release in theaters. Stores everywhere carried *Gremlins*-related products in the summer of 1984, making it look as if one simply just had to add water to generate a transmedia franchise overnight. Within the film itself, these little creatures transform into the malevolent gremlins of the film's title after tricking their owner into feeding them after midnight. Mayhem ensues as the gremlins propagate and terrorize the small town within the movie. Thus, if the narrative of *Gremlins* is an allegory for its own franchised existence, then this allegory suggests that the intentional, coordinated creation of consumer merchandise based on a film is something to be feared. From this perspective, licensed and branded toys, lunchboxes, backpacks, and so on, appear as evil spawn running rampant across the media and retail landscape of America.

Gremlins is just one example of franchised media production occurring in the 1980s. As Derek Johnson has shown, it was during this decade that franchising developed as a coherent strategy for media production, one that has continued unabated until today (2013: 6). As he states, the idea of franchising can be used within the context of film and media, "to explain the multiplied replication of culture from intellectual property resources" (2013: 6). From the 1980s onward, the mainstream American media industry has focused heavily on creating franchises, where an underlying intellectual property (IP) is manifested in any number of movies, television programs, toys, and other consumer products.

Given the apparent ubiquity of franchises in today's media culture, it is important to understand how these industrial practices came about historically. The recycling and reuse of existing IP goes back much farther than the 1980s. Remakes are as old as cinema itself, for instance, and film series and serials appear not long after that. Similarly, transmedia properties appear throughout much of the twentieth century (Santo, 2015: 6–10). Derek Johnson does not discount this history but, rather, shows how many of these practices came to be understood and industrially configured as "franchising" in the 1980s (2013: 21). Crucially, Johnson shows how the discursive, industrial, and cultural shift toward media franchising in the 1980s derives strongly from the franchising practices of the retail and restaurant industries, such as with McDonald's, in the 1960s (2013: 21, 33–41).

I do not want to reject any of the existing scholarship about media licensing or franchising that largely situates franchising as the business of the mainstream media industry. Instead, in this chapter, I aim to complement and further nuance our conception of media franchising by calling attention to another of its "generative precursors" (Santo, 2015: 9). Specifically, I argue that the textual recycling typical of marginal, independent film companies, which often traded in exploitation or horror movies, anticipated and had an impact on the franchising practices of the larger, mainstream Hollywood industry. To do so, I shall examine New Line Cinema's *Critters* film series (1986–92), which, like *Gremlins* before it, depicted hordes of little rascally creatures terrorizing a small community in a horror–comedy generic hybrid. The *Critters* films present a privileged case for understanding media franchising from the industrial and cultural "margins." As Johnson himself acknowledges, New Line Cinema innovated franchising through their sequalization of the *Nightmare on Elm Street* films in the 1980s and the *Teenage Mutant Ninja Turtles* movies in the 1990s (2013: 76). The *Critters* movies were produced alongside these other series, yet they never achieved the same popularity or financial success nor did they generate the transmedia blitz of *Gremlins*. Nevertheless, the *Critters* series demonstrates New Line's larger industrial ambitions and practices which largely resembled the franchise logic that was becoming a primary strategy of the major Hollywood studios during the same period.

New Line Cinema, bc—Before *Critters*

As a company, New Line Cinema is remarkable for at least two reasons: first, for its remarkable growth and financial success over a significant period of time, especially when compared with other independent companies operating during the same period; and second, for the diverse range of films, genres, and franchises that the company produced and/or distributed, which, over the life of the company, included such films as *Pink Flamingos* (John Waters, 1972), *House Party* (Reginald Hudlin, 1990), and the *Lord of the Rings* trilogy (Peter Jackson, 2001–03). Justin Wyatt has written a brief but critically insightful account of the company through to the mid-1990s, but he does not account fully for how New Line capitalized on eclecticism or the place of the horror film genre, including the *Critters* films, within its overall story (1998). In fact, based on the company's activities in the 1960s and 1970s, it would have been very hard to guess that New Line would release something like *Critters*, much less something so mainstream as the *Lord of the Rings* films.

New Line was founded by Robert Shaye in 1967, which was quietly announced in a tiny blurb at the bottom of a page in *Variety* ("Pictures: New Line Cinema Corp"). For the next several years, New Line operated as a small "non-theatrical" film distributor, primarily bringing foreign films and other unconventional fare to college campuses. The company released the Czech film *Martyrs of Love* (Jan Nemec, 1967) in 1969, for instance, and, in his review of the film, Vincent Canby noted, "The picture is the first release of New Line Cinema, a small independent company whose aim is to handle movies that other distributors wouldn't touch with a pole of any imaginable length" (1969). Shaye and the company used the oddness of its releases and the company's outsider status as assets, presenting itself in news articles and advertisements as a cutting-edge company that delivered movies more interesting to young audiences than those of Hollywood ("New Line Shaye Doubts Old-Line Showmen Re Kids"). The company programmed films from numerous countries, individually or as packages, and had its first minor "hit" with the Jean-Luc Godard film *Sympathy for the Devil* (1968), which made respectable money on campuses like Hunter College ("Display ad 361"), Cornell, Harvard, and the University of North Carolina (Canby 1970).

New Line continued to cultivate a "renegade" industrial identity when the company released Louis Gasnier's 1936 anti-marijuana exploitation film *Reefer Madness* in 1972 ("This 1936 Unexpurgated Classic from New Line

Cinema") and soon followed this with the notorious midnight movie *Pink Flamingos* ("Arty & 'Freak' Films Packaged to Theatre Trade by New Line; Sex, Too"). New Line's activities with these films and others during the mid-1970s, however, anticipate the company's more mainstream business and marketing practices in the 1980s. Specifically, during the 1970s, New Line displayed an ability to expand the range of its catalogue into new, marginal cinematic genres with dedicated fan bases. In this case, that meant continuing to release foreign art films (the company even released the film *Get Out Your Handkerchiefs* [Bertrand Blier, 1978], which won the Academy Award for Best Foreign Language Picture) while expanding into erotica, LGBT dramas, and more exploitation films ("New Line Cinema Forms Exploitation Film Unit").

Eric Schaefer has explained how, following World War II, foreign art films and American exploitation films occupied an overlapping social and cultural space as independent distributors and exhibitors traded in both types of film (1999: 330–7), while Joan Hawkins has discussed how the conflation of high and low genres, of art and exploitation cinemas, has continued through various sites of "paracinematic" movie culture (2000: 3–28). And, indeed, New Line was not alone in distributing a mixed slate of non-mainstream films in the 1970s; Roger Corman's New World Pictures, to take just one example, released a range of exploitation films, such as the "nurse" film *The Student Nurses* (Stephanie Rothman, 1970), the female prisoner movie *The Big Doll House* (Jack Hill, 1971), cheap monster movies such as *Piranha* (Joe Dante, 1978), and the low-budget sci-fi film *Death Race 2000* (Paul Bartel, 1975), at the same time that the company distributed foreign art films such as *Amarcord* (Federico Fellini, 1973), *Autumn Sonata* (Ingmar Bergman, 1978), and *Dersu Uzala* (Akira Kurosawa, 1975) (Cook, 2002: 328). The point is thus not that New Line Cinema was unique in the 1970s but rather that the company was in some ways typical of marginal, independent film distributors at the time that catered in any non-Hollywood product with market potential. New Line's activities in the 1970s would appear more exceptional, however, as the company later moved increasingly into the mainstream.

The other tactic that New Line used in the 1970s, and which had a significant impact on their practices in the 1980s and beyond, was its cultivation and reexploitation of IP in the form of film series and sequels. Following the success of *Reefer Madness*, New Line followed up by releasing the similar anti-drug film *Cocaine Fiends* (1937, released as *The Pace that Kills* in 1935) ("Motion Pictures Rated by the Code & Rating Administration"), and the anti-VD film *Sex Madness* (1938) ("Display Ad 37—no title"). Similarly, when *Pink Flamingos* proved to be a midnight movie hit, New Line continued releasing John Waters films, including *Female Trouble* (1974) and *Desperate Living* (1977), through the 1970s and beyond. Though none of these examples constitutes film series or sequels in a traditional sense, New Line helped organize them as such by marketing them in relation to one another over the course of their releases, such as in one advertisement for *Sex Madness* that stated, "From New Line Cinema, the gang that brought you 'Reefer Madness'" ("Display Ad 37—no title").

Perhaps the stronger example of New Line's nascent attempts at creating film series as protofranchises is their release of several *Street Fighter* films in the mid- to late 1970s. The market for East Asian martial arts action films was considerable in the early and mid-1970s, built significantly upon the success of the films starring Bruce Lee. As David Cook has written, "By early 1974, every major distributor but Fox and United Artists had picked up one or more 'chop socky' films, as *Variety* called them" (2002: 256). New Line exploited the development of this cycle of films when the company released the Japanese karate film *The Street Fighter* (Shigehiro Ozawa, 1974), starring Sonny Chiba, in 1974, which the small company even promoted with television advertising ("Pictures Grosses: 'Street Fighter' Bows with Heavy TV Drive"; "'The Street Fighter' is Set for 4-City Premiere"). Though there were signs that the market for "chop socky" films began to wane in the mid-1970s (Verrill, 1975), New Line released *Return of the Street Fighter* (Shigehiro Ozawa, 1974) in the fall of 1975 ("Display ad"). The sequel did more modest business than the original, yet it was also fairly successful ("'Fighter' Stout 27G, Philly; 'Tango' 4G"). The company followed these two films with yet another in the series, *Sister Street Fighter* (Kazuhiko Yamaguchi, 1974), in 1976 ("Miscellaneous"). By the time New Line released *The Street Fighter's Last Revenge* (Shigehiro Ozawa, 1974) in 1979, the company showed that it was dedicated to cultivating film sequels and series as a means of generating consistent revenues. Once again, though, New Line was not unique in this regard, as many cult action films were similarly franchised in this manner during the 1970s, such as with the multiple *Shaft* movies (Gordon Parks et al., 1971–73), and Hollywood studios released more mainstream film series, such as United Artists's James Bond films. Rather, this history suggests that New Line developed and used the strategy of reexploiting existing IP precisely because the company operated at the margins of the movie business and this strategy allowed the company to grow as time went on.

The Horror: Franchising Freddy

Perhaps more than any other genre or cycle, horror films appeared particularly prone to sequelization in the 1970s and 1980s. The blockbuster success of films such as *The Exorcist* (William Friedkin, 1973), *The Omen* (Richard Donner, 1976), and *The Amityville Horror* (Stuart Rosenberg, 1979), for instance, provoked the quick turnaround of sequels after only a few years of each of those films. Relatedly, Richard Nowell has detailed the industrial conditions that prompted the generation (and temporary decline) of the "teen slasher" film cycle that proliferated in the late 1970s and early 1980s, including films such as *Halloween* (John Carpenter, 1978) and *Friday the 13th* (Sean S. Cunningham, 1980) (2011a). Sequels were a definitive element of the slasher genre as it moved forward, with four additional *Halloween* and seven more *Friday the 13th* films appearing before the end of the 1980s. Scholar Paul Wells has gone so far as to refer to this sequelization as a kind of "McDonaldisation" of the horror genre where these cultural products have become routinized, rationalized, and evacuated of meaningful substance through the process of their being serialized for commercial gain (2000: 94).

By invoking McDonald's, Wells prompts us to consider how the sequelization of horror films in the 1970s and 1980s aligned with broader efforts to create film franchises by more mainstream media companies during the same period. Although Hollywood developed sequels to a number of blockbuster hits during this period, including such notable cases as the *Star Wars* (George Lucas et al., 1977–present), *Indiana Jones* (Steven Spielberg, 1981–present), and *Superman* franchises (Richard Donner et al., 1978–2006), the major studios largely refrained from producing or distributing horror films (Kendrick, 2014: 311). Thus, as James Kendrick notes, "the majority of horror films were therefore handled by smaller, independent production and distribution companies such as New Line Cinema," among others (2014: 311). This suggests that there were two parallel forms of reexploiting IP at the time and that the role of horror films and the impact of horror sequelization have been overlooked within the larger historiography of media franchising. New Line Cinema was in the middle of the historical transformation toward transmedia franchising within the media industry, just as the company was in the middle of its own transformation from a "marginal" into a "mainstream" media company in the 1980s, and the company's horror films were central to both of these transformations. Indeed, for Wells, New Line's *Nightmare on Elm Street* films epitomize the "McDonaldisation" of horror (2000: 94). But, whereas Wells suggests this results in a degradation of the genre, it seems more germane here to examine how such horror sequels bolstered New Line Cinema and helped innovate film franchising more generally.

New Line got its first experience handling horror cinema in the 1970s when the company handled a reissue of *Night of the Living Dead* (George A. Romero, 1968) in 1976 ("Pictures Grosses: 'Monsters' Sock $13,000, Cleve"), and catalogs for the company indicate that it continued to distribute the film until at least 1979 ("New Line Cinema Entertainment Catalog 4": 15). New Line also handled a rerelease of *The Texas Chainsaw Massacre* (Tobe Hooper, 1973) in 1980 after the original distributor, Bryanston, lost control of the film in a legal battle (Hughes, 1980). New Line made considerable money with the film, releasing it just as the teen slasher cycle, which was at least partially indebted to *Texas Chainsaw*, was reaching a critical mass ("'The Texas Chainsaw Massacre'"). The company fully entered the horror film market with three more releases in the early 1980s: an original production, *Alone in the Dark* (Jack Sholder, 1982), the sci-fi splatter film *Xtro* (Harry Bromley Davenport, 1983), and *The Evil Dead* (Sam Raimi, 1981), which found considerable success owing to the notorious reputation it earned prior to its theatrical release ("Pictures: New Line Picks Up Bloody 'Evil Dead'"). Thus, by the time New Line released Wes Craven's *A Nightmare on Elm Street* in November 1984, the company already had considerable experience in the low-budget horror film market and the young audiences these films typically attracted.

As history has shown, the first *Nightmare* film was something of a blockbuster, earning significant money at the box office and launching New Line into a new realm of industrial power. The company promoted the film heavily prior to and during its release. As Richard Nowell has argued, the wave of teen slasher films of the early 1980s was propelled by inventive marketing and advertising, especially poster ads, and he asserts that New Line, in particular, was innovative by "adopting the synergetic marketing tactics used for contemporaneous Hollywood blockbusters" (Nowell, 2011b: 65).[2] More particularly, New Line promoted the *Nightmare* films "as youth event pictures" (2011b: 65). As Nowell states, New Line was especially savvy in differentiating *A Nightmare on Elm Street* from previous teen slasher films by promoting the film as "female-friendly, middle-class-centered," and more similar to something like *Fast Times at Ridgemont High* (Amy Heckerling, 1982) than, say, *Halloween* (2012: 79–80). New Line thus treated the first *Nightmare* film as something bigger and better than other slasher films. The film earned $16 million in its first four months of release ("Scoreboard"), and the sequel, *A Nightmare on Elm Street 2: Freddy's Revenge* (Jack Sholder, 1985) was in theaters less than a year after the first film.

The *Nightmare on Elm Street* films thus transformed New Line Cinema at the same time that the company innovated on the conventional ways in which horror films were exploited in the market, specifically by mainstreaming and franchising these films. As Nowell asserts, "In order to increase the cultural visibility of its *Elm Street* property, New Line Cinema executed High Concept distribution practices. They included wide theatrical openings and intensive promotion, publicity, and licensing" (2012: 83).[3] That is to say, New Line sought to expand beyond the already common practice of sequelizing slasher and other horror films by turning *A Nightmare on Elm Street* into a transmedia franchise, particularly through the Freddy Krueger character.

New Line's 1987 annual report made this strategy explicit, saying that the company had recently completed business deals that had the "potential to significantly broaden the Company's operations," particularly through creating television spin-offs and licensed merchandise ("Annual Report 1987": 6). The report announced that the company's first television production would be *Freddy's Nightmares* ("Annual Report 1987": 6), an anthology horror series hosted by, and occasionally featuring, the Krueger character, somewhat resembling *The Twilight Zone* (1959–64) or *Tales from the Darkside* (1983–88). Crucially, the report also described how the company was increasingly engaged in licensing and merchandising its film properties, saying that this endeavor was "yielding a new and growing source of revenue" ("Annual Report 1987": 7). It detailed how Freddy Krueger had been successfully merchandised in the form of Halloween costumes, and that "model kits, *Elm Street* board games, hologram watches, skateboards, and a seven-foot blow-up Freddy figure are all now on the market" ("Annual Report 1987": 7). The Freddy Krueger costume, for instance, sold extremely well, and was reported to be the best-selling Halloween costume in 1987 (Willman, 1988). Additionally, there were video games on multiple platforms featuring the Freddy Krueger character, and there was even a 900 number phone service that allowed callers to pay and listen to messages from Freddy. *Nightmare on Elm Street* was no longer (simply) another slasher film series but rather a transmedia phenomenon.

What makes the *Nightmare on Elm Street* franchise especially remarkable is the way in which New Line managed to transcend the traditional market for teen slasher films and bring the Freddy character into a cultural mainstream. Much of this mainstream popularity was achieved by engaging child consumers through non-cinematic texts and commodities. Admittedly, similar merchandising occurred with Jason Voorhees from the *Friday the 13th* films as well as Leatherface from *A Texas Chainsaw Massacre*, including clothing, comic books, and video games. Yet none of these materials appears to have crossed over into mainstream culture in the way that Freddy Krueger did. As Nowell observes, the *Nightmare* franchise was consumed by preteen children, and this population was courted especially through merchandise and spin-off products (2012: 88). Ian Conrich has similarly noted how savvy marketing and merchandising of the Freddy Krueger character probably helped the character appeal particularly to child consumers (1997: 124–8). Thus, within the larger context of mainstream franchising practices among the larger media companies in the 1980s, New Line Cinema strived to "mainstream" *A Nightmare on Elm Street* into a multimedia franchise palatable to larger audiences, even child audiences.

Critters as Potential Franchise

Though New Line continued to release an eclectic slate of films in the mid-1980s, during this time the company capitalized upon the success of *A Nightmare on Elm Street* and reimagined itself as a company that could generate media franchises for mainstream audiences. *Critters* was their first attempt to do so. *Critters* (Stephen Herek, 1986) was developed in early 1985 as an original New Line Cinema production, with a planned budget of about $3 million, and with Robert Shaye acting as executive producer ("Pictures: 'Critters' First Pic in New Line's Coventure with RCA/Col Homevid"). The film was written by Dominic Muir and Steve Herek, who also directed the film, and shooting took place in the late summer of 1985 ("US News: New Line Shave's Pot of Gold After 'Elm Street' Success"). *Critters* served as an overt part of New Line Cinema's move toward attaining a more prominent position within the media industry, which involved a more robust slate of original productions and a larger scale for distribution. A story in *Screen International* detailed the company's efforts to publicize its new, larger status at the 1986 Cannes Film Festival and called particular attention to both *Critters* and *A Nightmare on Elm Street 2*, both of which played at the festival, as evidence of this growth ("Cannes: New Line Launches New Image with Bigger Production, Acquisition Slate"). Further, New Line developed *Critters* in tandem with the company's new partnership with RCA/Columbia Home Video for video distribution ("Pictures: 'Critters' First Pic in New Line's Coventure with RCA/Col Homevid"). By the mid-1980s, home video accounted for a large amount of revenue within the film industry, and many companies had secured similar partnerships for home-video distribution, or had even opened their own home-video divisions. In New Line's case, this deal was particularly important because the *Nightmare on Elm Street* films were so successful on home video. Further, because New Line was still a comparatively small, independent movie distributor, the partnership with RCA/Columbia Home Video indicates both the company's increasing engagement with the multimedia environment of the 1980s, as well as its hopes that *Critters* would expand beyond the conventional theatrical market as *A Nightmare on Elm Street* had.

New Line released *Critters* on April 11, 1986 on over six hundred screens, making it the company's largest theatrical release up to that point ("Pictures: New Line, Prepping Bow of New Apparatus, Adds Prod. & Branch Staffers"). The narrative of the film begins in outer space with the titular critters escaping from a prison on an asteroid. Two shape-shifting, alien bounty hunters track them to a small town on Earth. There, the film focuses upon the Browns, a family with a teenage daughter and preteen son, and Charlie, the Brown family's paranoid, alcoholic neighbor. The critters' ship lands near the Browns' farm, streaking through the sky and alarming everyone nearby. The creatures, which roll speedily across the ground, tumbleweed style, exit the ship and begin attacking everyone and everything in sight. The boy and his father stumble upon a mauled cow, for instance, and the critters also kill a police officer. We get our first direct look at the little toothy monsters when one of them attacks the father in the cellar. The creatures also attack the girl and kill her boyfriend in another bloody scene.

As the critters attack the Browns, various community members in town have their own altercations with the bounty hunters. Amid this mayhem, the film maintains moments of quirky humor, such as when one critter eats a firecracker and falls over, smoking, and another jumps into a toilet after catching fire. Similarly, the critters are occasionally depicted as more mischievous than deadly, such as when one comically stares down and bites the head off of an E.T. doll. Eventually, the bounty hunters arrive at the Brown farm and begin spectacularly killing off the critters. An especially large critter kidnaps the Brown daughter, but the son and Charlie save her at the last minute. The film ends with peace restored, but, in true horror film fashion, the last image of the movie promises a sequel by showing unhatched critter eggs in the barn.

Upon its release, *Critters* earned mixed reviews. A good number of these compared it with *Gremlins*, typically stating that *Critters* was the lesser film (Baltake, 1986; James, 1986; Wilmington, 1986). A critic at the *Boston Globe* stated, "'Critters' seems to want to do nothing more than rip off 'Gremlins' and half a dozen other sci-fi outings" (Carr, 1986b), for instance, while the critic for the *Philadelphia Inquirer* wrote, "What do you call a parody of a parody? Critters spoofs Gremlins, which itself spoofs E.T.—The Extra Terrestrial. That makes Critters a spoof twice removed" (Rickey, 1986). These comparisons to *Gremlins* are unsurprising, as that film had been released nearly two years earlier and had had lots of press. Though Dominic Muir asserted he wrote *Critters* before *Gremlins* came out (Carr, 1986a), it seems likely that, at the least, New Line was inspired by the financial success and cultural impact of *Gremlins* when it chose to produce and release *Critters*. In fact, New Line did nothing to avoid these comparisons with *Gremlins*; the company even drew these comparisons itself, as well as with other successful films and film series, in various marketing materials and public statements. In one news article, for instance, New Line executives referred to the film as a "thematic mix of 'Gremlins' and 'The Terminator'" ("Pictures: 'Critters' First Pic in New Line's Coventure with RCA/Col Homevid"). Though it is common practice for film distributors to associate their films with successful similar films, it is notable here that *Critters* was so firmly associated with *Gremlins* by the company as well as within the wider critical discourse.

In fact, *Critters* was wrapped up within, and contributed significantly to, a broader cycle of "little monster" movies appearing in the mid-1980s. Though initiated by *Gremlins*, this cycle was widely regarded as trash cinema. We see evidence of this in another review of *Critters*, where the critic for the *Los Angeles Times* wrote, "Two years ago we had 'Gremlins' . . . it was a good movie . . . Last year we had 'Ghoulies' . . . a stinker [that] cleaned up at the box office . . . And now we have another species: 'Critters'" (Wilmington, 1986).

Calling *Critters* "dumb" and "derivative," the author actually calls for a "temporary end" to films featuring little monsters of any variety (Wilmington, 1986). Aside from *Gremlins*, *Ghoulies* (Luca Bercovici, 1984), and *Critters*, other films in this cycle include *Munchies* (Bettina Hirsch, 1987), and *Hobgoblins* (Rick Sloane, 1988). In all of them, small, mischievous, or outright malevolent little monsters wreak havoc on a family or an entire community. Unlike *Gremlins*, however, which had a respectable production budget and got a major release by Warner Bros., all of the other little monster films were produced inexpensively and distributed by small, independent companies. The cycle, then, largely fell into the realm of cult or exploitation cinema, as a result of the films' industrial conditions and apparent derivativeness.

However much *Critters* may have been aligned with cult cinema, it appears that New Line hoped to emulate the success of *Gremlins* as a mainstream franchise, particularly drawing upon the company's previous experience with *A Nightmare on Elm Street*. Even before *Critters* was released, New Line situated *Critters* as a franchise-in-the-making by pairing it with the sequel to *A Nightmare on Elm Street* in publicity and advertisements. New Line took out a two-page spread in *Variety* in May 1985, for instance, which was aimed at promoting its films to international- and home-video distributors; the left-hand side was a full page ad for *A Nightmare on Elm Street: Part II*, and the right side was devoted to *Critters* ("Critters"). This cartoony advertisement prominently featured the gun-toting alien bounty hunters from the film standing above a pile of huddled critters with glowing eyes, and had the tagline, "They're loose, hungry and in pre-production" ("Critters"). Here, the company presents the two films side by side as though they are both related to the success of the first *Nightmare* film and as though they both somehow extend from that film. Similarly, New Line ran a two-page advertisement in *Variety* regarding the company's participation in the 1986 ShoWest convention, and highlighted *Critters* alongside details regarding the financial success of *A Nightmare on Elm Street: Part II* ("From 'L' New Line Cinema Distribution"). In combination, these ads suggested to distributors that *Critters* could soon be earning similar revenues as the *Nightmare* sequel, spawning another round of financial success.

Following its release, *Critters* was a financial hit, though more modest than *Gremlins* or New Line's own *A Nightmare on Elm Street*, earning $4.7 million in North American theaters (Thompson, 1987: 68). The timing of this success was important, as New Line Cinema became a publicly traded company in the fall of 1986, making an initial public offering through the investment bank Drexel Burnham Lambert, Inc. ("Securities Offering Calendar"). *Critters* became symbolically important in this context, and New Line used the film as evidence of its growing industrial strength and cultural significance. The company's 1986 annual report stated, "Our fiscal performance in 1986 achieved record levels, primarily as a result of the success of *Critters* and *A Nightmare on Elm Street 2: Freddy's Revenge*, in both the motion picture and home video marketplaces" ("Annual Report 1986": 1). And, indeed, the company had revenues of over $26 million in 1986, significantly smaller than the earnings of a major studio but remarkable for an independent distributor ("Annual Report 1986": 1).

The 1986 report tried to situate *Critters* as a franchise through a number of rhetorical tactics. It connected the success of *Critters* to that of the *Nightmare* movies and noted in particular the multimedia success, including home video, of both. Further, the report indicated that that these successes would generate future successes in the form of additional productions, particularly sequels for both *Critters* and *Nightmare on Elm Street*. The report mentioned specifically that *Critters 2* would open in November 1987, following the "popular" success of the first film ("Annual Report 1986": 1, 4–5), and the company's 1987 annual report similarly situated *Critters 2* as one of "two sequels to New Line box office hits," along with *A Nightmare on Elm Street 4* ("Annual Report 1987": 4). Throughout these documents, then, New Line Cinema treated the first *Critters* film as merely one instance in what was imagined to be an ongoing exploitation of a single IP. The reports reveal how New Line aimed to make sequels as a general production strategy, and positioned *Critters* alongside the *Nightmare* franchise as examples of this pursuit. Perhaps most overtly, a news article quoted a New Line marketing executive saying, "It would be nice if *Critters* turned out to be the same kind of franchise as *Nightmare*" (McDonagh, 1988). Here, the phrasing is precise and makes the logic clear: New Line aimed to be in the franchise business, and strove to make *Critters* as such.

New Line also had ambitions to license *Critters* into a *transmedia* franchise, just as it had done with *A Nightmare on Elm Street*. As noted above, the company's 1987 annual report boasted about its success in licensing the Freddy Krueger character into a variety of related texts and merchandise. Immediately following this passage, the report states, "Similar exploitation of items from *Hairspray* and *Critters* are now being explored" ("Annual Report 1987": 7). Appearing almost as an aside within the larger context of the report, this statement shows that New Line hoped to continue to exploit *Critters* in as many ways as possible, ideally as additional non-cinematic texts and products. And, although it does not appear that any *Critters* games, toys, or other merchandise were ever made at this time, this statement reveals that New Line sought to expand *Critters* into a franchise that included both sequels *and* other textual and paratextual manifestations of the IP. In any event, the company treated *Critters* as one of their important properties with mainstream potential, as indicated in an internal memo from 1988 that referred to both *A Nightmare on Elm Street* and *Critters* as "tried and true" successes in the company's efforts to make "teenage movies for a mass audience" (De Luca, 1988: 1).

Critters as Failed Transmedia Franchise

New Line released Mick Garris's *Critters 2: The Main Course* in April 1988, two years after the first film had arrived in theaters. The film's story follows from the first, with the critters back to terrorize the same small town. Several characters from the original also appear in the sequel, including the son of the Brown family and one of the bounty hunters. In general, the sequel tries to amplify all of the action and comedy found in the original film, as the townsfolk and the bounty hunters all work to destroy the monsters. There are also many similar gags related to the killing of critters, such as when one has the top of its hair blown off with a gun, making it bald, or when one explodes after biting into a tire and swallowing too much pressurized air. Perhaps the most remarkable of this film's "improvements" upon the first is that now the critters can cluster together as a single giant ball and roll around to smash people and destroy things.

Critters 2 received predictably mixed reviews, a number of which continued to compare it with *Gremlins* (Andrews, 1988; Harrington, 1988; Kehr, 1988). Perhaps more importantly, the film made nearly $2 million dollars in its opening weekend ("Pictures Grosses: Weekend Boxoffice Report"). Its revenues fell quickly in the following weeks, however, but the film did manage to earn around $3.8 million during its theatrical run ("Pictures: Boxoffice Roundup—'Heat' Raises B.O. Temps; 'Big' Brawny"). The movie was also not a big hit on video, as it ranked as the thirty-eighth best-renting title during its first two weeks of home-video release ("Video Retailing: Top Videocassettes Rentals"). Though not exceptional, *Critters 2*'s performance in theaters and on home video was certainly not a failure; indeed, the film could be considered a success within the context of other low-budget horror films.

Yet the film's earnings did raise questions about the potential continuation of the *Critters* series or finding mainstream popularity. A review from *Box Office* put it this way: "This klutzy sequel to the 'Gremlins' rip-off managed to gross only $3.3 million in its first two weeks, meaning that we've probably seen the last of those rotten old Crites" (Kozak, 1988). Nevertheless, following the film's theatrical run, New Line president Shaye stated, "'Critters' was less successful at the box office (than 'Elm Street'), but the greatest financial success of 'Critters' came through television and home video" (quoted in Lannon, 1988). Here, Shaye suggests how New Line strove to exploit the *Critters* property as widely as possible, in this case through emphasizing its success across multiple media venues. Even more pertinently, he stated, "There's equivalent interest for 'Critters 2.' If it's as strong as it seems, there might be a 'Critters 3.' But if we don't have an idea for it, there won't be one" (quoted in Lannon, 1988). Several things are important about this statement. First, one can surmise that New Line produced *Critters 2* with the hope that it would be successful across multiple media and different windows of release. Second, Shaye hedges as to whether the company will produce another film in this series, and hinges this decision upon the apparent demand for *Critters 2*, combining this also with creative issues. That is to say, there seems to be little question that New Line would produce another film in this series as long as there appeared to be clear financial reasons to do so and creative potential; the company would exploit and reexploit the underlying IP as long as it could. Third, and finally, there were *no* discussions at this point about licensing the *Critters* property in any way other than as another movie in the series. New Line no longer imagines "exploring" other ways of exploiting the IP as it had with the first film. The company may have treated *Critters* as a franchise but not any longer as a potential transmedia franchise, licensed across multiple texts and commodities, but simply as just another series of low-budget horror films.

This shift in the trajectory of the *Critters* franchise occurred within a specific context in the larger cinema culture and industry as well as within New Line's particular development as a company. First, a number of the other, low-budget "little monster" movies, as well as sequels to these films, were released in the later part of the 1980s, more firmly establishing this cinematic cycle as a low-budget "cult" genre. In addition to *Munchies*, which came out in 1987, both *Hobgoblins* and *Ghoulies 2* came out the following year. Also, somewhat similar horror films, featuring small monsters depicted through practical effects and dolls, like *Troll* (John Carl Buechler, 1986), *Child's Play* (Tom Holland, 1988), and *Puppetmaster* (David Schmoeller, 1989), were released around this time. At this point, then, films that featured small mischievous monsters, such as *Critters 2*, were more associated with cheap exploitation cinema than with mainstream, transmedia franchises such as *Gremlins*.

Second, although during the late 1980s, New Line continued to find success with *Nightmare on Elm Street* as both a film series and multimedia franchise, the company sought out other IP it might similarly exploit as crossover, mainstream franchises at the same time as it continued to expand the range of its business operations. This occurred when New Line got hold of the *Teenage Mutant Ninja Turtles* movie (Steve Barron, 1990), a turn of events that would once again elevate the company into a new echelon of the media industry and, simultaneously, more firmly establish New Line as a company that would nurture and exploit transmedia franchises. Having first appeared as a small, independent comic book, the Teenage Mutant Ninja Turtles entered mainstream media culture as action figure toys and an accompanying animated children's program that began airing in 1987. As Marsha Kinder has detailed, the Turtles rapidly became a pop culture phenomenon that crossed over multiple media platforms and consumer products, including video games, soundtrack albums, and lunchboxes; for Kinder, the *Teenage Mutant Ninja Turtles* epitomize a media "supersystem," where a specific figure or property becomes wrapped up in an intertextual, industrial network that spans across media and collectible merchandise, targets different consumer populations, and takes on the cultural impact of an "event" (1991: 121–3). The Turtles were, in other words, the very picture of a transmedia franchise in 1989.

As the cinematic manifestation of this franchise, the *Turtles* movie was produced in the summer of 1989 on a budget of around $12 million; notably, this movie was made outside the Hollywood studio system by the Hong Kong-based studio Golden Harvest, which was known for making martial arts films and other action movies aimed at the East Asian market ("Cannes 1989: Long Kong's show of Force"). Although Golden Harvest promoted the *Turtles* film heavily at the Cannes Film Festival that year, the company had difficulty securing a distributor for North America ("Golden Harvest Hawking 3 Pics; Pumping Prospects at 'Golden Club'"). New Line Cinema acquired the North American rights, as well as to any potential sequel, to the film in 1989 (Hazelton, 1989). New Line released the film in March 1990, in coordination with a deluge of Turtle-related products and merchandise, including "T-shirts, posters, caps, buttons, bumper stickers [and] trading cards" as well as cross-promotions with a wide range of restaurants, retailers, and other companies (Smith, 1990). The film itself was a major success, earning over $25 million in its first week of release ("The Hollywood Reporter Boxoffice"), and eventually earned over $133 million during the course of 1990 ("The Numbers Page: Top Twenty Boxoffice Performers"). This was New Line's biggest financial hit by far and the most successful independently distributed film up to that point.

There are several aspects of New Line's release of *Teenage Mutant Ninja Turtles* that are worth considering within the context of the company's efforts to create transmedia franchises in general and its production of the *Critters* series in particular. First, it is vital to note that New Line was *not* the production company behind *Teenage Mutant Ninja Turtles* and, even more importantly, that the company did not own the underlying IP. New Line's participation in the *Teenage Mutant Ninja Turtles* franchise was somewhat marginal, as the company was a licensee. Thus, as just one of many, many different companies that proliferated the Turtles through popular culture in myriad manifestations, New Line held a contingent relationship to the franchise as a whole. This is not to say that New Line did not coordinate its business practices with other aspects of the franchise or benefit from its network effects. Sandra Ruch, then copresident of New Line Marketing, put it this way:

> There were [*sic*] a tremendous amount of licensees already in place by the time we decided to take on this movie. New Line does not own the licensing rights, but we were able to contact all the licensees and try and do tie-ins specifically for the movie. Since we had people in place, what

we did was piggyback and capitalize on it and exploit it, if you will, even further. (Kelleher, 1990)

Here, we see New Line helping to integrate the *Turtles* film within an already existing franchise, and even more, to try to coordinate the multiple manifestations in such a way as to emphasize the importance of the film. Nevertheless, New Line came upon this franchise as a kind of add-on or tagalong, attaching itself opportunistically and strategically to the *Teenage Mutant Ninja Turtles* industrial and cultural network.

Insinuating itself so successfully within the *Turtles* franchise at this juncture was crucial for New Line, altering the fate of the company and, in turn, the way in which it would deal with the *Critters* property in the future. *Teenage Mutant Ninja Turtles* broke through as a mainstream, child-friendly, and decidedly *transmedia* franchise in a way that *A Nightmare on Elm Street* never could, even with its somewhat surprising iterations as a video game, television program, and children's Halloween costume. Even if *Turtles* wasn't New Line's property, the film did become a prominent and financially noteworthy element of that franchise, bringing new levels of industrial power and cultural legitimacy to the company. New Line remained an "indie" within the film industry, at least for the moment, but was now understood as a company that catered in transmedia franchises capable of crossing over into mainstream popular success.

What became of *Critters* in this context, particularly given that there were no indications that that property could be exploited as anything other than as a series of films, and that there were even questions about that possibility? Despite any misgivings the company may have had about the franchise following the performance of *Critters 2* at the box office, the company moved forward with the production of a sequel. In fact, the company was somewhat aggressive and actually set up productions for two sequels, *Critters 3* (Kristine Peterson, 1991) and *Critters 4* (Rupert Harvey, 1992), which were made back to back in the summer of 1991 ("Film Production"). One article, perhaps generously or perhaps mockingly, or both, likened New Line's double production to that of a mainstream franchise, *Back to the Future*, which also shot its second and third installments back to back (Pond, 1991). This comparison is noteworthy to the extent that New Line, like the producers of the *Back to the Future* films, had enough faith in the demand for this series to invest in multiple sequels all at once, not letting the success of the next determine the potential production of another. Moreover, it is notable that this strategy—which, though bold, also presents cost savings by eliminating the need to orchestrate an entirely new production—was used years later by New Line with the immensely successful *Lord of the Rings* trilogy (Thompson, 2007: 29–32).

But that is where the comparisons to *Back to the Future* or *The Lord of the*

Rings end, as *Critters 3* and *4* are decidedly low-budget, quirky, and campy films aimed at an audience of cult horror fans. *Critters 3*, and especially *4*, are notable for the very loose way in which their narratives connect to the previous films. Further, the production value for *Critters 4* appears especially reduced from the others, and the film also lacks any of the goofy humor of the previous films. *Critters 3* tells the story of a family who inadvertently brings critter eggs back to their apartment building while returning home from a vacation in the country. Once there, the film alternates between silly humor and lighthearted spookiness as the critters cause mischief and attack everyone they can while the residents of the building fight them off in a variety of humorous ways. Eventually, Charlie, one of the only characters to appear in all four *Critters* films, arrives and fights them off for good; yet the film concludes with a "To Be Continued" title, as a spaceship crashes into the building. The fourth film picks up immediately after this, yet the action takes place decades later, where Charlie and the critter eggs are now aboard a spaceship. As the crew of the ship deal with various technical problems, the critters hatch and kill them off. Interstellar officials, who appear just like the bounty hunters from the earlier films, arrive, but, instead of helping eradicate the critters, these officials become adversaries for other members of the ship's crew. Eventually, Charlie and two other crew members defeat the bounty hunters and escape from the critters on a different spaceship. The critters are defeated when their ship explodes, and Charlie and the crew fly away to safety.

Following the production of these films, New Line attempted to generate interest in them by screening *Critters 3* at both the American Film Market (AFM) and the MIFED film market in October 1991 ("New Line Cinema"), and *Critters 4* at the AFM in March 1992 ("From New Line Cinema"). It is unclear, however, if any deals were made at these markets. And, although New Line initially planned to release *Critters 3* in theaters (McCullaugh, 1991), ultimately both of these sequels were direct-to-video releases, with *Critters 3* coming out in December 1991 (Smith, 1991), followed by *Critters 4* in October of the following year ("New On Video"). From one perspective, we might take this as New Line dumping the *Critters* series into the cinematic dustbin, given the cultural status typically associated with straight-to-video movie releases at the time. And, indeed, one contemporaneous news article that examined the straight-to-video business asserted that "the B movies that get the most attention are inferior sequels or would-be imitations of hit films that played in theaters," citing New Line's video releases, and particularly *Critters 4*, as examples of this phenomenon (King, 1992). From this vantage point, we might see New Line as resigning *Critters 3* and *4* to the disreputable cultural zone already occupied by such films as *Ghoulies 2* or *Munchies*.

From another perspective, though, we might see the straight-to-video release of *Critters 3* and *4* as indications of New Line's savvy self-awareness as a growing company that had one foot in exploitation cinema and another in mainstream, transmedia franchises. Following the company's deal with RCA/Columbia Home Video from 1985, New Line formed its own home-video division, New Line Home Video, in the fall of 1990 (Sweeting, 1990). Prompted specifically by the immense success of the first *Teenage Mutant Ninja Turtles* movie on video, which was distributed on VHS by Live Entertainment (Goldman, 1990), the formation of New Line Home Video was clearly an attempt better to control all of the company's releases on this platform. Though RCA/Columbia ultimately continued distributing some New Line titles on video, as a kind of partnership with the two companies handling different aspects of home-video releases (Berman, 1991), the creation of the New Line Home Video division indicated the growing power and capabilities of the company and, perhaps most importantly, its desire and ability to control better the exploitation of its films and franchises across different media. With this deal in mind, New Line's release of *Critters 3* and *4* on video was actually a shrewd means of making the most of an existing IP *and* of an emerging aspect of their development into a mainstream, multimedia corporation. And, indeed, Ted Turner purchased New Line in 1994, making the company part of the much larger and diversified Turner Broadcasting System; in turn, New Line would become an element of Time Warner when the conglomerate purchased TBS in 1996.

Conclusion

In early 2018, *Variety* announced that the small, digital production unit within Warner Bros., called Blue Ribbon Content, would produce a new *Critters* television series, called *Critters: A New Binge* (Otterson, 2018). This series seemed designed for the new multiscreen, digital media landscape of the moment, as it was set to appear on Go90, a streaming service from telecom giant Verizon (Otterson, 2018). This new web series promised to continue the *Critters* franchise into the future, and possibly establish some small place for the *Critters* brand within contemporary Hollywood. Yet in June 2018, Verizon stated it would discontinue Go90 at the end of the following month, leaving the fate of *Critters: A New Binge* and other programming "yet to be determined" (Spangler, 2018).

This unsuccessful attempt to reboot *Critters* seems in keeping with the longer history of the franchise, with its moments of modest, opportunistic success, failed expansions, and spluttering disappearance. Ultimately, the *Critters* film series probably does not mean much within the grand scheme of contemporary transmedia Hollywood. Similarly, New Line Cinema might also appear somewhat marginal within the greater media industry, since it was radically downsized and incorporated as a branch of Warner Bros. in 2008, as part of the larger "Great Studio Pullback" away from independent cinema in that year (Tzioumakis, 2018). But New Line's development, from a tiny college-campus film distributor to the company behind *The Lord of the Rings* trilogy, is truly remarkable within the history of cinema. So much of this company's development depended crucially on the multiplied exploitation of distinct IP, often through the creation of transmedia franchises that transcended single media platforms to other cultural texts and products. But as history—and this chapter—has shown, New Line entered into these practices from a place of industrial and cultural marginality, as an independent company with an outsider status. If nothing else, the company had a particular gift in taking small, niche properties and transforming them into more broadly appealing texts and products. At one of the company's many crossroads, in fact, in 1990, president Bob Shaye gestured toward the company's mixed identity in comments prepared for a stockholders' meeting, stating, "our niche marketing philosophy has more viability than ever . . . and the selection of materials that have potential to cross-over to a broader audience for the big possible upside" ("Stockholders Meeting Comments", 1990).

Sitting literally and figuratively between *A Nightmare on Elm Street* and *Teenage Mutant Ninja Turtles*, the *Critters* series never reached that potential. Despite the allegory offered in *Gremlins*, one cannot make a transmedia franchise by just adding water; it takes hard work, strategy, and opportune timing. Whereas New Line managed to franchise Freddy into a broadly popular, multimedia figure, and cunningly took part in the already popular *Teenage Mutant Ninja Turtles* franchise, the company was unable to push *Critters* beyond the industrial and cultural space typically afforded to low-budget horror films; *Critters* was just one more "little monster" film series. Though New Line failed to spread *Critters* into toys and other merchandise, these films helped confirm the company's ability successfully to exploit a single IP multiple times, the very measure of success within the emerging conglomerate media landscape of the late 1980s. Thus, even in its limitations, *Critters* represents an instance of franchising at the margins of the media industry, and suggests how the franchising practices of the major conglomerates relate so closely to the exploitation tactics of small, indie companies such as New Line.

Notes

1. I use the phrase "transmedia franchise" in this essay to refer specifically to those media franchises that appear in multiple media forms and products. As Derek Johnson indicates, transmedia intertextuality and transmedia storytelling are neither the same as franchising nor are these phenomena inherently conjoined (2013: 30–3). This makes it all the more important, to my mind, to distinguish those franchises that do cut across media and products and those that appear primarily or exclusively within one cultural form, such as with film sequels or interrelated television series.

2. As Jennifer Holt notes, synergy has multiple, though related, definitions (2011: 3). In this context, we could draw on Derek Johnson's use of the word "synergy," which he states is a situation, "where the same content can dominate multiple markets and generate more value than the sum of its iterative parts" (2013: 67).
3. Justin Wyatt has defined Hollywood "High Concept" cinema as films that feature simple narratives, emphasize style, and work in congruence with marketing materials and related merchandising (1994: 1–22).

References

Andrews, Marke (1988), "It's Low-Budget Munchies as the Carnivorous Furballs Return," *Vancouver Sun,* 4 May, p. F8.
"Annual Report" (1986), University of Michigan Special Collections Library, New Line Cinema Papers—Box 2, Shaye Business Documents—New Line Cinema—Annual Shareholder Reports, 1986, 1988.
"Annual Report 1987" (1987), Wesleyan University Cinema Archives, John Waters Collection—Box 22, folder 19, "Film Projects, Distributors, New Line Cinema's Annual Report, 1987," New Line Cinema Corporation.
"Arty & 'Freak' Films Packaged to Theatre Trade by New Line; Sex, Too" (1973), *Variety*, 4 July, p. 6.
Baltake, Joe (1986), "'Critters': Hairy Little Toys Hanging Out," *Philadelphia Daily News*, 29 April, p. 54.
Berman, Marc (1991), "HomeVideo: New Line, RCA/COL Finalize Distrib Deal," *Variety*, 1 April, p. 22.
Canby, Vincent (1969), "Nemec's 'Martyrs of Love'," *The New York Times*, 4 February, p. 32.
Canby, Vincent (1970), "What Godard Hath Wrought," *The New York Times*, 29 March, p. 77.
"Cannes 1989: Long Kong's show of Force" (1989), *Screen International*, 6 May, p. 201.
"Cannes: New Line Launches New Image with Bigger Production, Acquisition Slate" (1986), *Screen International*, 3 May, p. 188.
Carr, Jay (1986a), "Caught Up with Critters," *Boston Globe*, 11 April, p. 44.
Carr, Jay (1986b), "The Good Bad and the Really Bad," *Boston Globe*, 27 April, p. 87.
Christensen, Jerome (2011), *America's Corporate Art: The Studio Authorship of Hollywood Motion Pictures*, Stanford: Stanford University Press.
Conrich, Ian (1997), "Seducing the Subject: Freddy Krueger, Popular Culture and the *Nightmare on Elm Street* films," in Deborah Cartmell, I.Q. Hunter, Heidi Kaye, and Imelda Whelehan, (eds), *Trash Aesthetics: Popular Culture and its Audiences*, London: Pluto Press, pp. 118–31.
Cook, David (2002), *Lost Illusions: American Cinema in the Shadow of Watergate and Vietnam, 1970–1979*, Berkeley: University of California Press.
"Critters" (1985), *Variety*, 8 May, p. 39.
De Luca, Jeff (1988), "Identifying New Line's Creative Niche," Memo, University of Michigan Special Collections Library—New Line Cinema Papers—Shaye Business Documents—New Line Cinema—Correspondence and Email, 1967–2008, 2 February.
"Display Ad" (1975), *Washington Post*, 10 September, p. B11.
"Display Ad 37—no title" (1973), *Washington Post,* 18 July, p. B14.
"Display Ad 361" (1970), *The New York Times*, 22 February, p. D16.
"'Fighter' Stout 27G, Philly; 'Tango' 4G" (1975), *Variety*, 24 September, p. 15.

"Film Production" (1991), *Variety,* 29 July, p. 10.
"From 'L' New Line Cinema Distribution" (1986), *Variety*, 29 January, pp. 20–1.
"From New Line Cinema" (1992), *Variety*, 24 February, p. 147.
"Golden Harvest Hawking 3 Pics; Pumping Prospects at 'Golden Club'" (1989), *Variety*, 10 May, p. 342.
Goldman, Kevin (1990), "'Ninja Turtle' Juggernaut Demolishes Box Office Record," *The Wall Street Journal*, 3 April, p. B1.
Harrington, Richard (1988), "More Carnivore Bore: The Lamentable Return of Critters", *Washington Post,* 4 May, p. D4.
Hawkins, Joan (2000), *Cutting Edge: Art-Horror and the Horrific Avant-Garde*, Minneapolis: University of Minnesota Press.
Hazelton, John (1989), "News: New Line Turns Turtle," *Screen International*, 12 August, p. 2.
"Hollywood Reporter Boxoffice, The" (1990), *Billboard*, 14 April, p. 48.
Holt, Jennifer (2011), *Empires of Entertainment: Media Industries and the Politics of Deregulation, 1980–1996*, New Brunswick, NJ: Rutgers University Press.
Hughes, Candice (1980), "'Chainsaw' Fight Nearly Over," *Atlanta Constitution*, 10 December, p. 8B.
James, Caryn (1986), "The Screen: 'Critters' from Space," *The New York Times*, 11 April, p. C20.
Johnson, Derek (2013), *Media Franchising: Creative License and Collaboration in the Culture Industries*, New York: New York University Press.
Kehr, Dave (1988), "'Critters 2: The Main Course' Is Manic, Eat 'Em Up Comedy," *Chicago Tribune (Pre-1997 Fulltext)*, 29 April, p. G.
Kelleher, Ed (1990), "New Line Readies Ninja Market Blitz," *The Film Journal*, 1 March, p. 8.
Kendrick, James (2014), "Slasher Films and Gore in the 1980s," in Harry M Benshoff (ed.), *A Companion to the Horror Film*, Hoboken, NJ: Wiley-Blackwell, pp. 310–28.
Kinder, Marsha (1991), *Playing With Power in Movies, Television, and Video Games: From Muppet Babies to Teenage Mutant Ninja Turtles*, Berkeley: University of California Press.
King, Thomas (1992), "B Films can Earn A-Plus as Video Hits," *The Wall Street Journal*, 28 July, p. B1.
Kozak, Jim (1988), "Reviews: 'Critters 2: The Main Course'," *Boxoffice*, 1 July, p. R-58.
Lannon, Linnea (1988), "They're Ba-ack: Sequels, Prequels, Never Equals," *Detroit Free Press (1923–1999)*, 22 August, p.18.
McCullaugh, Jim (1991), "HomeVideo: VSDA REPORT," *Billboard*, 27 July, p. 47.
McDonagh, Maitland (1988), "Nightmares Fuel New Line's Movie Dreams," *The Film Journal*, 91: 3, 1 April, p. 13.
"Miscellaneous" (1976), *Boxoffice*, 19 April, p. 6.
"Motion Pictures Rated by the Code & Rating Administration" (1973), *Boxoffice*, 24 September, p.10.
"New Line Cinema" (1991), *Variety,* 14 October, pp. A58, A59.
"New Line Cinema Entertainment Catalog 4" (1979), University of Michigan Special Collections Library, New Line Cinema Papers—Box 1, Shaye Business Documents—New Line Cinema—Catalogs, 1979–81.
"New Line Cinema Forms Exploitation Film Unit" (1974), *Boxoffice*, 6 May, p. 12.
"New Line Shaye Doubts Old-Line Showmen Re Kids" (1970), *Variety*, 11 February, pp. 4, 28.
"New on Video" (1992), *Austin American Statesman,* 16 October, p. 15.

Nowell, Richard (2012), "Between Dreams and Reality: Genre Personae, Brand 'Elm Street', and Repackaging the Teen Slasher Film," *Iluminace*, 24: 3, pp. 69–101.

Nowell, Richard (2011a), *Blood Money: A History of the First Teen Slasher Film Cycle*, New York: Continuum.

Nowell, Richard (2011b), "'Where Nothing is Off Limits': Genre, Commercial Revitalization, and the Teen Slasher Film Posters of 1982–1984," *Post Script – Essays in Film and the Humanities* 30:2, Winter, pp. 53–68.

"Numbers Page, The: Top Twenty Boxoffice Performers" (1991), *Boxoffice*, 1 February, p. 111.

Otterson, Joe (2018), "Blue Ribbon Content Sets Conan O'Brien Pet Comedy, 'Critters' Adaptation as go90 Series," *Variety*, 10 January, <https://variety.com/2018/tv/news/critters-series-conan-obrien-go90-1202659582/amp/?__twitter_impression=true> (last accessed 8 July 2018).

"Pictures: Boxoffice Roundup—'Heat' Raises B.O. Temps; 'Big' Brawny" (1988), *Variety*, 22 June, p. 4.

"Pictures: 'Critters' First Pic in New Line's Coventure with RCA/Col Homevid" (1985), *Variety*, 15 May, p. 31.

"Pictures Grosses: 'Monsters' Sock $13,000, Cleve" (1976), *Variety*, 11 February, p. 17.

"Pictures Grosses: 'Street Fighter' Bows with Heavy TV Drive" (1974), *Variety*, 18 September, p. 10.

"Pictures Grosses: Weekend Boxoffice Report" (1988), *Variety*, 4 May, p. 50.

"Pictures: New Line Cinema Corp" (1967), *Variety*, 13 September, p. 24.

"Pictures: New Line Picks Up Bloody 'Evil Dead'" (1983), *Variety*, 12 January, pp. 8, 46.

"Pictures: New Line, Prepping Bow of New Apparatus, Adds Prod. & Branch Staffers" (1986), *Variety*, 26 March, p. 28.

Pond, Steve (1991), "Summer Films Take a Fall," *Washington Post*, 7 June, p. C7.

Rickey, Carrie (1986), "Film: Balls of Fluff from Space in Kansas," *Philadelphia Inquirer*, 25 April, p. E8.

Santo, Avi (2015), *Selling the Silver Bullet: The Lone Ranger and Transmedia Brand Licensing*, Austin: University of Texas Press.

Schaefer, Eric (1999), *Bold! Daring! Shocking! True: A History of Exploitation Films, 1919–1959*, Durham, NC: Duke University Press.

"Scoreboard" (1985), *Philadelphia Daily News*, 7 March, p. 41.

"Securities Offering Calendar" (1986), *The Wall Street Journal*, Eastern edition, 22 September, p. 1.

Segers, Frank (1989), "Film: Cannes Festival News—Golden Harvest Reaped Film Sales, with 'Ninja Turtles' and "Force'," *Variety*, 31 May, p. 18.

Smith, Leo (1991), "Flicks, Film and Video File: Season's Rentals 'City Slickers' and 'Terminator 2' Are Due in Video Stores in Time for the Holidays," *Los Angeles Times*, 28 November, p. 3.

Smith, Stacy Jenel (1990), "Shell Shock," *Los Angeles Times*, 4 February, p. H26.

Spangler, Todd (2018), "Verizon Is Shutting Down Go90, Its Ill-Fated Mobile Video Service," *Variety*, 28 June, <https://variety.com/2018/digital/news/go90-shutting-down-verizon-1202860864> (last accessed 8 July 2018).

"Stockholders Meeting Comments" (1990), University of Michigan Special Collections Library—New Line Cinema Papers (Robert Shaye), Box 1—Shaye Business Documents—New Line Cinema—Correspondence and Email, 1967–2008, 1: 5.

"'Street Fighter, The' is Set for 4-City Premiere" (1974), *Boxoffice*, 23 September, p. 11.

Sweeting, Paul (1990), "New Line Forms Home Vid Unit with Ex-Vestron Execs at Helm," *Billboard*, 24 November, pp. 5, 86.
"'Texas Chainsaw Massacre, The'" (1980), *Variety*, 10 December, p. 16.
"This 1936 Unexpurgated Classic from New Line Cinema" (1972), *Variety*, 1 November, p. 27.
Thompson, Anne (1987), "The 12th Annual Grosses Gloss," *Film Comment*, 23: 2, p. 68.
Thompson, Kristin (2007), *The Frodo Franchise:* The Lord of the Rings *and Modern Hollywood*, Berkeley: University of California Press.
Tzioumakis, Yannis (2018), "10 Years After the 'Great Studio Pullback of 2008': American Independent Film Theatrical Distribution in the (Partly) Post-Hollywood Film Division Era," paper presented at the Media Industries Conference.
"US News: New Line Shave's Pot of Gold After 'Elm Street' Success" (1985), *Screen International*, 5 October, p. 12.
Verrill, Addison (1975), "Pictures: Chop-Socky Fad Shrinks, but Expect its Survival," *Variety*, 8 January, p. 8.
"Video Retailing: Top Videocassettes Rentals" (1988), *Billboard*, 19 November, p. 52.
Wells, Paul (2000), *The Horror Genre: From Beelzebub to Blair Witch*, London: Wallflower Press.
Willman, Chris (1988), "Freddy: No More Mr. Bad Guy?," *Los Angeles Times,* 11 September, p. 20.
Wilmington, Michael (1986), "Movie Review: Some More Whozits this Way Come in 'Critters'", *Los Angeles Times,* 3 May, p. C7.
Wyatt, Justin (1998), "The formation of the 'Major Independent': Miramax, New Line and the New Hollywood" in Steve Neal and Murray Smith (eds), *Contemporary Hollywood Cinema*, New York: Routledge, pp. 75–90.
Wyatt, Justin (1994) *High Concept: Movies and Marketing in Hollywood*, Austin: University of Texas Press.

PART II

VIDEO GAMES: SOFTWARE, HARDWARE, AND SPACE

4. THE HAPPIEST PLAYS ON EARTH: THEME PARK FRANCHISING IN DISNEYLAND VIDEO GAMES

Heather Lea Birdsall

> Experience Disneyland magic like never before! Fly, dance and dive into iconic attractions, share special moments with beloved Disney characters and take on challenging quests with the controller-free magic of Kinect for Xbox 360—and a touch of pixie dust. Kinect™: Disneyland® Adventures is your ticket to explore the timeless lands of Disneyland® park right in your living room.
>
> Microsoft's *Kinect Disneyland Adventures* webpage[1]

Introduction

In 2010, in lieu of a traditional handheld controller, Microsoft released the Kinect, an input device for its Xbox 360 video game console that sensed the gamer's body in three-dimensional space. With it, the video-game industry took a tentative (though ultimately ill-fated) step toward a new level of immersive and embodied home gaming. The following year, Microsoft Studios published *Kinect Disneyland Adventures* (2011) for the Xbox 360, which used the Kinect sensor to allow the player to move kinetically in real space in order to move a virtual avatar through a mimetic recreation of the Disneyland Park in Anaheim, California.[2] The above epigraph, a quotation taken from the official Xbox website for *Kinect Disneyland Adventures* (*KDA*),[3] points to the many issues at play in a game that sits on the convergent crossroads of physical theme parks, virtual video games, and media franchises. It combines references to immersion via digital technology ("controller-free magic"), Disney franchises ("beloved Disney characters"), embodied theme park experiences ("Fly, dance and dive into iconic attractions"), the blurring of geographical, physical, and virtual boundaries ("explore the timeless lands of Disneyland® park right in your living room"), and intellectual property (as the multiple trademark symbols indicate).

Though *KDA* represents a unique moment in the development of Disney park video games, video games have digitally recreated Disney theme park space for nearly three decades, from home consoles to mobile devices. Since 1990, video games have brought Disneyland into the home, digitally freeing the parks from their material limitations as built, physical spaces. In much the same way that the ABC television series *Disneyland* (1954–58) used the televisual medium to free the parks from their geographical constraints and reach consumers in the home, Disneyland video games capitalize on their medium's embodied and interactive nature to further this project of virtual/physical dissemination. Similarly, these games use interactive play to guide users and future park guests toward the desired modes of interaction with, and consumption of, the park franchise. Just as Walt Disney used the theme park and television to synergistically promote one another, the Disneyland video games use media to cross-promote their different media franchises. As the parks themselves facilitate consumption of other Disney franchises, video games based on or in the parks reflect and repeat this in a reciprocal fashion. The parks promote media franchises, but media, in this case video games, also promote Disneyland as a media franchise itself.

As Derek Johnson has explained, media franchising is characterized by "the multiplied replication of culture from intellectual property resources" (2013: 6). Building on this definition, I argue that Disneyland can be read not only as its own distinct media franchise manifested through multiple incarnations within media (for example, television, theme parks, movies, and video games) but also as a superfranchise, or "a franchise organizing and enabling other franchises," as it comprises (though not exclusively) component media texts and franchises (Buhler, 2017: 21). As a superfranchise, Disney is able to assert authorship over a number of disparate franchise intertexts, ranging from their own original franchises (such as *Pirates of the Caribbean*) to those of their acquired subsidiaries (such as Marvel or Pixar) as well as externally licensed IP (such as *Avatar*). As in the case of the licensing of *Star Wars* for Star Tours foreshadowing the eventual acquisition of Lucasfilm, incorporating these franchises under the greater Disney umbrella through the parks themselves often precedes Disney's actual acquisition of these properties. Properties within the Disneyland superfranchise may be identified primarily as part of the Disneyland franchise (for example, Space Mountain) or may occupy a more nebulous identity both within and without a park context (for example, The Haunted Mansion attraction and film). In each case, by bringing various properties together, the parks reinforce a more cohesive brand identity for Disney.

The Disney parks have been franchised in a number of ways. The original Disneyland park in Anaheim was initially franchised through the eponymous television series *Disneyland*—the park itself was the subject of the television series, while both the series and the park also synergistically featured and promoted Disney's other franchises. Merchandise, too, was capitalized on in these early stages to build and disseminate the Disneyland franchise. In later years, Disneyland was physically replicated across the globe; there are currently eleven Disney theme parks, six of which are modeled after Disneyland, beginning with the creation of the Magic Kingdom at Walt Disney World in Orlando, Florida in 1971, and eventually leading to Disneylands in Tokyo, Paris, Hong Kong, and Shanghai. Disneyland Tokyo is the only physical park that has been franchised under a licensing agreement. Like retail or food merchandising, the park including the layout and attractions, is licensed from The Walt Disney Company—but is owned and operated by The Oriental Land Company (Fisher, 2017). Each of these Disney parks can be read as a text itself, together comprising the "Disney Parks" franchise, but these iconic "Disneyland" parks reinforce the core identity of that franchise as the original and most recognizable iterations of a Disney park. Disneyland has also been franchised through feature films with park-based movies (such as *The Haunted Mansion* [Rob Minkoff, 2003] and *Tomorrowland* [Brad Bird, 2015]). The 2012 release of "Thrills and Chills," a four-movie DVD collection of films based on Disney parks attractions, shows the conceptual linkage of park attraction-based films as "family favorites [that] bring the magic from the parks straight into your home!"[4] In the case of the *Pirates of the Caribbean*, park space-based films have even been established as their own successful franchises. Disneyland video games serve this larger project, too, while expanding it beyond the physical confines of the parks themselves.

Though analog (built) and digital (video game) theme parks may seem ontologically at odds with one another, video games are particularly apt tools for franchise creation for what are ultimately similarly ludic and interactive virtual and physical park spaces. In other words, the parks themselves and their digital incarnations are interactive spaces cocreated through their active use. The players (or guests) themselves constitute the virtual and physical parks through the negotiated and participatory construction of meaning within these game worlds. As Johnson explains, it is the design as well as the "emergent creative use" of franchise worlds by communities of users that "allows franchising to be understood in terms of its creative structures, but also in terms of the dynamic creative subjectivities and strategies that emerge from shared use of worlds" (2013: 115). Guests actively manifest Disneyland's built park spaces by inhabiting them.[5]

Though the immersive construction of these spaces seeks to guide the guest experiences, it is not always successful[6] and is rather a process of negotiation that produces an experience somewhere between control and free play.[7] This is arguably similar to the tension between free play and linearity in many "open-world" video games (such as the *Grand Theft Auto* franchise [Rockstar Games, 1997–present]). In his discussion of audience participation in the Indiana Jones Epic Stunt Show Spectacular at Disney's Hollywood Studios, J. P. Telotte observes how their immersion "emphasizes the deceptive power that both [movies and theme parks] wield, how they are equally able to sell us their illusions" (2008: 138). He further notes how such immersion "practically celebrates our participation in that deception, our complicity in the bargain being offered for our consumption, ultimately calling into question just who wields the 'control technology' here" (Telotte, 2008: 138). Disneyland guests cocreate the park experience, and this interactive element of active creation is also present in Disneyland's virtual game versions and thus in the Disney park franchise itself. As Dusenberry has explained with regard to the use of Disneyland in *Disney's Epic Mickey* (Disney Interactive Studios, 2010):

> By using players' knowledge of Disney and allowing them to make choices about how they will participate with the different media embedded in the game, *Disney Epic Mickey* highlights the potential of film, TV, and game convergence to create meaningful spatial and temporal stories, where each medium adds complexity to the player's experience with the franchise and where nostalgia is used to initiate and sustain long-term engagement with a broad, transmedia story. (2013: 184)

Dusenberry's argument points to the significance of choice and agency as well as individual experience and nostalgia in the active creation of story experiences.

In promising to bring the Disney park experience into the player's home, these games show a persistent impulse, if not strategy, to further the Disneyland franchise by bridging the experiential gap between the physical theme park and the virtual (and reproducible) screen. This recalls what Lynn Spigel has identified as early television's power to provide "technological solutions to distance" (1992: 7). Like television, these games attempt to address a significant obstacle to theme-park franchising: overcoming the physical specificity of the theme park as media object. In other words, to franchise the Disney parks, which are experience-based, Disney must work out how to deliver the park experience outside of the park gates. Despite repeated attempts and technological advances, however, these games ultimately fail to capture the nature of the Disney park experience. They may invoke typical park behaviors and sights, and often do serve as compelling advertisements or nostalgic trips down Main Street/memory lane, but they can only approximate or invoke the park experience, rather than truly replicate or create it. The rerelease of *Kinect Disneyland Adventures* as *Disneyland Adventures*, with the loss of the "immersive" Kinect technology, emphasizes just how large this experiential gap is and the inherent challenges of using digital technology in an attempt to bridge it. Tracing the history of Disney park-based games and apps, and considering other ways in which the parks have been "gamified," reveals an ever-deepening trend of using digital game modalities to expand the Disney parks beyond their physical limitations as a means by which to establish and further them as a supermedia franchise.

Using select case studies from the history of Disneyland video games, this chapter traces how Disney has used video games for park franchising to understand how and why these games elicit the experience of being in a Disney park.[8] While there are other video games that connect in some way to the parks—whether they use loose interpretations of park space as settings for narrative storylines such as the *Kingdom Hearts* franchise (Square-Enix, 2002–present) or are based on films derived from rides like *LEGO Pirates of the Caribbean: The Video Game* (Disney Interactive Studios, 2011), this discussion is limited to focus on games that recreate the park space and in which navigation of the park space is a central experience of the game.[9] In particular, I explore video games that recreate the site-specific geography/architecture and invoke specific interactive guest behaviors of a Disney theme park. That is, I am interested in games whose core focus and attraction are inhabiting park space and reproducing park experiences. Disney's impulse to recreate the embodied park experience, and to transmit it into our homes, pockets, and beyond reveals how the company uses digital technology to dissolve physical park boundaries.

Out of the Television and Into the Home: Disney and *Disneyland*

For Disney, the origins of the theme park are inextricably linked to the idea of bridging gaps between media. Disneyland originated as a new kind of media/park convergence, beginning with the very idea of the newly conceived "theme park" itself. As J. P. Telotte has noted, "The first step in developing Disneyland—and its subsequent offspring—involved creating a kind of entertainment hybrid, an amalgam of the amusement park and movie experience" (2008: 119). Ontologically, in Disneyland, Walt Disney combined the embodiment of the amusement-park form with the narrative immersion of the movies to create the "theme park," a new kind of media experience that has since been expanded on and replicated throughout the world. As Telotte (2004), Christopher Anderson (1994), and Jennifer Gillan (2015) have demonstrated, however, Disneyland also developed out of the television industry.[10]

In 1954, ABC signed a deal with Disney, whereby the network promised $2 million for a fifty-two-week series with a seven-year option for renewal.

More importantly for the park, ABC agreed to purchase a 35 percent share in what would be Disneyland for $500,000, providing Walt Disney with the funds needed to finance his vision (Anderson, 1994: 141). The series, produced by Walt Disney Productions under the *Disneyland* title from 1954 to 1958,[11] took up the anthology format, allowing Disney to use the program synergistically to promote several different aspects of the burgeoning Disney media empire. The *Disneyland* television series was something of a mash-up that combined preexisting content (including the studio's back catalog of short cartoons), promotional segments for new studio releases, and sections of (or even entire) episodes devoted to showcasing the new theme park, all tied together with Walt as the host and personal guide through the world of Disney (Telotte, 2004: 64–79). The use of the series as a vehicle for promoting existing Disney content and franchises was such a central focus for *Disneyland* that the "only completely new programming" in its first season was the three-part *Davy Crockett* serial and Walt Disney's hosted lead-ins (Telotte, 2004: 27).[12] *Davy Crockett*—later expanded into films, merchandise, and tie-in theme-park appearances—proved an early lesson for Disney in how to use the combination of television, movies, and theme parks to capitalize on (and later to build) a successful transmedia franchise (Telotte, 2004: 31–5).

While *Disneyland* quite literally brought Disneyland into existence through the solid financial backing of ABC, Anderson argues that "television's figurative representation of Disneyland actually called the amusement park into existence," bringing the possibility of such a place into the minds of future park guests (1994: 134–5). Even before the park was finished, the program raised consciousness about this new phenomenon—the theme park—and, as Gillan has demonstrated, showed viewers how to *think* about Disneyland (2015: 227–35). As Karal Ann Marling and Kathy Merlock Jackson have noted, the show's structure divided *Disneyland* based on the park's lands of Frontierland, Tomorrowland, Adventureland, and Fantasyland, "corresponding with the studio's cinematic genres: its signature animation, as well as action adventure, the Western, and science fiction" (Jackson 2011: 21). This fused park structure *and* navigation to the park's conceptual roots in Disney media through the television. The *Disneyland* television series conceptually brought the park into existence for the viewer while bringing the park virtually into their home through the television set.

The show simultaneously brought the viewers *to* Disneyland, however, as in *Disneyland*'s first episode, "The Disneyland Story," where scale models, maps, and aerial footage are used virtually to transport home viewers to the still-unfinished park (Figure 4.1). Following these maps and aerial shots, the camera pans through, in Walt's words, a "quarter-inch-to-the-foot scale model of Disneyland" to guide audiences through the mock park. As Telotte notes, "[T]he most curious element of 'The Disneyland Story' episode is precisely

Figure 4.1 "The Disneyland Story."

this insistence on blurring boundaries" between television and park (2004: 76). As Walt himself says as he introduces the park by way of an aerial map, "Disneyland the place and *Disneyland* the TV show are all part of the same" (Telotte, 2004: 10). This is accomplished, in part, by using the televisual medium to fluctuate between the simultaneous movement inward, from the park into the home, and outward, from the home into the park. The concept of flying the viewer into the park space was later brought to life in a 1958 episode, "An Adventure in the Magic Kingdom," in which Walt quite literally asks an animated Tinker Bell to fly guests down the freeway (after sprinkling the screen/audience with pixie dust) and into the park for a visit (Gillan, 2015: 215). While destabilizing virtual and physical boundaries, the bird's-eye view, scale model, or map also orients the viewer to park space and primes them to navigate that space. Disneyland video games would later employ this televisual technique for similar purposes.

More than simply presenting the parks to viewers, *Disneyland*'s blurring of these spatial boundaries also functioned as a means by which Disney could compel viewers to become visitors. Through the television series, the park was made into a narrative setting that would become activated when viewers actually travel to the park to take their parts in the "inhabitable text" of the park. According to Anderson:

> A trip to Disneyland—using the conceptual map provided by the program—offered the family viewer a chance to perform in the Disneyland narrative, to provide unity and closure through personal experience, to witness the "aura" to which television's reproductive apparatus could only allude. (1994: 152–3)

With regard to the idea of "aura," Walter Benjamin has argued that, "even the most perfect reproduction of a work of art is lacking in one element: its presence in time and space, its unique existence at the place where it happens to be" (2007: 220). The lack of this aura in the televisual representation of the physical park, combined with a narrative to be inhabited by the viewer, created a compelling impetus for viewers to become guests. Through the combination of park and television, Disney developed a template for using different media forms as a method by which to expand the reach of its brand and to disseminate a distinct physical location to a potentially limitless number of household television sets. This replication and dissemination are precisely the goals of franchising.

In addition to presenting the park conceptually as an "inhabitable" narrative to be activated by a park visit, the *Disneyland* television series also helped to condition viewers in park behavior. In show segments focused on the park space itself, viewers are presented with not only how to think about the park but what to do there. In the "An Adventure in the Magic Kingdom" episode, as the aerial camera flies over the freeway, the *Disneyland* narrator and "voice of Disneyland" Dick Wesson notes that "if it wasn't for Tink, we'd be driving to the Disney studio party [in Disneyland] in those cars down there," a remark that reinforces the idea of a magical Disney experience, while also showing the future guest's reality of using the freeway to access Disneyland, which also taps into the park's reliance on a mid-century burgeoning car culture.[13] As the camera tours around the park, it shows—and Wesson tells—viewers what to do there and thus *how* to use park space; in Frontierland alone, viewers are impelled to "warm up [their] six shooters," "take an excursion on the Mark Twain riverboat," and "explore caves" on Tom Sawyer's island.

Television, however, is not the only medium the Walt Disney Company has turned to as a vehicle for disseminating the experiences of its theme parks and compelling viewers to become visitors. As technologies advanced in the second half of the twentieth century, video games offered a new opportunity to bring the parks into the home and virtually to transport new guests to the parks. In her discussion of Disney-affiliated sitcoms and Disney Channel shorts, Gillan notes how, like the *Disneyland* television series, shows such as ABC's

Modern Family (2009–present) or *The Middle* (2009–18) "became paratexts for an experiential relationship to the park, which would begin when television viewers became site visitors" (2015: 235). Similarly, video games became another method by which Disney created an experiential relationship to the park in the domestic space via gaming consoles and computers. Like these television shows, video games "provide narratives through which Walt Disney desired viewers to experience the park. They shape viewers' perceptions of what they would see at the park when they did make a site visit" (Gillan, 2015: 235). As Dorene Koehler notes, games like *KDA* and *Epic Mickey* "return the experience of the park to the living room, utilizing the merchandizing [*sic*] and mythic power of the park as a place to both sell games and, like the *Disneyland* and *Mickey Mouse Club* shows of the past, turn the television into a magic window into the park" (2017: 163). In this way, video games represent a digital continuation of the project begun by the *Disneyland* television series of the 1950s to bring the parks into the home and cultivate desired viewer/guest relationships to them as a distinct media text.[14]

Because video games are inherently more interactive than television, however, requiring players to make decisions and affording them a degree of agency, the effect of this perceptual training is heightened as players become invested, through the reward mechanism of the video game, in conforming not only to the targeted perceptions of the park but also to the desired active *behaviors* of park interaction. Video games *require* players to act, to participate kinetically with their hands and bodies rather than just watch, and with the presence of the disembodied yet digitally manifest player (typically communicated visually through the presence of an avatar), games can train the player in how to behave in the physical environment of the theme park through the virtual space of the digital park on the screen. In other words, these simulated behaviors can then be reenacted "in real life" when players visit the park. As Michael Z. Newman has observed, early video games were characterized as a form of interactive, participatory television. According to Newman, "this idea of manipulable television, of the audience as active agents making graphics move around the television set, implies a transformation not only in the address of media to audiences, but also in the orientation of audiences to media" (2017: 66). A textual analysis of Disneyland video games demonstrates that, like the *Disneyland* television series, these games allow Disney to shape how its parks are experienced by educating potential future guests in not only how to think of, but how to *embody*, the space of the park.

For previous park guests, Disneyland video games can reinforce desired park behaviors, but they can also refresh and enhance memories of actual park visits. In particular, these games have the potential to invoke the nostalgia of past park experiences, to reshape memories in ways that emphasize preferred park interactions, and to reward the player with the satisfaction of already being familiar with the layout and attractions, having enacted the behaviors typical of the park experience. In her analysis of the dystopian version of Disneyland in *Disney Epic Mickey*, Dusenberry argues that "selecting Disneyland as a model for the game world triggers existing nostalgia in players familiar with Disney *and* manufactures nostalgia in players who understand gameplay but are unfamiliar with the game's transmedia links" (2013: 185–6). This gratification is probably enhanced for frequent park visitors, particularly for annual passholders who typically take pride in their park knowledge and experience—for them, these games can reinforce specialized inside knowledge. Disney's preferred methods of reception (how to conceive of the parks) and of behavior (how to act in them) are communicated through these virtual spaces by training future guests as they interact with the game interface. Simultaneously, the parks are brought into the home, disseminating the park space and enabling more frequent interaction with the parks as a media franchise.

Disneyland Goes Digital: Early Disney Parks Video Games

The replication of Disney parks through video games dates back to the 1990 release of *Adventures in the Magic Kingdom,* developed and published by Capcom for the Nintendo Entertainment System (NES), which first brought an interactive digital Disneyland into the home. At the time *Adventures in the Magic Kingdom* was released, the NES dominated the North American home-video gaming market, having revitalized it after the industry crash of 1983. By 1991, more than thirty-three million homes had NES consoles (Kent, 2001: 46). When *Adventures in the Magic Kingdom* came out in 1990, the conditions were prime for using this new, widespread personal technology as a vehicle for bringing the parks into the home.

In *Adventures in the Magic Kingdom*, the player navigates a rudimentary eight-bit overworld map of the Magic Kingdom using an avatar personalized with his or her name.[15] Similar to the way in which the model and park map situate the viewer in the *Disneyland* television series, the digital map clearly situates the gameplay within the park space by approximating the actual hub-and-spoke layout of a Disney Magic Kingdom Park (Figure 4.2). The title locates the player within *a* Disney Park but not necessarily a specific one—while "Magic Kingdom" is indeed the official name of the Walt Disney World, Florida version of Disneyland park, other "Disneyland" parks, and typically Disneyland in Anaheim park, are also colloquially referred to as the "Magic Kingdom." The multiple franchised physical "Disneyland" parks (at the time of release only in Anaheim, Orlando, and Tokyo) are thus all potentially and simultaneously invoked by the identifiable but non location-specific game map.

The ostensible object of *Adventures in the Magic Kingdom* is to collect the six silver keys that unlock the castle for Mickey's big parade. In doing

Figure 4.2 *Adventures in the Magic Kingdom.*

so, the gameplay involves three modes of engagement with the Disney parks: navigating the park space, answering trivia questions, and "experiencing" the attractions. These different behaviors correspond to desired modes of interaction with the "real" Disney parks. Players are virtually engaged in an approximation of what it is like physically to inhabit and interact with the park space. The attractions in *Adventures in the Magic Kingdom* are themed minigames with gameplay mechanics akin to other NES games. For example, the Autopia level mimics top–down vehicle racers like *Road Fighter* (Konami, 1985), the Haunted Mansion and Pirates of the Caribbean level recalls scrolling platformers like *Super Mario Bros.* (Nintendo, 1985), and Space Mountain is similar to space simulator shooters like *Star Voyager* (Acclaim, 1986). An obvious obstacle to recreating the park experience, especially in eight-bit form, is how to convert the built ride experience into a two-dimensional video game. The solution here, as in later games, is to translate rides into established video game modalities, a process often resulting in the creation of game narratives distinct from the actual attractions in the physical park. As M. J. Clarke has explained in relation to television video-game tie-ins, "licensed games typically translate their licenses into preexisting genres and formats" (2013: 108). As the technology progresses, we can see this borne out to a certain extent—later games bear an "ever-increasing fidelity" to their source material, but "are constructed and played in a manner common to video games" (2013: 109).[16]

While game progression requires passing these minigame levels, the player cannot complete the game without also correctly answering a number of trivia questions about Disney history, which speaks to another mode of interaction with the park space: the cultivation and demonstration of franchise knowledge and fandom. Questions vary from basic facts about the parks (such as "Which is the largest Disney Magic Kingdom?") to trivia on early Disney television series (such as "What was the name of the girls' ranch in the 'Spin and Marty' serial on the 'Mickey Mouse Club'?") and Disney films (such as "What is the name of the King of Hearts from 'Alice in Wonderland'?").[17] In essence, the gameplay trains the player as a Disney fan; players are either rewarded for possessing the kind of knowledge that only a Disney devotee might know off the top of the head (pre-Internet) or taught the correct answer if wrong. Trivia also easily translates into gameplay, particularly when technology limits the ability to translate more accurately the park experience.

Through gameplay, the player is both exposed to and educated in a number of desired behaviors required for progression through the game and which also constitute engagement with the Disneyland franchise. Focus on attractions as the most desired destinations in the theme parks is emphasized through their privileged position as the "exciting" part of the game, as the visual and gameplay mechanics of these are more dynamic than simple map navigation or trivia selection. This recalls how Disney Imagineers use attractions as "weenie," or "visual magnets" that help circulate guests through physical park space (Sklar, 2015).[18] Because the object of the game is to track down and collect a set of keys, even the collection of items and protopurchasing behaviors are encouraged (or, rather, required). The creation of a narrative storyline in which players are the catalyst ("Maybe you can find the six keys for us!"), however simple, implicates players as generative of these behaviors despite the fact that they may not have a great degree of agency in the game. At the parks, guests are guided through the space but ultimately decide how the "play" progresses. Even in a relatively early video game, with a basic premise and simple gameplay, desired forms of thinking about *and* behaving in the park are both taught and reinforced as a means of bringing the park experience into the home while also encouraging the player to become a Disneyland fan.

In 1996, Disney Interactive—the studio's video-game division established in 1994—released *Walt Disney World Explorer*, a CD-ROM application that represented a continuing effort to use digital media to further the Disney Parks franchise through digital Disney parks. Though less a game than an interactive map of Orlando's Walt Disney World, *Walt Disney World Explorer* nevertheless used digital video game technology to fulfill many of the same functions as *Adventures in the Magic Kingdom* and even the *Disneyland* television series to bring the park into the home and teach the player how to inhabit park space and relate to the Disneyland franchise. According to its tagline, *Walt Disney World Explorer* let users "[e]xperience the sights, sounds and magic of Walt Disney World!" It functioned as a virtual tour guide, advertising the park's features for potential guests interested in learning more about the park.

Walt Disney World Explorer also taught users preferred ways of experiencing the park. Like *Adventures in the Magic Kingdom*, it reinforced brand knowledge through trivia facts, behind-the-scenes videos, and even a timeline of Walt Disney World history. Similar, too, was *Walt Disney World Explorer*'s incorporation of collection, perhaps the most traditionally game-like feature of the application. Instead of keys, the focus was on collecting "Hidden Mickeys", which are images or symbols of Mickey, typically in the form of the simplified three-circle head-and-mouse ears silhouette icon, with a long tradition of being hidden by Imagineers in the Disney theme parks and even in the company's animated films as a kind of inside joke with the fans. Among Disney parks fans, there is an established tradition of collecting and exchanging knowledge and sightings of the locations of Hidden Mickeys, such as with the fan-run website hiddenmickeys.org. The application thus not only trained users as parkgoers but also initiated them into fan practices. This helps to establish and maintain a fanbase for the parks that is distinct from, though deeply interconnected with, the company's other media franchises.

In 2005, Walt Disney Parks and Resorts launched *Virtual Magic Kingdom*, developed in conjunction with the Sulake Corporation, as part of its "Happiest Celebration on Earth" promotional campaign to celebrate the fiftieth anniversary of Disneyland. A massively multiplayer online game, *Virtual Magic Kingdom* brought fans together in a digital space to experience the park in a virtual representation of a Disney-style theme park. *Virtual Magic Kingdom* resonated enough with players that after Disney shut it down in 2008, a devoted community of fans on myvmk.com recreated the game. While *Virtual Magic Kingdom* contains lands and attractions based on real park features, the game design still uses a bit of artistic license. The park space is less a total mimetic simulation than an approximation of a Disney park. While its map is relatively faithful to Disneyland's, its park layout lacks the kind of photographic and cartographic accuracy of later games. Instead, identifiably accurate elements, such as the Jungle Cruise tableau of safari guides treed by an angry rhino, or the isometric rendition of the flying-over-London room in Peter Pan's Flight, indicate that this is a "Disney park."

Like *Adventures in the Magic Kingdom* and *Walt Disney World Explorer*, *Virtual Magic Kingdom* also recreates the Disney park experience by invoking typical park behaviors. Minigames serve as attractions, and collecting and purchasing are elicited through the inclusion of special virtual pins and badges acquired through gameplay and even the opportunity to "shop" in the game (Lukas, 2008). The connection between virtual park and physical park was further solidified with the implementation of in-park quests that rewarded users for visiting the real Disneyland and Walt Disney World.[19] Guests would print out quests at home, and log into the in-park VMK kiosk to answer questions and earn both virtual and physical prizes (Knight and Knight, 2006: 153). With the integration of in-park quests with the otherwise strictly at-home gaming experience, *Virtual Magic Kingdom* reinforced a core purpose: to motivate users to become guests and to teach them how to interact with the parks once they do. The virtual game space and the physical park space are blended, and the actions taken in one (answering questions on a park kiosk) affects the other (receiving virtual in-game prizes). The game literally translated virtual behaviors into actual physical behaviors in the Disney parks. Like the games discussed above, *Virtual Magic Kingdom* brought the Disney parks into the home via a home computer and Internet connection but it also allowed the player to have a "home" in the virtual park. Players are given virtual private rooms, themed to different Disney park spaces, which can be decorated with items bought and collected in the virtual Magic Kingdom. The park is brought into the home, while the "home" is brought into the (virtual) park, completing the reciprocal cycle from one kind of park space to the other.

(Kinect) Disneyland Adventures

It wasn't until the development and release of *Kinect Disneyland Adventures* in 2011 that digital graphic and interface technology allowed Disneyland park space to be realistically recreated with a high degree of accuracy, combined with a play mechanism that attempted to bridge the divide between the actual parks, which are fundamentally dependent on physical experience, and the screen experience of video games. Preceding games used the more traditional interfaces of a controller or a keyboard and mouse as the point of interaction. Coming soon after the introduction of Microsoft's Kinect Technology, *KDA* was the first Disney park game to utilize the new body-as-controller motion-capture technology of Microsoft's Kinect virtually to adapt the embodied experience of the physical park and bring it into the home.[20] While it shares many of the same impulses as earlier Disney Parks game texts, *KDA* harnesses the newly available technology to blur further the boundaries between the virtual and the physical by prompting players to move around in, interact with, play in, and consume park space using their bodies. The kinetic theme park became more accessible at home through this new technology whose interface integrates the player's own body and voice.

Released in November 2010, Microsoft's Kinect was initially wildly successful, selling eight million units in just over a month (far surpassing the company's five million unit estimate), with utopian promises to revolutionize at-home console gaming (Morris, 2011). The Kinect is a "depth camera," that creates an image through infrared light to detect where objects are in space, so that the player's body and voice act as controller (Borenstein, 2012). In terms of play experience, it is similar to Nintendo's popular Wii interface, with its motion-sensitive controllers, while going one step further to erase the interface by eliminating the use of handheld controllers during gameplay. The removal of a handheld input device is meant to heighten the immersive gaming experience into a new kind of embodied gaming, turning the static experience of sitting with a controller into the more kinetic gaming experience of interacting with one's whole body while standing.

The packaging for *KDA* impels players to "[u]se your body, voice—and a touch of pixie dust—to explore the timeless worlds of *Disneyland* park in *Kinect: Disneyland Adventures*," and a pack-in insert announces that, with Kinect games, "you are the controller." *KDA*'s body/sensor input interface privileges the body—not just the hands/eyes—as the site of interaction. To start, players must hold their hand out in front to navigate the game menu where they create an onscreen avatar to serve as their proxy in the virtual park space. Character interactions are instigated through bodily movements as well, as players must wave their hand to interact with characters onscreen. Other virtual character interactions are triggered by their "real-life" proxies: players can hold their hands out to ask for autograph, bow to dance, hold their arms out to the side for a hug, or high-five characters (Figure 4.3). These physical gestures result in mini in-game cutscenes, where the onscreen avatar will hug,

Figure 4.3 Virtual character interactions in *Kinect: Disneyland Adventures*.

dance, or high-five the digital character. This produces a mirroring effect, with the player's physical movements virtually duplicated as the onscreen avatar synchronously recreates the player's "real-life" gestures and postures.

Enhancing the immersive mechanics of the Kinect, where a player's physical full-body movements correspond more directly to the embodied park experience, *KDA* also capitalizes on the increasingly realistic graphics enabled by the seventh-generation Xbox 360 console. As Clarke has observed of television-based licensed games, "[W]on by the technological advances of sixth- and seventh-generation gaming consoles and compelled by the cultural virtue of photorealism, licensed game makers maximize connection with the on-air series through the strict digital translation of recognizable visual elements" (2013: 108). Similarly, realistic fidelity to park space, including layout, scale, and overall appearance, reinforces the game's connection with the "real-life" Disneyland. The game's use of photography in gameplay reflects this tendency toward photographic realism. Some quests prompt the player to collect photographs with characters using the virtual in-game camera—photographs are of the player's avatar and the character, though the player is typically prompted to copy the character's pose with their body, thus matching their avatar's pose with the character's. Similarly, designated photo spots direct players to take virtual "photographs" of the virtual space (Figure 4.4). Most interestingly, the Kinect camera automatically takes photographs of the real-life player playing the game at points in the minigame play, a feature recalling the analog ride photos taken on Disneyland attractions such as Space Mountain or Splash Mountain.

Figure 4.4 A designated photo spot in *Kinect: Disneyland Adventures.*

The new Kinect technology was promoted as a step toward greater immersion and more authentic embodiment, and this kind of discourse characterizes the language used in promotions of *KDA*. The game trailer on the DisneyGames website carries the tagline: "Experience Disneyland magic like never before," and the girls in the trailer exclaim, "I wish there was a Disneyland in our house!" (Walt Disney Company). The game promises, therefore, to bring a heightened Disneyland experience "like never before" into "our house" and, according to some reviewers, the game succeeded. Reviewer Jack DeVries of *IGN*, a self-confessed "Disneyland nut," remarked, "*Kinect Disneyland Adventures* recreates the Magic Kingdom so well that it sometimes surpasses Disneyland itself" (2011). In many ways, that is the role *KDA* fulfills—a faithful virtual recreation of Disneyland that players can navigate and virtually "inhabit." Players can simulate being in the park itself by collecting photographs, autographs, and pins while shopping for virtual souvenirs in various stores around the park. Players can also "ride" simulated rides; certain attractions are included as non-playable cutscenes that simulate the first-person perspective of riding an attraction, such as Dumbo the Flying Elephant, the King Arthur Carrousel, or the Mad Tea Party.

Simulation is only part of the game, however. Beyond these simple ride simulations and following the form of other Disney park video games before it, *KDA* also includes playable minigame "attractions." These attractions are frequently loosely based on real attractions, most often with new storylines and content.[21] For example, part of the narrative of the second "chapter" of the Matterhorn minigame is that yetis have kidnapped Goofy; players help rescue him with "real-life" throwing motions that cause their avatars to throw snowballs at the monsters.[22] While the physical Matterhorn attraction at Disneyland does include Audio-Animatronic Abominable Snowmen, it can hardly be said to have more than a rudimentary narrative, much less one that somehow involves Goofy. Moreover, the physical spaces of these playable rides in *KDA* are generally not faithfully recreated, putting them in sharp contrast with the larger park space of *KDA*, which approaches the uncanny in its adherence to actual park topography. While the Matterhorn *KDA* minigame does feature a toboggan ride similar to the Matterhorn roller coaster, the virtual Matterhorn is much more fantastical, and game-like, than its physical counterpart. In the first "chapter" of the minigame, players actively steer the in-game bobsled by leaning from side to side while, in the third, players "ski" by leaning, pumping their arms to go faster, and jumping to jump in-game (Figure 4.5). Similarly, in the Pirates of the Caribbean minigame, scenes from the ride become "playable" by players, such as the famous "prison dog" scene: in the ride, this scene is a tableau that guests drift by but, in the game, players can throw fruit (with their arms) at the starving pirate prisoners. In the pirate-ship fight scene in *KDA*, the player is the one actively ducking, jumping, and swordfighting, in contrast to

Figure 4.5 The Matterhorn minigame in *Kinect: Disneyland Adventures.*

the animatronic figures on the ride. In a sense, these minigames could arguably feel *as or more* embodied than the actual rides, with a somewhat greater degree of actual physical agency (for example, moving your body instead of having your body moved for you).[23]

According to Clarke, minigames "flesh and fill out licensed games," as they can function as "play sequences that briefly diverge from the dominant visual design and game mechanics, often in the form of a narratively contextual puzzle" (2013: 110). In *KDA*, minigames diverge from the primary narrative of a visit to Disneyland—and its realism—in order to capitalize on the Kinect technology by foregrounding new mechanics and narratives for the rides. With its combination of simulations and games, *KDA* is more of a hybrid than some of its predecessors. It combines the questing game structure of *Adventures in the Magic Kingdom* and *Virtual Magic Kingdom* with the lifelike simulation of *Walt Disney World Explorer*.

The various functions of *KDA* help shape the potential future park experience for gamers and thus how they experience Disneyland as the core text of the Disney parks franchise. *VentureBeat* critic Joe Sinicki cynically, but insightfully, calls *KDA*, "a game that basically serves as a commercial" (2011). While he is correct in one sense, that *KDA* serves a function similar to the *Disneyland* television series in that it is meant in part as a promotional tool on behalf of the parks, there is a deeper function of the game: subtly to train the player in how *physically* to inhabit park space through the more embodied Kinect technology. Like the Disney park video games before it, *KDA* exploits player agency inherent in video gaming, and the increased sense of agency implied by the physical movement necessary to interact with the Kinect sensor,[24] to reinforce the training of desired behaviors for using park space: what to do at and how to experience Disneyland.

As in earlier games, KDA reinforces brand knowledge through virtual park guides who impart rare Disney facts when the player walks by them or engages them with a wave (again, virtually duplicated as the onscreen avatar mirrors the player's real wave). Even the background chatter of the other virtual "guests" that populate the park train the player in typical Disneyland discourse.[25] They prompt the player to marvel at the environment, as atmospheric NPCs marvel, "Can you believe the detail they put into this place?" This line in particular seems to be a winking reference to both the physical Disneyland and its laboriously mimetic virtual corollary. Another "guest" asks "Does anyone know what time it is?," suggesting that Disneyland is a place of another time, an atemporal fantasy space where all relation to the "real" world becomes irrelevant in relation to the fun everyone is having there.[26] This dialogue, too, is an integral part of the experience of Disneyland, and even supplies future guests with some standard lines—and thoughts—for when they actually visit the park.

Other activities prompted by *KDA* include seeking out characters for meet-and-greets. Waving while in proximity of a character (indicated by a sparkling ring area) enables the player to interact with that character. Once engaged with a character, players use physical movements to produce further interactions; for example, players can bow to trigger a dance sequence between the character and the player's onscreen avatar, hold both hands out in front to ask for an autograph, hold both arms out to the side for a hug, or high five for a high five. When prompted to pose for a photo with a character, players are encouraged to match their bodies with the character's onscreen pose. Most quests in *KDA* are given to the player by the characters. Character quests often revolve around collecting items, sometimes using in-game tools such as cameras, magic wands, and blasters that other characters give the player. Some of these more fanciful quests do not have literal analogs to park behavior (apart from Disneyland guests' affinity for buying and carrying lightsabers). Other quests, however, very clearly do have real counterparts typical of a visit to the physical theme park. In the game, characters will ask the player to collect autographs from, or take specific photographs with, other characters or to buy specific items from the in-game store, which the player must purchase using coins collected primarily from playing the attraction minigames.

In addition to character-driven quests, *KDA* also prompts players to replicate inpark behavior by collecting pins from the attraction minigames. This ties into a common guest behavior at the physical parks, a fan phenomenon officially referred to as "Disney Pin Trading." Players unaware of pin trading, which tends to be practiced more among park enthusiasts rather than casual tourists, are given a practical primer on the ins and outs of pin trading as they seek special-edition versions of pins in the game, which mirrors real-life pin traders' quests for the rare and desirable. The game also includes another inpark behavior, souvenir shopping, with a proliferation of virtual shops in which players can purchase clothing, costumes, autograph books, and photo albums, versions of which are also available in those actual shops in the physical Disneyland. This ties back to similar shopping gameplay in *Virtual Magic Kingdom*. Since purchasing these items requires ingame coins collected by walking around the park or by playing the minigames, the need for large amounts of coins to complete the character quests can result in minigame gameplay that becomes merely a means to collect coins and purchase items. In *IGN* reviewer Jack DeVries's opinion, "Between the Hidden Mickeys, pins, costumes, autographs, and photos, there's a ton of stuff to collect and discover within the park. Most of it comes very naturally through the game, eschewing the needless collectible quests in favor of a more organic progression" (2011). In addition to being more or less covertly disguised through game mechanics, however, consuming and purchasing are also openly promoted by the industrial discourse surrounding the game. Frontier's own *KDA* website lists "[c]ollect Disney-themed items and purchase popular souvenirs" as being among the "Features" of the game (Frontier Developments). Though gamers tend to bemoan "collect-a-thons," the use of the game as a consumer-training device is less masked than touted as an essential, fun, and meaningful part of the Disneyland experience in *KDA*.

Like *Virtual Magic Kingdom*, *KDA* includes a feature by which players can take action in the "real" world that crosses over into the virtual game space. For *Virtual Magic Kingdom*, this was the completion of quests in the parks that award virtual prizes in the game. For *KDA*, the mechanism is inverted: players could purchase plush toys of Mickey, Minnie, Donald, and Goofy at Toys "R" Us that they would then scan into the game for additional character quests. Instead of focusing on the quest as the primary activity, as in *Virtual Magic Kingdom*, the purchase of consumer goods is the primary activity for *KDA*'s real-life quest component. Ultimately, *KDA* brings the park into the home while also educating the player/park guest in not only how to think about the park, like the *Disneyland* television series before it, but also in how to experience park space, as both a part of the game and as a consumer.

It is significant, however, that this is done here through the Kinect technology. Promotional texts, from Disney's games website, Microsoft's *KDA* page, the Frontier Developments *KDA* page, and even the packaging inserts for the physical Xbox game disc all contain a variation on the same language—as the insert pamphlet reads, "Kinect™: Disneyland® Adventures is your ticket to the magical world of Disneyland park through Kinect." The carefully crafted message here is that it is the new, more immersive and embodied Kinect technology that allows the player to experience Disneyland. Cutting-edge digital technology is touted as a means of access to authentic immersive experiences. With *KDA*, Disney, faced with a changing technological climate, uses cutting-edge consumer gaming technology to bring the most analog of phenomena (the theme park) into the digital age.

In the years since the original release of *KDA*, Microsoft has discontinued manufacture of, and support for, the Kinect sensor and games (Wilson 2017).[27] It was not game over for *KDA*, however, which was remastered and released for Xbox One and Windows 10 as *Disneyland Adventures* (sans Kinect) in 2017. The elimination of the Kinect sensor strips *Disneyland Adventures* of the whole-body interaction that had made *Kinect Disneyland Adventures* a uniquely compelling attempt at imitating the park experience by bridging the virtual/physical screen divide through the player's physical body. Some critics and gamers praised the non-Kinect port for its ease of use and 4K remaster, as Ian Morris's review subtitle suggests: "All the fun of Disney, with none of the nonsense of Kinect" (2017). Others pointed to the awkwardness of converting a game founded on Kinect controls to traditional handheld controllers; for example, John Elliott of *XboxAddict.com* observed that "as a former Kinect game it doesn't translate well into controller based gameplay, as it feels like you are playing *Dance Dance Revolution* or *Rock Band* on a controller" (2017). In citing other kinetic games predicated on an embodied experience, Elliott underscores how a controller-based adaptation of *KDA* is fundamentally at odds with its purpose.

KDA, like the Disney Parks games before it, ultimately failed to capture more than an instructive approximation of Disney park space but, because of the Kinect technology, it came much closer than previous games in capturing both the motion and the emotion of the lived park experience. The loss of that particularly haptic experiential quality negates what gave the game its impact and what could be thought of as a redeeming quality in a game that still pales in comparison to a "real" theme-park experience. As with earlier console and computer games, the park space is again navigated via controller or keyboard and mouse. The choice to remaster it in spite of the abandonment of the Kinect technology leaves a game where players go through the motions of a trip to Disneyland, with little motion at all.

Conclusion: Disney Parks as Game Space

Recently, Disney has begun pushing further into the mobile-game space as a means of extending their park franchise outside of its physical bounds, as well as outside of the home. Since 2009, Disney has released a number of official Disney Parks apps, including informational guide apps, shopping apps, and photo manipulation apps that have brought mobile digital technology into park space.[28] These apps further smooth over the gap between physical and virtual space using a mobile—rather than home—experience. From home video games that trained guests on how to engage virtually with park space, Disney has moved to mobile apps that blur these boundaries to provide a more synchronous engagement between the virtual and the physical space, and mobile games that actively gamify its physical parks.

In 2016, Disney partnered with developer Gameloft to release the mobile theme-park creator game app, *Disney Magic Kingdoms*. The game charges players with reconstructing the "Magic Kingdom" after a takeover by Disney villains. Players rebuild the parks bit by bit by collecting characters, building park components, and sending characters on quests to earn magic and create happiness. Just as the mobile game itself recreates the park in the hands of the player, the player also actively reconstructs the park through gameplay. Moreover, the player's active creation of happiness and collection of Disney magic emphasizes the player's interactive role as a constitutive component of the park experience. Like earlier Disney video games, *Disney Magic Kingdoms* is designed around the typical hub and spoke map, eliciting the experience of navigating Disney's parks. As players build rides, earn parades, and interact with characters by tapping to give virtual "high fives," they also enact typical park behaviors. In using mobile devices as a gaming platform, *Disney Magic Kingdoms* opens up the possibility of playing the game while in the park itself, further collapsing the boundaries between these spaces. Beyond mobile gaming, this move to dissolve physical contexts also fits into a larger, more diffuse move by Disney to break down the boundaries between the physical and virtual with the increasing gamification of park space through digital technologies.

There are a number of ways that park space has operated as game space. Most broadly, the parks themselves operate analogously to open-world (as opposed to linear) video games, where progression through space and order of gameplay is determined by the player. More concretely, examples of smaller attempts at making the park spaces more game-like have abounded in recent years. Some attractions resemble video games, such as Buzz Lightyear Astro Blasters, an attraction that combines the dark ride with the "shooter" game genre via interactive ride vehicles equipped with laser guns. At the end of the ride, guests may email souvenir photographs to themselves that digitally combine on-ride photos of the guests in action with a virtual cartoon game space. Other attempts have brought more overt elements of game play into the parks, such as with Walt Disney World's "Sorcerers of the Magic Kingdom" or "A Pirate's Adventure—Treasures of the Seven Seas," where guests complete missions, becoming the protagonist of their own game narrative within the park.

Scott Bukatman has discussed how the structure of Disneyland mimics the structural architecture of computers, a useful tool for understanding how theme parks work on a systematic scale and how the visitors, or "users," function within that system. In his "terminal theme park," the users are integrated into a technologically informed park system of nodes organized by function and kind (rides are "files," and lands are "folders"). At Disneyland, Bukatman argues, "the computer becomes a site of bodily habitation and experience in the theme parks—a technological interface so effective that most users are unaware of the interface at all" (2003: 26). The 2013 introduction of Magic Bands, RFID-enabled wristbands that serve a multitude of functions, including as a digital key, entry ticket, credit card, and FastPass, can perhaps be considered a manifestation of Bukatman's theory.[29] Now equipped so as to become a digital part of the park, guests' natural "play" in the park space can be customized via interactive environments and props that read the bands (while data about their use of park space can be gathered, of course) (Barnes, 2013).

In 2018, the *Play Disney Parks* app was released, which represents the next step in the gamification of park space. Leading up to its release, Disney promised that "guests can play in the parks like never before using this soon-to-launch mobile app to access exclusive experiences at both [US] locations through activities that interact with the surrounding environment" (Smith, 2018). This app demonstrates Disney's continued focus on using digital video-game media to combine virtual and physical play spaces by dissolving the spatial boundaries of the theme park.[30] Players are now invited to play while inside the park. The GPS-enabled app has ride-specific games that unlock only by waiting in ride queues. This "play" thus redirects down time in the park in ways that are controlled by Disney and reabsorb the players/guests in the Disney experience.[31] Appropriating from video-game culture, in-app achievements may be unlocked by riding select attractions. *Play Disney Parks* also fosters environmental interactivity as it uses Bluetooth technology to activate special effects in the park itself (Carey, 2018). Other functions within the app, including land-specific trivia quizzes and Apple Music-integrated playlists, work both inside and outside of the parks. *Play Disney Parks* allows players to begin their park experience before they enter the gates, to continue it after they leave—seamlessly to carry it everywhere. According to Dan Soto, Vice-President, Digital Experience, Disney Parks and Resort Digital, "This app allows our guests to feel a powerful and emotional connection to their favorite parks in a whole new way, which is always our goal" (Hayner, 2018).

In what promises to take this gamification of park space even further in terms of dissolving boundaries and embodied game space, the upcoming Star Wars: Galaxy's Edge lands at Disneyland and Walt Disney World also promise a more game-like interactive park experience, as guests actions will reportedly be tracked, allowing them to build reputations and "become a Star Wars character" (Goldman, 2017). According to Soto, these new lands represent a "next-level experience" (Carey, 2018). In an early testing of Project Natal (the first incarnation of what would become Microsoft's Kinect), *Gizmodo*'s Mark Wilson and Matt Buchanan effusively praised the spectacular new potential of the technology:

> Project Natal is the vision of gaming that's danced through people's heads for decades—gaming without the abstraction of controllers, using your body and natural movements . . . seeing it, *feeling* it in person, makes me want to believe that this what the future of gaming looks like—no buttons, no joysticks, no wands. The only thing left to get rid of is the screen, and even that'll happen soon enough. (2009)

Microsoft's now-defunct Kinect notwithstanding, both Disney's *Play Disney Parks* app and Star Wars: Galaxy's Edge suggest further moves toward this disappearing interface, continuing the progression of bringing video games together with actual park space where even if the screen is still there, at least you can stow it in your pocket while you continue your game in the park itself.

Notes

1. The blurb for the 2017 remastered, non-Kinect rerelease of *Disneyland Adventures* for Xbox One and Windows 10 reads: "Disneyland Adventures allows children and Disney fans of all ages to explore Disneyland park in 4K Ultra HD and HDR, enjoy immersive adventures based on 18 popular attractions, engage in 100 challenging quests, and interact with 35 beloved Disney characters. Disneyland Adventures is your ticket to the magical world of Disneyland right in your living room." In this description, the physical references to flying, dancing, and diving are conspicuously absent, gone with the now-defunct Kinect sensor.
2. *Kinect Disneyland Adventures* was externally licensed by Disney and developed by Frontier Developments for release by Microsoft Studios.
3. "Official" according to game developer Frontier's own webpage for *Kinect Disneyland Adventures* (http://disneyland.frontier.co.uk/).
4. The collection features *The Haunted Mansion, Tower Of Terror* (D. J. MacHale, 1997), *Mr. Toad's Wild Ride* (Terry Jones, 1996), and *The Country Bears* (Peter Hastings, 2002).
5. While Disney's efforts are certainly intended to control and guide as much of the experience as possible, it is important to note that guests are not simply naive objects of manipulation but have an impact on the creation of the park as narrative through their own knowledge and choices. They constitute the variables and the missing element of the narrative as they add not entirely controllable organic elements, including those of play, memory, and nostalgia, as well as knowing participation, to the park space.
6. An oft-cited example is early guests' failure to understand that they were supposed to be Snow White in the Snow White and her Adventures attraction, leading to a redesign where the Snow White figure was added in to the ride (Barrier, 2007: 258).
7. See Jess-Cooke (2010) for a discussion of immersion as control.
8. While Disney parks games released exclusively outside of the United States exceed the scope of the present study, further analysis of such games could help to expand

the present discussion, particularly into how Disney parks video games work in transnational contexts. Examples include *Adventure of TOKYO DisneySEA* (Konami, 2001) or *Mickey no Tokyo Disneyland Daibōken* ("Mickey's Great Adventure in Tokyo Disneyland," Tomy, 1994), which contain levels based on park attractions but little representation or independent navigation of park space and whose park map serves primarily as a menu screen locating in which land the coming level/attraction is set.

9. See Dusenberry (2013) for an analysis of nostalgia and space in *Disney Epic Mickey*.
10. Telotte complicates this discussion by arguing that Disneyland actually lies at the convergent point of film, television, and theme park: "a key element of that intersection is a shift in image technology that it heralds, the movement from early television's dominantly live or kinetoscoped image to the filmed image" (2008: 1).
11. The series was renamed and reconceived throughout the next half-century, most notably for the development and depiction of the park as *Walt Disney Presents* (1958–61) and *Walt Disney's Wonderful World of Color* (1961–69). See Telotte (2004) for a concise but cogent analysis of the series.
12. See Gillan (2015): 206–11 for a discussion of the use of Walt's persona as a form of brandcasting.
13. See Marling (1992): 174–7 for a discussion of Disneyland's relationship to automobile culture.
14. In Disney parlance, a park visitor or company customer is referred to as a "guest." Similarly, Disney employees are referred to as "cast members," which helps to frame the media "show" aspect of the park experience.
15. Although the title, *Adventures in the Magic Kingdom*, and the cover art, which features Cinderella's Castle from the Magic Kingdom Park in Orlando, seem to indicate that the game takes place at Walt Disney World, the game's map is more reminiscent of the Disneyland Park in Anaheim. The phrase "Magic Kingdom" is perhaps meant to refer to Disney theme parks more generally.
16. Though Clarke is talking specifically about video games licensed from television series here, his argument is also borne out in these Disney parks games.
17. According to the game, the answers are "Tokyo Disneyland," "Little King," and "Circle H."
18. "Imagineer" is a Disney portmanteau of '"magination" and "engineer" used to designate the creative workers who design and build Disney parks and attractions.
19. Disney would also bridge the virtual and physical using video game technology with its *Disney Infinity* series (Disney Interactive Studios, 2013–15) where physical toy figures bring characters and playsets into the digital game world.
20. *Kinect Rush: A Disney–Pixar Adventure* (Microsoft Studios, 2012) followed KDA and employed similar aesthetics and mechanics. Though not set in a Disneyland park, but rather a fictional "Pixar Park," *Kinect Rush* arguably served similar superfranchise functions as it brought together Disney and Pixar properties, some of which also appear in Disney parks.
21. Similarly, Walt Disney World attractions inspired the racetracks in *Walt Disney World Quest: Magical Racing Tour* (Eidos Interactive, 2000). The game's intro recalls the *Disneyland* television series, with the player "flying" into the Magic Kingdom and down Main Street U.S.A. Players use a stylized map of the resort to select levels.
22. Goofy and his son Max encounter "the legendary Bigfoot" in the 1995 animated film *A Goofy Movie*, though the yetis in *KDA*, with their long white fur, red eyes, sharp fangs, and snowy environment, recall the much more frightening Abominable Snowmen from Disneyland's Matterhorn attraction.

23. Though some of the more interactive attractions like the Buzz Lightyear Astro Blasters or Davy Crockett's Explorer Canoes do elicit more active physical participation.
24. The "ghost physics" and "semi-embodied" nature of interfaces such as the Kinect have complicated implications for notions of agency and immersion. For further discussion of this, see Gregersen and Grodal (2009).
25. As evidenced by Disney terms such as "guest," "Imagineer," and "weenie," esoteric Disney language is an integral part of Disney park culture.
26. The game's fashion and technology are arguably also somewhat temporally nebulous. Human characters are dressed simply and generically in styles that look like they could range from the 1990s through the 2010s. In terms of technology, some atmospheric NPCs carry cameras, and Mickey gives the player a "magical camera" to complete photo quests, not a smartphone as one might expect.
27. According to Alex Kipman, creator of the Kinect, and Matthew Lapsen, GM of Xbox Devices Marketing, the decision to discontinue manufacture and support for the Kinect primarily resulted from Microsoft's efforts to compete with Sony's PlayStation 4 console, both in terms of lowering the price and as a refocus on traditional gaming in an attempt to recapture the "hardcore gamers" who rallied around the PS4 and away from the experimental interface of the Kinect which had been bundled with the Xbox One since its launch in 2013. See Wilson (2017).
28. These mobile apps include informational park guides such as *Mobile Magic* (2009), *Disneyland Explorer* (2012), *My Disney Experience* (2012), *MyMagic+* (2013), and *Disneyland* (2015) and interactive brochures like *Disneyland Explorer* (2012). Complementing these park information apps are a series of peripheral apps, including *Shop Disney Parks* (2015) for park merchandise and the photo-manipulation app *Show Your Disney Side* (2016) that further gamified the park experience.
29. Magic Bands are used in conjunction with Disney's *MyMagic+* app.
30. This trend is also supported by Disney's recent—and costly—park-wide WiFi upgrades. See Martin (2018).
31. *Play Disney Parks* is probably an appropriation of existing guest behavior, as it has become common to see line queues full of guests playing mobile apps such as Ellen Degeneres's *Heads Up!* as a way to pass the time during long waits.

References

Anderson, Christopher (1994), *Hollywood TV: The Studio System in the Fifties*, Austin, TX: University of Texas Press.

Barnes, Brooks (2013), "At Disney Parks, a Bracelet Meant to Build Loyalty (and Sales)," *The New York Times*, 7 January, <https://www.nytimes.com/2013/01/07/business/media/at-disney-parks-a-bracelet-meant-to-build-loyalty-and-sales.html> (last accessed 22 April 2018).

Barrier, Michael J. (2007), *The Animated Man: A Life of Walt Disney*, Berkeley and Los Angeles: University of California Press.

Benjamin, Walter (2007), "The Work of Art in the Age of Mechanical Reproduction," in Harry Zohn (trans.) and Hannah Arendt (ed.), *Illuminations*, New York: Schocken Books.

Borenstein, Greg (2012), *Making Things See: 3D Vision with Kinect, Processing, Arduino and MakerBot*, Sebastopol, CA: O'Reilly Media.

Buhler, James (2017), "Branding the Franchise: Music, Opening Credits, and the (Corporate) Myth of Origin," in Stephen C. Meyer (ed.), *Music in Epic Film: Listening to Spectacle*, New York and London: Routledge.

Bukatman, Scott (2003), *Matters of Gravity: Special Effects and Supermen in the 20th Century*, Durham, NC and London: Duke University Press.

Carey, Bridget (2018), "Turn Your Phone into a Magic Wand and Kill Time at Disney," *CNET*, 17 July, <https://www.cnet.com/news/play-disney-parks-app-games-while-waiting-in-line/> (last accessed 10 August 2018).

Clarke, M. J. (2013), *Transmedia Television: New Trends in Network Serial Production*, New York and London: Bloomsbury Academic.

DeVries, Jack (2011), "Kinect Disneyland Adventures Review: The Happiest Kinect Game on Earth," *IGN*, 18 November, <http://www.ign.com/articles/2011/11/18/kinect-disneyland-adventures-review> (last accessed 30 July 2018).

Dusenberry, Lisa K. (2013), "Epic Nostalgia: Narrative Play and Transmedia Storytelling in *Disney Epic Mickey*," in Gretchen Papazian and Joseph Michael Sommers (eds), *Game On, Hollywood!: Essays on the Intersection of Video Games and Cinema*, Jefferson, NC: McFarland & Company, Inc., pp. 183–96.

Elliott, John (2017), "Staff Review of Disneyland Adventures (Xbox One)," *XboxAddict*, 8 December, <http://www.xboxaddict.com/Staff-Review/14026/Disneyland-Adventures.html> (last accessed 23 July 2018).

Fisher, Marla Jo (2017), "Tokyo Disneyland Resort Will Undergo Huge $2.7 Billion Expansion," *The Organge County Register*, 30 November, <https://www.ocregister.com/2017/11/30/tokyo-disneyland-resort-will-undergo-huge-2-7-billion-expansion/> (last accessed 23 July 2018).

Frontier Developments, "Kinect Disneyland Adventures," *Frontier Developments*, <http://disneyland.frontier.co.uk/> (last accessed 25 April 2015).

Gillan, Jennifer (2015), *Television Brandcasting: The Return of the Content–Promotion Hybrid*, New York and London: Routledge.

Goldman, Eric (2017), "Disneyland's Star Wars Land Will Let You Control the Millennium Falcon and Choose to Help the Resistance or First Order," *IGN*, 15 April, <http://www.ign.com/articles/2017/04/15/disneylands-star-wars-land-will-let-you-control-the-millennium-falcon-and-choose-to-help-the-resistance-or-first-order> (last accessed 22 April 2018).

Gregersen, Andreas and Torben Grodal (2009), "Embodiment and Interface," in Bernard Perron and Mark J. P. Wolf (eds), *The Video Game Theory Reader 2*, New York: Routledge, pp. 65–84.

Hayner, Chris E. (2018), "Play Disney Parks App Gamifies Your Disney Vacation," *GameSpot*, 1 July, <https://www.gamespot.com/articles/play-disney-parks-app-gamifies-your-disney-vacatio/1100-6460025/> (last accessed 23 July 2018).

Jackson, Kathy Merlock (2011), "Synergistic Disney: New Directions for Mickey and Media in 1954–1955," in Kathy Merlock Jackson and Mark I. West (eds), *Disneyland and Culture: Essays on the Parks and Their Influence*, Jefferson, NC: McFarland and Company, Inc., pp. 19–28.

Jess-Cooke, Carolyn (2010), "Sequelizing Spectatorship and Building Up the Kingdom: The Case of *The Pirates of the Caribbean*, Or, How a Theme Park Attraction Spawned a Multibillion-Dollar Film Franchise," in *Second Takes: Critical Approaches to the Film Sequel*, Carolyn Jess-Cooke and Constantine Verevis (eds), Albany: State University of New York Press, pp. 205–24.

Johnson, Derek (2013), *Media Franchising: Creative License and Collaboration in the Culture Industries*, New York and London: New York University Press.

Kent, Steven L. (2001). "Super Mario Nation," in Mark J. P. Wolf (ed.), *The Medium of the Video Game*, Austin: University of Texas Press, pp. 35–48.

Knight, Michael and Trisa Knight (2006), *Fodor's Disneyland and Southern California with Kids*, New York: Fodor's Travel Publications.

Koehler, Dorene (2017), *The Mouse and the Myth: Sacred Art and Secular Ritual of Disneyland*, Bloomington, IN: Indiana University Press.
Lukas, Scott A. (2008), *Theme Park*, London: Reaktion Books Ltd.
Marling, Karal Ann (1991), "Disneyland, 1955: Just Take the Santa Ana Freeway to the American Dream," *American Art* 5:1/2, winter–spring, pp. 168–207.
Marling, Karal Ann (1997), "Imagineering the Disney Theme Parks," in *Designing Disney's Theme Parks: The Architecture of Reassurance*, Karal Ann Marling (ed.), New York: Flammarion, pp. 29–177.
Martin, Hugo (2018), "Why Theme Parks Are Spending Millions to Give Guests Free Wi-Fi," *Los Angeles Times*, 20 July, <http://www.latimes.com/business/la-fi-theme-park-wifi-20180720-story.html#> (last accessed 23 July 2018).
Microsoft (n.d.), "Disneyland Adventures," *Microsoft*, <http://www.microsoft.com/en-us/p/disneyland-adventures/9n6z8dqxsqwh> (last accessed 30 July 2018).
Morris, Chris (2011), "Microsoft Kinect Sales Blast Predictions," *Variety*, 5 January, <http://variety.com/2011/digital/news/microsoft-kinect-sales-blast-past-predictions-1118029819/> (last accessed 30 July 2018).
Morris, Ian (2017), "Disneyland Adventures Xbox One Review: Back to the Magic Kingdom," *Everybody Plays*, 1 December, <https://www.everybodyplays.co.uk/review/Disneyland-Adventures-Xbox-One-Review-Back-to-the-Magic-Kingdom/2657> (last accessed 23 July 2018).
Newman, Michael Z. (2017), *Atari Age: The Emergence of Video Games in America*, Cambridge, MA: The MIT Press.
Sinicki, Joe (2011), "Review: Kinect Disneyland Adventures Is the Most Unique and Most Disappointing Kinect Game Yet," *VentureBeat*, 15 November, <http://venturebeat.com/2011/11/15/kinect-disneyland-adventures-is-both-the-most-unique-and-most-disapointing-kinect-game-to-date/> (last accessed 30 July 2018).
Sklar, Martin (2015), *One Little Spark!: Mickey's Ten Commandments and the Road to Imagineering*, Glendale, CA: Disney Editions.
Smith, Thomas (2018), "All-New Play Disney Parks App Coming to Disneyland Resort and Walt Disney Resort This Summer," *Disney Parks Blog*, 18 April, <https://disneyparks.disney.go.com/blog/2018/04/all-new-play-disney-parks-app-coming-to-disneyland-resort-and-walt-disney-world-resort-this-summer/> (last accessed 20 April 2018).
Spigel, Lynn (1992), "Installing the Television Set: Popular Discourses on Television and Domestic Space, 1948–1955', in Lynn Spigel and Denise Mann (eds), *Private Screenings: Television and the Female Consumer*, Minneapolis and London: University of Minneapolis Press, pp. 11–46.
Telotte, J. P. (2004), *Disney TV*, Detroit: Wayne State University Press.
Telotte, J. P. (2008), *The Mouse Machine: Disney and Technology*, Chicago: University of Illinois Press.
Walt Disney Company (n.d.), "Kinect Disneyland Adventures," *Walt Disney Company*, <http://games.disney.com/kinect-disneyland-adventures-video-game> (last accessed 23 July 2018).
Wilson, Mark (2017), "Exclusive: Microsoft Has Stopped Manufacturing the Kinect," *Fast Company*, 25 October, <https://www.fastcompany.com/90147868/exclusive-microsoft-has-stopped-manufacturing-the-kinect> (last accessed 23 July 2018).
Wilson, Mark and Matt Buchanan (2009), "Testing Project Natal: We Touched the Intangible," *Gizmodo*, 3 June, <http://gizmodo.com/5277954/testing-project-natal-we-touched-the-intangible/> (last accessed 30 July 2018.

5. "NOW *THEY'RE* PLAYING WITH POWER!": NINTENDO'S CLASSICS AND FRANCHISE LEGACY MANAGEMENT

Matthew Thomas Payne

Introduction

For fans and critics alike, Nintendo's NES (Nintendo Entertainment System) Classic Edition represents some of the company's best and worst tendencies. A miniature replica of their famous 1985 home console containing thirty titles, the NES Classic was such a massive hit that retailers' limited inventories sold out almost immediately. This popular retro-styled microconsole was not long for this world, however, with Nintendo ending its production run in only five short months.[1] The company's subsequent production of a Super NES (SNES) Classic Edition—a remake of their 1991 sixteen-bit system, containing twenty-one games—was met with similar success, selling over four million units as of early 2018 (Makuch, 2018). Nintendo has announced plans to produce more Classic Editions; a promise that has done little to quell the fears of a suspicious fan base that has become inured to the company's long history of chronic underproduction (Covil, 2016; Frank, 2017). If genres represent tacit social contracts between producers and audiences, then the NES and SNES Classic Editions (or simply "Classics" for this chapter's purposes)—both the limited lineups of titles and their conspicuous underproduction—reveal a fairly one-sided relationship (Buchanan, 2006: 73–5; Feurer, 2010).

It may be tempting to dismiss these microconsoles as mere nostalgic novelties; easy-to-make playthings created to profit from an already loyal fan base. That is not wrong. But it isn't the whole picture either. The NES and SNES Classics should be thought of as constituting their own subfranchise genre for the ways they embody a highly selective version of Nintendo's past. Their curatorial design elevates selected games and ways of playing as worthy of canonizing and remembering. Furthermore, the devices underscore notions of quality and prestige that have long been vital to the brand identity. Nintendo's notoriety as a fountainhead of playful design is integrally tied to its past—one that begins with its genre of classic games.

Genres are not static textual collections but are dynamic categories that emerge from the process of categorization itself (Mittell, 2017: 81–3). Building on work in film and television studies, game scholars have adapted genre theory for crafting descriptive taxonomies (Wolf, 2001), for highlighting the form's medium specificities (Apperley, 2006), and for tracking how generic expectations evolve over time (Arsenault, 2009). As the case of Nintendo's Classics illustrate and as this chapter argues, there is utility in thinking about how producers organize constellations of their so-called "classic" games, how titles achieve canonical status, and how industrial strategies engender unequal relationships between producers and consumers. What Nintendo deems as classic is based less on the games' textual features than on contemporary business exigencies; namely, its classic genre is a calculated assessment of how packaging Nintendo's past successes might serve its present and near-future projects.

This is why Nintendo did not equivocate when it stated that the NES Classic was never intended to be a long-term commitment. This chapter contends that it was never a long-term commitment because it was never truly intended for consumers. The NES and SNES Classic Editions were instead created by Nintendo for Nintendo. Critical genre analysis enables us to move beyond simple categorization to identify non-gameplay functions served by underproducing non-networked consoles[2] with a fixed roster of games, and how amplifying demand for these devices through manufactured scarcity is less about fostering player nostalgia than about codifying corporate gaming histories for the purposes of crafting a prestige product. As television scholar Jason Mittell notes, "Genres are more than just any category. They must be culturally operative within a number of spheres of media practice, employed by critics, industries, and audiences" (2004: 10). The present question is, how does Nintendo go about discursively constructing a classic genre for itself, one that is not necessarily "true" in any objective sense but nevertheless functions as if it were the case?[3]

On a more personal note, this chapter is an opportunity to return to one of my first research projects. In 2008, I published a piece in Zach Whalen and Laurie Taylor's *Playing the Past* anthology on the Jakks Pacific's line of "Plug it in and Play TV games" (Payne, 2008). That essay focused on the ways these electronic toys collapsed and effaced elements of gaming history, representing a divide between how media producers and gaming communities remember gaming's past.

I'd like to use this chapter as an opportunity to revisit my thinking about the commodification of cultural memory from the perspective of franchise management. Specifically, I'm shifting focus from thinking about how retro-styled microconsoles represent diverging points of view among producers and players as it concerns nostalgia and video-game history, to examining the ongoing strategies of a media giant to cement its own industrial history with its retro-styled devices. Jakks Pacific licensed a handful of Atari's and Namco's libraries when it remediated those 1970s and 1980s games into its plastic, third-party, all-in-one devices. In contrast, Nintendo is both the hardware manufacturer and the publisher of over half the NES Classic titles and two-thirds of the Super NES Classic games. When one thinks of Nintendo, one typically thinks of its decades-old franchises, like *Super Mario Bros.* (1985–present), *The Legend of Zelda* (1986–present), *Metroid* (1986–present), *Punch-Out!!* (1984–present), and *Star Fox* (1993–present). Thus, unlike Jakks Pacific's more removed handling of Atari and Namco titles, which does not offer a fulsome view into those publishers' pasts, Nintendo is critically concerned with deploying its marquee franchises, memorable characters, and associated intellectual property (IP) originating from other publishers (for example, Konami's *Contra* and *Castlevania* or Capcom's *Mega Man* and *Street Fighter II*) to solidify its central place in gaming history.

Nintendo isn't using its Classics to engage in some grand historical mystification. Their objective is simpler: to reaffirm those gameplay elements, fictional realms, and popular characters that Nintendo deems foundational to its brand. The Classics are a communications management tool first, a retro gaming device second. Media scholars have analyzed and emphasized how franchise logics operate for a range of transmedia properties and brands. For instance, narrative is often privileged as the foundational worldbuilding element, providing multiple storytelling directions for various authors across time. Henry Jenkins discusses this with regard to *The Matrix* (2006) and *The Walking Dead* (2013) franchises, whereas Kristen Thompson discusses similar points in her analysis of *The Lord of the Rings* universe (2007). Derek Johnson, however, points to the creative context of world-sharing wherein valued IP becomes a shared set of creative resources upon which multiple creators from different industries might draw (2013). If Jenkins and Thompson emphasize "worldbuilding" in their work, and Johnson advocates "world-sharing" in his analysis, then Nintendo's Classics might be best thought of as being "world-reinforcing." The Classics are an ad hoc means of ensuring brand legacy across eras, platforms, and titles. Indeed, by closely examining these routinely underproduced microconsoles—including their hardware and interface design, lineup of games, and popular characters—and their supporting materials, we can understand how Nintendo manages its Classic Editions with the goal of being remembered as an integral player (so to speak) of the classic gaming experience for its current and near-future ventures.

Commodifying Gaming Nostalgia

This chapter's title plays on Nintendo's well-known NES marketing slogan from 1986, "Now You're Playing with Power," a tagline that was revived for the launch of the NES Classic. During its original deployment, it hailed young players with the promise of technologically advanced games and hardware. As Stephen Kline, Nick Dyer-Witheford, and Greig de Peuter note in *Digital Play: The Interaction of Technology, Culture, and Marketing*, this slogan also "tacitly conveyed the corporate confidence of an enterprise that had remade and dominated the interactive games sector through much of the 1980s" (2003: 109). And why wouldn't Nintendo be confident? It was North America's best-selling toy in 1988, and only two years later there was a Nintendo in a third of American homes (Kline et al., 2003: 111). For a number of years, the words "Nintendo" and "video games" were virtually synonymous.

The company's marketplace successes were not owed to game design alone. Key to this process was, and remains, Nintendo's wildly successful branding efforts. Kline et al. contend that Nintendo was the first to brand the video-game market where their players became the "Nintendo Generation" and where retail aisles were transformed into "Nintendo Worlds." Nintendo's marketing strategy has shifted emphasis over the years. Though the company succeeded in selling the original NES to children in the 1980s, competition with the likes of Sega and Sony meant that Nintendo in the mid-1990s had to recalibrate its messaging to target teens. In the 2000s, Nintendo recalibrated its messaging yet again to market their Nintendo Wii console to families and to players who didn't identify with the "gamer" label (Boyer, 2009).[4] Nintendo has remained consistently open to innovating with new technologies and peripherals over the years, and to selling to customers on those novel experiences. Some gadgets served to amplify sales, even if they were rarely utilized; consider the NES's light-gun, Robotic Operating Buddy (or ROB), and Power Pad flood-mat controller. Even commercial flops, such as the Virtual Boy and Wii U platforms, underscore Nintendo's willingness to experiment in ways that differentiate it from its more risk-averse competitors (Boyer, 2009). (2018's cardboard-based Nintendo Labo "maker" kit is another example of this playful design tradition.) Nintendo has demonstrated a historical willingness and desire, through its multiple platforms, peripherals, and aggressive marketing efforts, to colonize multiple spaces of play. Moreover, if the "colonizer is Nintendo itself, and the colonized the children who play the games," as Kline et al. argue, then what's past is truly prologue (2003: 124–5). With the Classics we are witnessing the colonization of gaming memory through retail canonization.

Those familiar with Nintendo know that it is no stranger to commodifying nostalgia. Indeed, cynics and fans alike (which are not mutually exclusive groups) regularly lament Nintendo's overreliance on a handful of decades-old in-house franchises and how it regularly repackages and recycles these titles.[5] This practice of repackaging sometimes manifests textually as remastered versions of older games, such as *Super Mario All-Stars*, a 1993 SNES game that combines the original *Super Mario Bros.* trilogy (Nintendo, 1985–90) with *Super Mario Bros.: The Lost Levels* (Nintendo, 1986) in a single collection. And sometimes this tendency manifests extratextually, such as the small museum in New York City's Nintendo World Store, or with Nintendo's partnership with comic publisher Dark Horse Books on a three-part compendium of lore for *The Legend of Zelda* series.[6] But for those looking to replay games of their youth or to play a previous generation's titles for the first time, the most reliable platform has been Nintendo's Virtual Console.

Steve Cuff and Christopher Terry note that the company's "Virtual Console"—which launched with the Nintendo Wii in 2006—is a digital download service that provides "access to a carefully curated selection of video games from the 1980s and 1990s" (2017: 20). The electronic sell-through service has since been expanded to the 3DS and the Wii U. In September of 2018, Nintendo launched a monthly subscription service—"Nintendo Switch Online"—which brought classic NES titles to the Switch (McWhertor, 2018). (The move to a paid monthly subscription model follows the lead set by Microsoft and Sony.) Speaking to the Virtual Console's power to reshape cultural memory, Cuff and Terry state:

> On the Virtual Console, Nintendo's history of heated competition with these former rival companies is scrubbed clean and each digital title is assimilated into the broader Nintendo universe. By funneling a breadth of titles from multiple platforms and numerous video game console generations under Virtual Console umbrella, Nintendo commodifies not only Nintendo nostalgia, but also video game history in a broader sense. The Virtual Console stands as an example of the Sisyphean task of attempting to recreate the experience of playing a video game on modern technology in a manner which lives up to the lofty expectations dictated by player nostalgia. (2017: 21)

Making the Virtual Console accessible across platforms represents a nexus of corporate strategies for Nintendo's ongoing commodification of nostalgia. Technologically, it allows players to access these games across time—accounting for the planned obsolescence of its console hardware—and across play spaces (taking advantage of the mobility of the 3DS and Switch, and to a lesser extent that of the Wii U).[7] Financially, players wanting to play the same titles across platforms would need to repurchase the same game for every new console they acquire; there is currently no sharing of software across platforms. This electronic sell-through channel also prevents "brick and mortar" retailers of second-hand games, such as GameStop in North America or GAME in the United Kingdom, from profiting from this form of commodified nostalgia. Finally, discursively, the fact that each platform since 2006 has featured a Virtual Console with a growing library signals Nintendo's continued interest in curating and commodifying its own vision of gaming history. That is, with each instantiation of the Virtual Console, Nintendo offers one more reminder that they were once the world's biggest game supplier, constituting 20 to 30 percent of the software market in the late 1980s, and that it has zero interest in ceding this notoriety or mindshare to any other entity (Kline et al., 2003: 117).

"The Classics"

The Classics are at once the inevitable progression of the commodification of gaming nostalgia *and* a radical break from business as usual. On the one hand, the Classics are a more forceful expression of the Virtual Console; they are virtual nostalgia made material. Because they cannot be officially altered, their game rosters will not fluctuate with the vagaries of the licensing agreements and marketplace demands that determine software appearing on the Virtual Console (Cuff and Terry, 2017: 28). (It bears noting that this has not stopped modders from hacking the consoles to insert complete libraries and other features [Machkovech, 2017].)

Moreover, the Classics have been crafted to look and play like the originals. While the original NES was designed to resemble another piece of grey audio-video equipment akin to a videocassette recorder, its miniaturization evokes the Japanese *kawaii* aesthetic, or "culture of cute." The core physicality, the static game list, and the plastic, wired controllers lend these microconsoles a sense of living-room permanence absent from the Virtual Console.

On the other hand, because the Classics are non-networked devices that can neither be patched nor updated, they are technologically quaint. This is also no doubt part of their charm. Yet it is hard to understand why they would be of any long-term value aside from their discursive power as a celebratory, anniversary artifact. After all, Nintendo cannot track consumers' play choices with such devices nor can they extract more money from players via micro-transactions. Cuff and Terry note of the Virtual Console that it "is not a true reflection of the past so much as it is a distortion which represents Nintendo's ideal past brought to life in the present" (2017: 28).

I'd like to extend this argument with a caveat. There is not some true version of gaming history against which the Classics ought to be judged, just as there is no true version of a gaming genre. Any history that media companies or game historians narrate is necessarily contingent and incomplete. We can still ask, however, what *kind* of history do these particular commodities reveal? What do the production, design, and marketing choices say about Nintendo's priori-

Figure 5.1 The NES Classic Edition replicates the look and feel of the original console and controller while updating the gameplay experience by including a set roster of games and hardware connections for modern displays.

ties? Just as the Virtual Console offers but a blurry picture of retro gaming, the Classics appear to say more about Nintendo's current and future plans than they disclose about its previous accomplishments.

The Classics stress an authentic (that is, company-sanctioned) retro-gaming experience. The gaming press largely praised the NES Classic, with journalists' foremost complaint being the inclusion of a single controller with a comically short cable (Stein, 2016). The SNES classic was similarly well-received, with Nintendo having seemingly heard the initial round of criticism; it included two controllers with longer cables (Dornbush, 2017). Both units connect to monitors via HDMI (high-definition multimedia interface) connections and can be powered by USB ports.

The Classics have been crafted with casual and former hardcore gamers in mind (Juul, 2009). They share similar menus, featuring their gaming libraries in rotating carousels. They offer three display options—CRT filter, 4:3, and "Pixel Perfect"—allowing gamers to choose from a variety of "looks" for rendering their eight- and sixteen-bit games on modern displays. Both units allow users to save mid-game progress with multiple suspend points. The SNES even goes a step further by including a rewind feature that enables one to undo a mistake—a particularly helpful feature for the more demanding action and platforming titles.

If the modernized hardware and interface design grant Nintendo a way of updating the past, then the limited game rosters and their starring characters

Figure 5.2 The NES Classic Edition features thirty games available for users to select from a scrolling menu.

provide a glimpse into how these devices discursively reify Nintendo's history for its current and near-future projects. The NES Classic has thirty titles; the SNES has twenty-one. This is a mere 4 percent and 3 percent of their respective libraries.[8] Compare this with similar all-in-one devices; the "Sega Genesis Flashback" boasts of eighty-five games and the "Atari Flashback 8 Gold" sports 120 games.[9] Why the disparity among these one-time rivals? It isn't a hardware limitation or a licensing issue for Nintendo.[10] Indeed, Nintendo's relative restraint guards against accusations of "shovelware"—the practice of adding content without regard to its quality, fit, or utility—that have hobbled their competitors, and this selectivity likewise underscores the company's rhetoric of quality control that has long been credited with having saved the industry in the 1980s.

The Classics also support Nintendo's broader IP ecosystem—in particular character-rich franchises like *Super Smash Bros.* (1999–present) as well as their lineup of collectible amiibo figures—by showcasing many of their characters' origins. The popular *Super Smash Bros.* fighting series features brawlers that appear in over half the games on the NES Classic and nearly two-thirds of the titles on the SNES Classic.[11] The Classics' games often present characters' beginnings, origin stories that have been largely omitted from the non-narrative *Smash* universe. The microconsoles introduce or refamiliarize players with Nintendo's marquee characters in an easy-to-use device, one that seeks to win back fans and gain new ones as the company transitions from the failed Wii U to the Switch.

The Classics also extratextually support Nintendo's NFC-enabled amiibo figures that can unlock digital content for select 3DS, Wii U, and Switch games. The fact that the non-networked Classics cannot communicate with amiibo does not mean they are not of value. As is the case with the *Smash* franchise, the Classics fill in the narrative and gaming backstories for numerous amiibo.[12] Furthermore, because many collectors never use their figures to unlock digital content, the non-connectivity between the Classics and the figures may not be perceived as much of a shortcoming, especially for gamers who prioritize collecting Nintendo's physical products above accessing digital content (Peckham, 2015). The Classics are even seen by Nintendo as a means of driving consumer interest in the Switch. The first slide in Nintendo President and Managing Director Tatsumi Kimishima's "Corporate Management Policy Briefing" given in early 2018 states that their basic strategy is "expanding the number of people who have access to Nintendo IP" (Nintendo Co. Ltd, 2018). Later, Kimishima notes: "We plan to continue selling NES Classic Edition and Super NES Classic Edition this year. We view them as an opportunity to garner interest in Nintendo Switch from those who have not interacted with video games in a long time, or ever" (Nintendo Co. Ltd, 2018: 11).

Perhaps also aimed at those who have not played Nintendo games in some time are the hardbound companion books *Playing with Power* (Rocha and von Esmarch, 2016) and *Playing with Super Power* (Haley and Marie, 2017). Published by Prima Games, an imprint of Random House and formerly the largest game-strategy guide publisher, these books offer additional, paratextual insights into how the company wants its games and its brand to be remembered. Produced to look like cartridges (or what Nintendo called "game paks"), these large coffee-table-style books feature level overviews, hardware descriptions, and reprinted content from *Nintendo Power* magazine (1988–2012). For Kline et al., two of Nintendo's most important marketing innovations in the 1980s were *Nintendo Power* magazine and the call-in lines where players could seek help from game counselors (Kline et al., 2003: 120). It is notable, if predictable, that *Playing with Power* and *Playing with Super Power* contain reprinted material from their house organ, including level walkthroughs, gameplay tips, and cheat codes from counselors. *Playing with Super Power* differs importantly from its predecessor in its inclusion of more ephemeral, supporting materials such as concept art, marketing materials, and fan art. The fan-authored material—including images of cosplay, interviews with speedrunners, and other expressions of fan appreciation—is one of the few instances where Nintendo recognizes the productivity of its users in relation to its classics.

Playing with Historical Power

Nintendo filed trademark paperwork with the European Union Intellectual Property Office on July 18, 2017 that included images of an N64 controller

(Sarkar, 2017). Game journalists have since been speculating that an N64 Classic could be in the works. It remains unclear at the time of writing if there will, in fact, be an N64 Classic. It is also uncertain if there will be other Classic iterations; the Game Boy Classic, the GameCube Classic? What is clear is that Nintendo will continue to use projects such as their Classics to brand themselves and their legacy franchises as a site of historical and cultural value while maintaining a tight control on contemporary projects. To wit, Nintendo's need for control extends into contemporary online spaces as it regulates how gamers create levels in *Super Mario Maker* (Nintendo, 2015) and how their games are streamed on sites such as YouTube and Twitch.tv (Witkowski and Manning, 2017).

The battles between game producers capitalizing on their IP versus gamers preserving their own versions of game history with ROMs and emulators—points I explore in my previous work (Payne, 2008)—remains a real concern for historians, archivists, and gaming communities more than ten years later (McFerran, 2018). What the case of the Classics makes clear is that the contestation over gaming's past is not only a push and pull between producers and audiences. Nintendo strategically deploys a variety of textual and extratextual efforts that help to overdetermine a constellation of games that might be thought of as "classic" and thus worthy of collecting and preserving. In the process, Nintendo positions itself as the corporate centerpiece of retro-gaming history, thereby generating prestige status that can be wielded for current and future ventures. The brandcentric subgenre formation of classic NES and SNES games is not about narrating the gaming past; it is about shaping the narratives of past games and past gaming experiences. We can see this in President and Chief Operating Officer of Nintendo of America Reggie Fils-Aimé's Foreword to *Playing with Super Power*, wherein he reflects on his personal experiences with the Super NES and its cultural legacy. He writes:

> What people remember most is the gameplay—how these adventures made people feel. Super NES helped build Nintendo's reputation for surprising people and putting smiles on their faces, something we still strive for today. Not only were these games great for their time, but almost all of them still hold up today. Every time we re-release one of these games, either on the Virtual Console service or in some other format, we're overwhelmed by the positive fan response. The love is still there. While nostalgia certainly plays a part, people keep going back for more because of the engrossing experiences that Super NES games deliver. The Super NES console generation continues to influence the video game industry, from the iconic controllers to the deeper stories the console made possible. A generation of game developers grew up on the Super NES, and it left an impression. Ask any modern game designer to name a favorite game, and invariably one or more Super NES games are in the mix.

> Whether you remember those games, or haven't yet checked them out, the Super NES era is a story worth telling. (Haley and Marie, 2017: 2)

". . . made people feel," "smiles on faces," "love is still there," "engrossing experiences," "iconic controllers," "a story worth telling": Fils-Aimé weaves together the language of affect, design, and memory in making the case for the console's rich historical legacy. The SNES may be a sincerely personal memory for Nintendo of America's leader but it is also undeniably strategic. The NES and SNES platforms attain some mythical status only when discourses organize them in a way where that label becomes culturally salient. Moreover, genre categories and prestige labels gain their saliency and legitimacy through multiple, historically specific discursive practices (Mittell, 2004: 12). The physicality of the Classics, their select lineup of games and interface design, the limited production cycles, and supporting paratexts all represent attempts by Nintendo to canonize a textual body as classic.

Nintendo Power is a revealingly apropos magazine title given the aims of this analysis. Power, after all, is about control. But control in this case is not about a magazine gifting players with insider information that will empower their gameplay; rather, it is about a media firm's attempts to shape how users remember and replay the past today with the goal of managing in-house franchises into the future.

Notes

1. Nintendo announced plans to bring back their NES Classic Edition in June 2018 (Byford, 2018).
2. The NES and SNES Classic Editions do not have internet connectivity and therefore cannot be updated with code patches or be expanded with additional content. Most contemporary consoles and handheld gaming devices require an internet connection to access their features.
3. Or, as Mittell (2004) later states, "Our goal is analyzing generic discourses is not to arrive at a genre's 'proper' definition, interpretation, or evaluation, but to explore the material ways in which genres are culturally operative" (p. 14).
4. It is worth noting that former Nintendo President Satoru Iwata described their willingness to take risks as a "blue ocean strategy"; to try the untried, to look for new things (Williams, 2013).
5. There is no shortage of opinion pieces when it comes to Nintendo's relationship with its in-house franchises, especially *The Legend of Zelda*, *Mario Kart*, and *Metroid*, to name but a few. There are those who would recommend doubling down on the current strategy of exploiting established intellectual property (Ars Staff, 2004). And there are those who would guard against allowing nostalgia to guide Nintendo's future plans (Biggs, 2017; Statt, 2015).
6. The *Legend of Zelda* trilogy includes *Hyrule Historia*, *Arts and Artifacts*, and *Encyclopedia*.
7. For a broader discussion of how planned obsolescence operates in the video-game industry, see Cruea, 2017.

8. There are 714 official NES titles and 721 SNES titles; these numbers do not reflect home-brewed titles.
9. At the time of writing, game publisher SNK announced plans to launch a "Neo Geo Classic Edition" to celebrate its console's fortieth anniversary (McFerran, 2018).
10. There is no difference in the performance as the Classics use the same inexpensive components (Strickland, 2017).
11. *Smash Bros.* characters appear in sixteen of thirty NES Classic games (53 percent), and in fifteen of twenty-one SNES games (71 percent).
12. Style note: amiibo is the singular and plural form of the word.

References

Apperley, Thomas (2006), "Genre and Game Studies: Toward a Critical Approach to Video Game Genres," *Simulation & Gaming*, 37:1, pp. 6–23.

Ars Staff (2014), "How We'd Save Nintendo," *Ars Technica*, 11 May, <https://arstechnica.com/gaming/2014/05/how-wed-save-nintendo/> (last accessed 30 May 2018).

Arsenault, Dominic (2009), "Video Game Genre, Evolution and Innovation," *Eludamos: Journal for Computer Game Culture*, 3: 2, pp. 149–76.

Biggs, John (2017), "Nintendo and the Limits of Nostalgia," *Tech Crunch*, 4 May, <https://techcrunch.com/2017/05/04/nintendo-and-the-limits-of-nostalgia/> (last accessed 30 May 2018).

Boyer, Steven (2009), "A Virtual Failure: Evaluating the Success of Nintendo's Virtual Boy," *Velvet Light Trap*, 24, pp. 23–33.

Buchanan, Ian (2006), *Fredric Jameson: Live Theory*, London: Continuum, pp. 73–5.

Byford, Sam (2018), "Nintendo Is Bringing Back the NES Classic on June 29th," *The Verge*, 13 May, <https://www.theverge.com/2018/5/13/17350686/nintendo-nes-classic-mini-re-release-date> (last accessed 30 May 2018).

Covill, Max (2016), "The NES Classic and the History of Nintendo's Product Scarcity," *Paste*, 28 December, <https://www.pastemagazine.com/articles/2016/12/the-nes-classic-and-the-history-of-nintendos-produ.html> (last accessed 30 May 2018).

Cruea, Mark D. (2017), "Business Models, Planned Obsolescence, Externalities: Examining the Virtual Hand of the Video Game Industry," in Casey B. Hart (ed.), *The Evolution and Social Impact of Video Game Economics*, Lanham, MD: Lexington Books, pp. 31–45.

Cuff, Steve and Christopher Terry (2017), "Nintendo's Retro Revolution: Commodified Nostalgia and the Virtual Console," in Casey B. Hart (ed.), *The Evolution and Social Impact of Video Game Economics*, Lanham, MD: Lexington Books, pp. 15–30.

Dornbush, Jonathon (2017), "SNES Classic Edition Review," *IGN*, 27 September, <http://www.ign.com/articles/2017/09/27/snes-classic-edition-review> (last accessed 30 May 2018).

Feuer, Jane (2010), "Genre Study and Television," in Robert C. Allen (ed.) *Channels of Discourse, Reassembled: Television and Contemporary Television* 2nd edition, Chapel Hill, NC: University of North Carolina Press, pp. 138–60.

Frank, Allegra (2017), "Nintendo's Long Been Hostile to Anyone who Wants to Buy Classic Games," *Polygon*, 13 April, <https://www.polygon.com/2017/4/13/15294568/nes-classic-edition-discontinued-nintendo-switch-virtual-console> (last accessed 30 May 2018).

Haley, Sebastian and Meagan Marie (2017), *Playing with Super Power: Nintendo Super NES Classics*, Roseville, CA: Prima Games.

Jenkins, Henry (2006), *Convergence Culture: Where Old and New Media Collide*, New York: New York University Press.

Jenkins, Henry (2013), "*The Walking Dead*: Adapting Comics," in Ethan Thompson and Jason Mittell (eds), *How to Watch Television*, New York: New York University Press, pp. 373–81.

Johnson, Derek (2013), *Media Franchising: Creative License and Collaboration in the Culture Industries*, New York: New York University Press.

Juul, Jesper (2009), *A Casual Revolution: Reinventing Video Game and their Players*, Cambridge, MA: The MIT Press.

Kline, Stephen, Nick Dyer-Witheford, and Greig de Peuter (2003), *Digital Play: The Interaction of Technology, Culture, and Marketing*, Montreal: McGill-Queen's University Press.

McFerran, Damien (2018), "SNK is Releasing a 'Neo Geo Classic Edition' to Mark its 40th Anniversary," *Nintendo Life*, 19 April, <http://www.nintendolife.com/news/2018/04/snk_is_releasing_a_neo_geo_classic_edition_to_mark_its_40th_anniversary> (last accessed 30 May 2018).

McFerran, Damien (2018), "The Retro Gaming Industry Could Be Killing Video Game Preservation," *Eurogamer*, 18 February, <http://www.eurogamer.net/articles/2018-02-09-the-retro-gaming-industry-could-be-killing-video-game-preservation> (last accessed 30 May 2018).

Machkovech, Sam (2017), "Super NES Classic Hacks Are Now Oh, So Easy to Pull Off—You Can Even Add Features," *Ars Technica*, 9 October, <https://arstechnica.com/gaming/2017/10/super-nes-classic-hacks-are-now-oh-so-easy-to-pull-off-you-can-even-add-features/> (last accessed 30 May 2018).

McWhertor, Michael (2018), "Nintendo Switch Online Service Will Launch with 20 NES Games," *Polygon*, 7 May, <https://www.polygon.com/nintendo/2018/5/7/17329728/nintendo-switch-online-classic-nes-games-lineup> (last accessed 30 May 2018).

Makuch, Eddie (2018), "SNES Classic Sells 4 Million Units," *Gamespot*, 31 January, <https://www.gamespot.com/articles/snes-classic-sells-4-million-units/1100-6456489/> (last accessed 30 May 2018).

Mittell, Jason (2004), *Genre and Television: From Cop Shows to Cartoons in American Culture*, New York: Routledge.

Mittell, Jason (2017), "Genre," in Laurie Ouellette and Jonathan Gray (eds), *Keywords for Media Studies*, New York: New York University Press, pp. 81–3.

Nintendo Co., Ltd (2018), "Corporate Management Policy Briefing," <https://www.nintendo.co.jp/ir/pdf/2018/180201_2e.pdf> (last accessed 30 May 2018).

Payne, Matthew Thomas (2008), "Playing the Déjà-New: 'Plug It in and Play TV Games' and the Cultural Politics of Classic Gaming," in Zach Whalen and Laurie Taylor (eds), *Playing the Past: History and Nostalgia in Video Games*, Nashville, TN: Vanderbilt University Press, pp. 51–68.

Peckham, Matt (2015), "14 Things Nintendo President Tatsumi Kimishima Told Us," *Time*, 3 Decmber, <http://time.com/4131306/nintendo-kimishima-interview/?xid=homepage> (last accessed 30 May 2018).

Rocha, Garitt and Nick von Esmarch (2016), *Playing with Power: Nintendo NES Classics*, Roseville, CA: Prima Games.

Sarkar, Samit (2017), "Nintendo Trademark Suggests Nintendo 64 Classic Edition," *Polygon*, 20 July, <https://www.polygon.com/2017/7/20/16003876/n64-classic-edition-nintendo-trademark> (last accessed 30 May 2018).

Statt, Nick (2015), "Nintendo's Future Still Lies in the Glory Days of Gaming," *CNET*, 14 March, <https://www.cnet.com/news/nintendo-future-still-lies-in-the-glory-days-of-gaming/> (last accessed 30 May 2018).

Stein, Scott (2016), "Nintendo NES Classic Edition Review: Mini NES Classic Is Your Childhood in a Brilliant Little Box," *CNET*, 11 November, <https://www.cnet.com/products/nintendo-nes-classic-edition/review/> (last accessed 30 May 2018).

Strickland, Derek (2017), "SNES Classic Edition Guts Are Identical to NES Mini," *TweakTown*, 27 September, <https://www.tweaktown.com/news/59296/snes-classic-edition-guts-identical-nes-mini/index.html> (last accessed 30 May 2018).

Thompson, Kristin (2007), *The Frodo Franchise:* The Lord of the Rings *and Modern Hollywood*, Berkeley: University of California Press.

Williams, Mike, (2013), "Nintendo's Iwata Doesn't Fear Failure," *GamesIndustry.biz*, 10 October, <https://www.gamesindustry.biz/articles/2013-10-10-nintendos-iwata-doesnt-fear-failure> (last accessed 30 May 2018).

Witkowski, Emma and James Manning (2017), "Playing with(out) Power: Negotiated Conventions of High Performance Networked Play Practices," *Proceedings of DiGRA 2017*, <http://digra2017.com/static/Full%20Papers/112_DIGRA2017_FP_Witkowski_High_performance_Play.pdf> (last accessed 30 May 2018).

Wolf, Mark J. P. (2001), "Genre and the Video Game," in Mark J. P. Wolf (ed.), *The Medium of the Video Game*, Austin: University of Texas Press, pp. 113–34.

6. FROM CINELUDIC FORM TO MISE-EN-GAME: THE LUDIFICATION OF CINEMATIC STORYWORLDS IN THE *STAR WARS* VIDEO GAMES

Andreas Rauscher

At the beginning of *Star Wars: Episode VII—The Force Awakens* (J. J. Abrams, 2015) Resistance pilot Poe Dameron (Oscar Isaac) teaches First Order deserter Finn (John Boyega) how to fly a TIE fighter: "Use the toggle on the left to switch between missiles, cannons and mag-pulse. Use the sight on the right to aim. Trigger is to fire." These instructions would have been perfectly suited for a tutorial flight in the video game *Star Wars: TIE Fighter* (LucasArts, 1994). The scene plays out almost like a reversal or, to stay within the terminology of video games, a reverse engineering, of the term *remediation,* defined by media scholars Jay David Bolter and Richard Grusin as "the formal logic by which new media refashion prior media forms" (2000: 273). The phenomenon of *remediation* is associated with the process of *immediacy* hiding the presence of the medium in favor of transparency, and the opposing strategy of *hypermediacy* that makes the user aware of the medium itself (Bolter and Grusin, 2000: 272). The *immediacy* of the *X-Wing* and *TIE Fighter* series (LucasArts, 1993–99), whose games merge realistic flight simulation and science-fiction combat, turns into *hypermediacy* by being reminiscent of a video-game tutorial within the *mise-en-scène* of *The Force Awakens*. In contrast to the transparent effect of *immediacy*, *hypermediacy* as a form of *remediation* refers to "a style of visual representation whose goal is to remind the viewer of the medium" (Bolter and Grusin, 2000: 272).

The references in *The Force Awakens* are not directed towards any real fighter plane but to the fictional physics of the TIE fighter. Of course, we probably remember those fighter jets from the first *Star Wars* trilogy—consisting of *Star*

Wars: Episode IV—A New Hope (George Lucas, 1977), *Star Wars: Episode V—The Empire Strikes Back* (Irvin Kershner, 1980) and *Star Wars: Episode VI—Return of the Jedi* (Richard Marquand, 1983)—and maybe we have a physical impression from owning a toy model. But the puzzling aspect of the scene between Poe and Finn is that we may even have memories of flying a *TIE Fighter* in the video game. The effect of *hypermediacy* in *The Force Awakens*, that a scene is reminiscent of video-game situations and tropes, results from what could be called the cineludic form. There are several moments in the more recent *Star Wars* films that look like remediated scenarios from the vast library of video games based on the franchise: the Millennium Falcon's obstacle course run through crashed Star Destroyers and labyrinthine mines in *The Force Awakens* already had appeared in the *Rebel Assault* CD-ROM games (LucasArts, 1993, 1995). Luke Skywalker's last stand as a force projection at the end of *Star Wars: Episode VIII—The Last Jedi* (Rian Johnson, 2017) is reminiscent of strategy games such as *Star Wars: Empire at War* (LucasArts, 2006) that also feature prominent characters from the films as a support with special skills activated for several rounds that allow the player's troops to escape during an Imperial attack.

The remediation of ludic structures and situations provides a key aesthetic element to Lucasfilm and Disney's new *Star Wars* films. They are integrated more seamlessly by directors J. J. Abrams and Rian Johnson than in George Lucas's earlier, prequel trilogy (1999–2005). Instead of employing the full canvas of digital possibilities as the prequels did, the new episodes use practical effects and on-location shooting to recall the first trilogy. But they also demonstrate an awareness of aesthetic choices and resist the temptation to stage retro scenarios or to provide fan service. The approach to action sequence choreography and cinematography in *The Force Awakens* and *The Last Jedi* is aware of its own stylistic traditions. This approach can be compared to neo-noir films such as *Body Heat* (Lawrence Kasdan, 1980), *L.A. Confidential* (Curtis Hanson, 1997) and *The Black Dahlia* (Brian De Palma, 2006) referencing classical film noirs such as *Double Indemnity* (Billy Wilder, 1944). Though neo-noir films, and games such as *L.A. Noire* (Rockstar Games, 2011), have the full spectrum of color grading and virtual illumination at their hand, they apply only the low-key aesthetics of classical noir. The same conscious decisions can be found in the third *Star Wars* trilogy, comprising—at the time of writing—*The Force Awakens* and *The Last Jedi*. In this regard, they differ from the aesthetic of CGI excess that had been applied in the prequels and resulted in an interesting but uneven mixture of *Looney Tunes* cartoon, video-game set piece, *Flash Gordon* serial and 1950s Douglas Sirk melodrama. *Star Wars: Episode I—The Phantom Menace* (George Lucas, 1999), *Star Wars: Episode II—Attack of the Clones* (George Lucas, 2002) and *Star Wars: Episode III—Revenge of the Sith* (George Lucas, 2005) featured game-like scenes, from the pod race in *Episode I*, to the side-scrolling, platform-like chase through a droid factory in *Episode II*, up to the animated opening long take of *Episode III* following a Republic fighter through space battles. But their exposure as set pieces made them look almost like isolated cutscenes instead of being integrated into the cinematic flow. In this regard, the integration of cineludic forms in *Episode VII* and *Episode VIII* can be compared to the more organic integration of scripted events in video games and ludic choices as an alternative to cutscenes and quick-time events.

In the context of replay, the redundant elements hearkening back to the first trilogy in the latest episodes, such as the attack on Starkiller Base in *The Force Awakens* recalling the Death Star battle from *A New Hope*, can be considered from a perspective of knowing ludic variation instead of simple redundant narrative repetition of the plot. These elements focus on different actions from the original sequences, and they contribute to the rich cinematic texture of *The Force Awakens* and *The Last Jedi*. These films do not reboot or retell their predecessors; rather, they replay certain game-like standard situations from the original trilogy in a new context. In *The Force Awakens*, the saved game in the Millennium Falcon's chessboard (a game referred to as "Dejarik" in the franchise's canon) that arouses Chewbacca's interest could be interpreted as a key symbol for the whole sequel trilogy. Just like the vintage game of Dejarik being reactivated for another round, the galaxy's setting has been reloaded to continue the game featuring the original characters that had been delegated to other media for several decades.

This chapter will discuss the concept of cineludic forms as a way of meaningful play and as an aesthetic approach toward the synergies between video games and films. Instead of focusing on the technological convergence in the production of films and video games or the narrative complexity of media convergence, the synergies between films and games are considered from a ludic perspective resulting in the reconfiguration of cinematic forms. At first, the cineludic form will be discussed on the microlevel of closed game scenarios that correspond to the traditional concept of goal-driven repeatable game structures. Like many arcade games from the 1980s, the early *Star Wars* video games—produced by Atari from 1983 to 1985 for arcade machines and television consoles—convert set pieces from the films (such as the Battle of Hoth from *The Empire Strikes Back*) that are quite adaptable to procedural structures in the sense of *ludus* (that is, games with a clear winning structure). Later game titles in the franchise, such as the Super Nintendo Entertainment System's *Super Star Wars* platform games (JVC, 1992–94) and the CD-ROM-based *Star Wars: Rebel Assault* (LucasArts, 1993), combine several standard situations to create a linear cinematic experience mirroring the films' structures like a theme park attraction.

In a second step, the cineludic form will be examined on the macrolevel of genre structures that enable players to leave the hard rails of the earlier arcade games. Games such as the *Jedi Knight* (LucasArts, 1995–2003) and *X-Wing* series adapt the setting of the galaxy far, far away to the demands of different game genres. In this regard, the video games continue the strategy of genre crossover established by the film series, which has taken cues from Akira Kurosawa's samurai epics, spaghetti Westerns, World War II dogfight films, and science-fiction serials. The same process can be found in the *Star Wars* video games, which have, for example, combined first-person-shooter mechanics with role-playing elements, and have used the framework of a flight simulator to conduct an exploration of the Dark Side not yet featured in the films. The configurations associated with participatory culture—which invites fans to appropriate and interpret popular franchise culture in their own idiosyncratic way—and with playful game structures correspond to the idea of free role play in the sense of *paidia* (that is, games as play without a set goal, in the tradition of Roger Caillois) (Egenfeldt-Nielsen et al., 2008: 27).

Finally, in a third step, the ludification of the storyworld as a set of possibilities and as an interconnected mental map of scenarios and playgrounds will be discussed. Referring to the idea of "playing with *Star Wars*" coined by game scholar James Newman (2017: 148), the chapter concludes with further perspectives on a playful approach toward transmedia worldbuilding that can, at the same time, provide a platform and archive for cultural and historical perspectives.

Ludonarrative Death Star Battles and Other *Star Wars* Standard Situations

The early years of academic game studies were dominated by a productive, albeit sometimes a little fierce, discussion about the meaning of narrative in games. Ludologists from the Center for Computer Games Research at Copenhagen demanded that games should be discussed on their own terms, an idea comparable to the insistence of film studies on finding its own vocabulary beyond adapting discussions more suited to literature, media or theater studies.[1] Since the expected Imperial full-frontal attack on the new discipline did not happen, the role of the phantom menace was delegated to narrativists, a more extreme form of narratologists that consider everything to be a story, even the experience of riding elevators and purely abstract games like *Tetris* (Various, 1984–present).

In a well-known discussion during the height of the controversy about the contradictions between ludic and narrative structures, game scholar Jesper Juul proclaimed that the story of *Star Wars* from the 1983 Atari arcade video game could not be understood (2001). Media scholar Henry Jenkins replied that, "[T]he *Star Wars* game may not simply retell the story of *Star Wars*, but it doesn't have to in order to enrich or expand our experience of the *Star Wars* saga" (2004: 121). Jenkins regards the *Star Wars* arcade game in relation to the larger structure of transmedia storytelling. In a discussion of the *Matrix* films (The Wachowskis, 1999–2003), their accompanying video games, comic books, animated short films and other building blocks to a larger network of worldbuilding, Jenkins defined *transmedia storytelling* as consisting of "stories that unfold across multiple media platforms, with each medium making distinctive contributions to our understanding of the world, a more integrated approach to franchise development than models based on urtexts and ancillary products" (2006: 293). In "Game Design as Narrative Architecture" (2004), he names four types of narrative—evocative, enacted, embedded, and emergent—-that can very well be combined with transmedia storytelling structures that also include contributions from fan fiction and other forms of participatory culture not being limited to the official canon established by the copyright holders.

The idea of transmedia storytelling corresponds to current concepts from adaptation studies such as Linda Hutcheon's *Theory of Adaptation* (2007) that take into account alternatives to a central urtext. Instead of transferring a single text from one medium to another, certain characters, settings, and situations can also be adapted. In this regard, transmedia storytelling provides an alternative to traditional linear narratives in suggesting the idea of a patchwork featuring several storylines. In the process of worldbuilding, those different story blocks can create a structure similar to a theme park or an adventure playground that represents not a single linear narrative but several set pieces from a franchise that can be visited as self-contained places. Audiences can use different points of entry without being restricted to a certain pathway and can even rearrange items in a way very similar to the configuration of game spaces.

Juul is quite right with regard to the early arcade games and titles, such as *Star Wars Episode I: Racer* (LucasArts/Nintendo, 1999), that translate only a few set pieces from the films into repeatable game mechanics. In contrast to later video-game adaptations such as *Super Star Wars* and the *LEGO Star Wars* series (Eidos Interactive et al., 2005–present) that try to reconstruct the films' broader dramatic structure in their level architecture, the early arcade games are much closer to board games and theme-park attractions. The arguments in the discussion between Juul and Jenkins seem to focus on two very different levels. Juul, on the one hand, focuses on the very foundational elements of game design; Jenkins, on the other hand, talks about the larger structure of *Star Wars* as an open-ended mythological patchwork adapting different media and cultural experiences. Both approaches are helpful in discussing decisive elements of the *Star Wars* experience throughout the last forty years. Juul's focus on games illustrates how key film scenes such as the Death Star battles, the escape from Hoth, and the pod race can be turned into intuitive game structures. Jenkins's idea of transmedia storytelling, in contrast, is essential to understanding how *Star Wars*—with an ever-expanding universe originating from novels, comics, and video games—created several narrative kernels and satellites in the 1990s that are no longer tied to a linear structure but that can, instead, be navigated like the sites in a theme park. Nevertheless, both approaches focus on structural, rather than aesthetic, aspects. So, in retracing the creation of cineludic forms on the microlevel, let us take a closer look at the first trilogy and its relation to game concepts.

The mission briefing before the Death Star battle in *A New Hope* is staged like a design pattern for future level instructions from the *X-Wing* games. The point-of-view shots from the X-wing's cockpit defined the perspective for generations of space shooters to come, and the failed first Death Star trench run looks like an instructive demonstration of how *not* to blow up a space station. On a structural level, the Death Star battle provides a template for future game design. It features all the formal elements that Tracy Fullerton lists in *Game Design Workshop* (2008): objectives (for example, to blow up the Death Star before it can fire at the Rebel base on Yavin 4), procedures (for example, use an X-wing fighter and its laser weaponry), rules (for example, one attempt to hit the exhaustion shaft per trench run), resources (for example, limited number of shields), conflict (for example, a family feud which, at this point of the story, has not yet been recognized to be one), boundaries (for example, limited to three attack waves in the Death Star battle, and an inability to go into hyperspace with the *X Wing* series), and outcome (for example, only relevant with regard to the high score; after one Death Star blows up, the player does not have to wait until *Return of the Jedi* for another to appear).

From a media-specific perspective, the sequence in *A New Hope* remains rather committed to the language of classical cinema by intercutting among Leia Organa (Carrie Fisher), C-3PO (Anthony Daniels) and the Rebel high command in their threatened base on Yavin 4; Grand Moff Tarkin (Peter Cushing) in the Imperial command station on the Death Star; and the dog fights between the Rebels and the Empire representing the next step on Luke Skywalker's (Mark Hamill) hero journey. It does not employ the artificial canvas of self-conscious cinematic devices associated with postmodern cinema found in the cinema of Luc Besson or of Quentin Tarantino. When Han Solo (Harrison Ford) and Chewbacca (Peter Mayhew) come to the rescue, the scene follows traditional, Classical Hollywood narrative patterns such as the trope of the last-second rescue, employed in everything from the silent cinema of D. W. Griffith to Steven Spielberg's *Indiana Jones and the Temple of Doom* (1984). Nevertheless, the concept of pastiche also informs the first *Star Wars* trilogy and its relation to video games. In preparing to film the sequence, Lucas, as shown in the documentary *Empire of Dreams* (Kevin Burns and Edith Baker, 2004), edited together several fight scenes from World War II aerial fight films such as *The Dam Busters* (Michael Anderson, 1955). Comparable to later animated storyboards, the visual effects were arranged by Industrial Light and Magic with a new camera system. The action emulated in video games from Atari's *Star Wars* to the *X-Wing* series, therefore, is not based on the simulation of actual realistic aerial fights but rather on the larger-than-life aesthetics of classical adventure films.

Many early space shooters, such as *Space Invaders* (Taito/Midway, 1978) and *Gradius* (Konami, 1985), took inspiration from the popularity of space action-adventure films, and used the concept of attack waves, at first increasing in speed and, later on in side-scrolling shooter games, in numbers and size. The Atari *Star Wars* video game follows this concept, presenting a first stage against Imperial TIE fighters guarding the Death Star. A casual narrative reference to the film is included by Darth Vader's TIE fighter being indestructible, as the villain needs to be around for the next episode even if players do not encounter him in the games. In the second level, players have to cross the battle station's surface guarded by armed defense towers. The third level uses vector-based graphics to recreate the Death Star trench run with increasing numbers of obstacles that require evasive maneuvers.

This aspect of the cineludic form that translates cinematic *mise-en-scène* arrangements into a ludic mise-en-game can also be found in several other sequences from Lucas and Spielberg films: from the opening sequence of *Raiders of the Lost Ark* (Steven Spielberg, 1981), which served as a template for action-adventure games like *Tomb Raider* (Eidos Interactive/Square Enix, 1996–present) featuring hidden traps, puzzle-solving skills, and rolling boulders that appealed to the player's reaction skills, to the mine car chase in *Indiana Jones and the Temple of Doom* (Steven Spielberg, 1984) that informed the film's 1984 Atari arcade game adaptation, as well as a difficult cartoon chase from Nintendo's *Donkey Kong Country* (1994). Beyond the Death Star battle, the first *Star Wars* trilogy created several iconic scenarios built upon a game-like structure. The Hoth sequence from *The Empire Strikes Back* consists of a countdown and provides an original game mechanic in attacking the Imperial AT-AT walkers by wrapping a line fired by a harpoon from a snow speeder around their legs. The scene has been adapted in several *Star Wars* games, from *Star Wars: The Empire Strikes Back* (Parker Brothers, 1982) to *Rebel Assault, Star Wars: Shadows of the Empire* (Nintendo/LucasArts, 1996) and *LEGO Star Wars*. The maneuver to bring down the AT-ATs has even been referenced by pop-culture connoisseur Peter Parker (Spider-Man) in *Captain America: Civil War* (Anthony and Joe Russo, 2016) and, just like his reference to the *Alien* series in *Avengers: Infinity War* (Anthony and Joe Russo, 2018) it turns out to be the right move—even if all the metareferences annoy Tony Stark (Robert Downey Jr).

Game scholar Hans-Joachim Backe writes about the *The Empire Strikes Back* Atari 2600 game:

> It referenced the film on an audio-visual level, and it even succeeded in reenacting an iconic filmic sequence in a recognizable way – a remarkable achievement considering the crude graphics available at the time. What is more, its game mechanics – which are an extrapolation of *Space Invaders* (1978) – correlate perfectly with the key passage in the film, where the relentlessly advancing Imperial troops have to be detained long enough to allow for an escape from the planet Hoth. (2015: 93)

The *Star Wars* films defined set pieces corresponding to film historian Tom Gunning's concept of the *cinema of attraction* (2000). But, in contrast to Gunning's traditional film studies perspective only recognizing a return to the distinct exposure of effects and sensations of early non-narrative cinema, the Death Star battle, the escape from Hoth, and the chase through the asteroid field also establish a ludic structure that goes beyond the narrative content and adheres to Juul and Fullerton's ideas about game mechanics. The audio-visual attractions on display become game-specific tasks and obstacles that are signaled on a semantic level and turned into a ludic syntax. Instead of simply advancing the plot, this syntactic structure strongly emphasizes the excitement and originality of the moves and maneuvers taken to overcome the obstacles. The form of the *mise-en-scène* becomes even as important as the story develops, but the syntax nevertheless does not resolve into pure spectacle; instead, it articulates a ludic progression in cinematic terms. Players adapt the simple narrative moves into turns within a game system compatible with both analog and digital game design. It is no coincidence that Indiana Jones's passage through the temple of doom, the Death Star battle, and the escape from Hoth have all been turned into video games and into board games with differing degrees of complexity. The elements Juul deems to be characteristic of classic game structures (for example, variable valorization of outcome, negotiable consequences, rules, player attachment, and player effort) (2005: 36) can be applied to the smallest units of the cineludic form found in the *Star Wars* and *Indiana Jones* films. Media scholar Stefan Hall names "identifying core cinematic referents" as one of the essential building blocks of *Star Wars* video games (2017: 95). But one important aspect of those building blocks that extends beyond the cinema of attractions or the pure cineludic fun of the early arcade and board games is that they can be combined in a ludic montage.

Just as the juxtapositions of sequences created small narrative units in early cinema and extended the traditional chase scenes found in slapstick comedies and *The Great Train Robbery* (Edwin S. Porter, 1903) into the dramatic narratives of classical cinema, the game scenarios of *Star Wars* and their changing sets of rules and mechanics could be combined into multilinear sequences that provided the structure for pen-and-paper role-playing games (RPGs), as well as video games featuring different phases of gameplay. In the development of the *Star Wars* films and video games, the set pieces developed into stock scenes that informed both audiences and players about what steps to take next in the battle against the Empire or the Rebel Alliance. The train robbery in the spin-off *Solo: A Star Wars Story* (Ron Howard, 2018) not only recalls Edwin S. Porter's first Western from 1903 but also parallels the level design and mechanics (that is, the game syntax) from the *Jedi Knight* series. The semantic character of approaching Imperial probot droids, a presence still mysterious and threatening in the opening shots of *The Empire Strikes Back*, as the prelude to the first attack wave has made its way back to the screen via video games as well. *Return of the Jedi* already employs the syntax established by level structures in games. To attack the second Death Star with Lando Calrissian (Billy Dee Williams) and the Rebel fleet, the shield generator on the forest moon Endor first must be destroyed. Atari's isometric arcade action-game tie-in *Return of the Jedi* (1984) combines both levels, the assault on the shield generator and the following Death Star battle, in a sequence of challenges.

Railway through the Galaxy between Theme Park and Playground

The video games related to the *Star Wars* franchise not only refer to the films in a complex way by integrating their aesthetic into the gameplay and filling out narrative blanks in the background story but they also fulfill the requirements of cineludic forms in a paradigmatic way, contributing to the larger structure of the *Star Wars* transmedia galaxy by arousing curiosity in further characters and locations, while simultaneously achieving a balanced combination between cinematic and game-specific aesthetics. In the arcade games, the gameplay and the different attack patterns are organized by an invisible rail structure comparable to a theme-park attraction or a very long tracking shot. In *Video Game Spaces* (2008), media scholar Michael Nitsche writes about the spatial organization of hard-rail shooter games:

> The world might appear to be accessible but can be navigated only in the confines of a very limited set track. Because this track's layout might not be clear initially to the player, this kind of approach includes an element of discovery. It is a guided journey along which the individual points are important. (2008: 175)

The next step in the development of *Star Wars* video games initiated by the introduction of CD-ROMs in the early 1990s was the simulation of game worlds that continued beyond the settings known from the films.

The desire to go beyond the frame of the cinematic image became a defining quality of LucasArts productions, reflected in Ron Gilbert's innovative

Figure 6.1 *Super Star Wars* (JVC, 1992).

adventure game *The Secret of Monkey Island* (LucasArts, 1991), which was inspired by Disney's Pirates of the Caribbean theme-park attraction (Deutsches Filminstitut et al., 2015: 48). The same promise of navigating a cinematic universe informs *Star Wars* video games, such as *Rebel Assault* and *Shadows of the Empire*, that featured improved graphics and film sequences stored on CD-ROMs. In contrast to the straightforward film adaptations featured in the Atari arcade games and the SNES platform games, these early 1990s titles featured new planets, characters and scenarios.

Like the novels around *Heir to the Empire* by Timothy Zahn and the comic book series by publisher Dark Horse Comics, released from 1991 onwards, these games reignited public interest in the rather underused franchise after two movies, a television series on the *Ewoks* (ABC, 1984–86) and the animated series *Star Wars: Droids* (ABC, 1985–86). In contrast to the complex pen-and-paper RPG system developed by West End Games in the late 1980s, video games such as *Rebel Assault* and *Shadows of the Empire* made sure that players would not miss any attractions, which included exclusive, new footage,

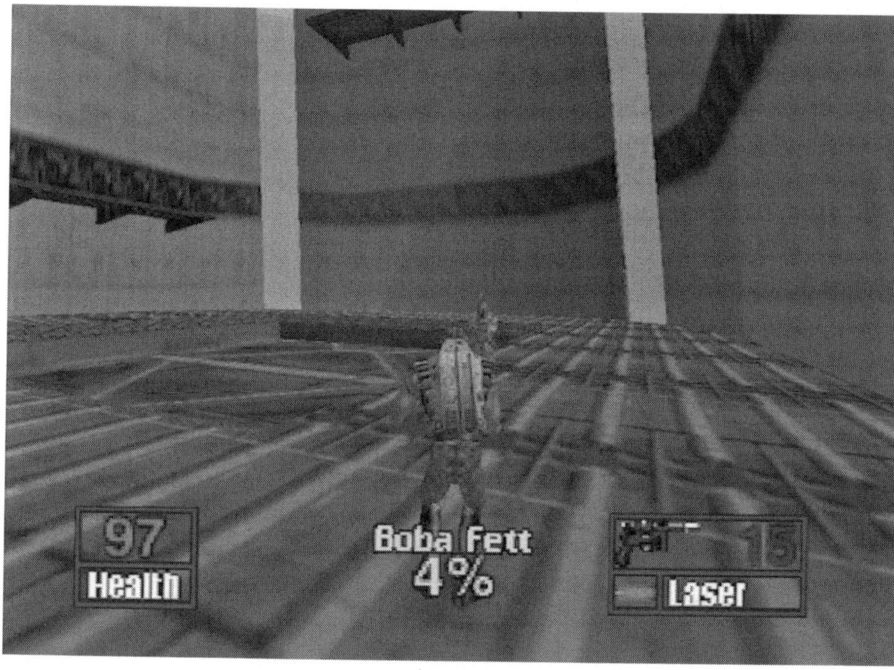

Figure 6.2 *Star Wars: Shadows of the Empire* (Nintendo/LucasArts, 1996).

by providing only one pathway through the game scenario. Comparable to the Disney theme park attraction *Star Tours*, which had opened in 1987, and the earlier *Star Wars* arcade games, *Rebel Assault* and *Shadows of the Empire* followed an invisible rail track. Well-known stock scenes such as the battle of Hoth and the chase through the asteroid field appeared in both games, but their scenarios also featured new situations that, like the risky maneuvers onboard the *Falcon*, would reappear in a slightly varied form in *The Force Awakens* and *The Last Jedi*. Looking back on the early CD-ROM games, some elements almost appear to be a test run for the ultimate Kessel run absolved in *Solo: A Star Wars Story*.

Video games such as *Shadows of the Empire* allowed players to enter places like the galactic capital Coruscant that had been introduced in the novels by Zahn and others. *Shadows of the Empire* combined hard-rail flight game, third-person shooter, and racing mechanics in a linear ludic montage that bridged *The Empire Strikes Back* and *Return of the Jedi* by focusing on the search for Han Solo who has been frozen in carbonite and taken away by bounty hunter Boba Fett to collect the money promised by Jabba the Hutt. According to Newman, the event-like character of *Shadows of the Empire* and its accompanying novel, comic books, soundtrack and action figures represented "the first concerted *Star Wars* transmedia project and [. . .] unlike many tie-in projects (such as the numerous expanded universe novels) which were typically set in distant times and spaces set apart from the canonical movie narrative and worlds, *Shadows of the Empire* was located directly between *The Empire Strikes Back* and *Return of the Jedi*" (2017: 145).

Shadows of the Empire takes a crucial step away from the cross-media promotion that informed the early video games accompanying the films in the first trilogy, and towards transmedia storytelling. In this regard it followed the concept proved and tested in the comic books, novels, and RPGs that at first explored only worlds mentioned in the films before later developing completely new settings and cultures. In contrast to the stories of the *Expanded Universe*, the video game featured only one point of view, from smuggler and Han Solo replacement Dash Rendar. Further stories were told in the accompanying comic-book series focusing on the celebrity deathmatch between Boba Fett and other bounty hunters such as IG-88 and Dengar. These characters shown in a few shots of *The Empire Strikes Back* turned into fan favorites. The vast discussions of if and how Boba Fett escaped from the Sarlacc pit, as shown in *Return of the Jedi*, indicate his exceptional status and demonstrate the ability of the *Expanded Universe* to turn a sideshow attraction into a main character. His potential escape has been featured in the short-story collection *Tales of the Bounty Hunters* (1996) as well as several fan films such as *Escape from the Dune Sea* (Oliver Thompson, 2015). So it seems to be a logical response to the fans that LucasArts featured the bounty hunter and his associates prominently, especially in the visual transmedia incarnations of *Shadows of the Empire*.

The standalone novel by Steve Perry mentions Boba Fett's fight against his former fellow bounty hunters only *en passant*. Since those are rather interesting on a visual and ludic level, the video game pits the player against several of them. The story told exclusively in the novel focuses on Leia, Lando, Luke and Chewbacca infiltrating the Black Sun crime syndicate controlled by new villain Prince Xizor. The several storylines, Dash Rendar's pursuit of the bounty hunters, Leia's rescue mission on Coruscant, and the standoff between Boba Fett and his former colleagues are spread across several media that can be combined in different ways. The concept of ludic montage informs not only the video game but also seems to be an essential part of the whole *Shadows of the Empire* experience. In a RPG, players pick only those characters that arouse their interest. In the *Shadows of the Empire* transmedia event, fans follow only the media-specific storylines and game plots that cover their preferred elements of the *Star Wars* universe. Some readers enjoy the comic-book-specific duel of the bounty hunters, others prefer the space-opera-like narrative of the novel, while video-game players can traverse the setting by taking on Boba Fett themselves. The narrative architecture of *Shadows of the Empire* is quite reminiscent of the early James Bond formula that philosopher

Umberto Eco and film scholar James Chapman have compared to a game of chess: "different characters (pieces) make certain predetermined moves [. . .] By the time that academics had started taking [author and creator Ian] Fleming seriously, James Bond had transcended his origins as the hero of a series of popular novels and had become nothing less than a cultural phenomenon" (Chapman, 2000: 5). Similar observations could be made upon the situation of *Star Wars* in the mid-1990s. With the "Special Edition" rerelease of the first trilogy in 1997, *Star Wars* turned from a cult phenomenon of the late 1970s and early 1980s into a well-received classic. But, in contrast to the films that were regarded as a cinematic form in itself, the so-called *Expanded Universe* organized the cultural phenomenon into different forms and segments. Besides expanding the known diegetic cosmos, *Shadows of the Empire* employed many strategies typically reserved for a comic-book series or a RPG campaign. Since it was obvious from the beginning that the Rebels could not save Han Solo prior to the events already shown in *Return of the Jedi*, enjoyment results from variations of the cineludic forms associated with *Star Wars* on multiple tracks spread across several media. The concept of *Shadow of the Empire* was repeated twenty years later with *Star Wars: The Force Unleashed* (LucasArts, 2008), which bridges the events between *Revenge of the Sith* and *A New Hope*. Nevertheless, the event release of video games which filled in the gaps between the films represents only one line of *Star Wars* gaming. Other game genres left the laid-out tracks of hard-rail level design reminiscent of a theme park ride to enable ludic configurations and explorations rather than scripted attractions.

Configurations of Worldbuilding

In contrast to lackluster video games based on successful films—such as British software studio Ocean's adaptations of very different films *Beverly Hills Cop* (Martin Brest, 1984) and *Robocop* (Paul Verhoeven, 1987) into similar 2-D platformer games—the *Star Wars* video games moved beyond the theme-park-like rail tracks of linear game concepts and articulated an invitation to play that encouraged exploration and socialization in and about the galaxy far, far away. The *X-Wing* series focused on strategy-related gameplay while also providing what Jenkins would term "embedded narrative" by investing players in the larger surroundings of the *Star Wars* galaxy (2004: 126). In the first *X-Wing* game, the mandatory Death Star battle, its recurrence almost comparable to rituals such as the annual cattle drive in the European Alps in the spring, is featured once again. But it is only encountered by the player *en passant*, while the several other missions indicate a storyworld—or rather a series of connected adventure playgrounds—much larger in scope than the few glimpses given in the first three films.

The *principle of minimal departure*, discussed by narratologist Marie-Laure

Ryan (2012), that a fictional story world is imagined in relation to our own, is put to innovative use in how the *Star Wars* games leave the tracks of the rail shooter games. The games relate to our knowledge about the imaginary world that we gathered from the films, comic books, and novels. As soon as the stolen plans to the Death Star arrive and have to be handed over to the next Rebel commando in *X-Wing*, twenty-five years before the background story was presented in the film *Rogue One: A Star Wars Story* (Gareth Edwards, 2016), it was highly motivating that the events of *A New Hope* depended on the successful completion of the mission given to the player. The same effect can also be experienced in *Dark Forces* (LucasArts, 1995), a first-person shooter resembling *Doom* (Id Software, 1993). This first title in the successful *Jedi Knight* franchise combining first-person-shooter and RPG elements and moving beyond the attack of the *Doom*-clones (for example, *Heretic* [Raven Software, 1994]) through improving a few elements of the original game (for example, the option to look up and down), the inclusion of the ambivalent mercenary Kyle Katarn, and the addition of a new stormtrooper design that would appear years later in *Rogue One* and establish another perspective on well-known events from the films. But, in contrast to Jenkins's notion of additive comprehension, which adds further layers of complexity to our understanding of a storyworld, the narrative interest covers only one possible level of involvement.[2] Because the narrative outcome is already known, there are stronger motivations to get involved in the backstory of the stolen Death Star plans from a ludic and a structural point of view. This kind of involvement may best be described by the term "ludonarrative," discussed by game designer Clint Hocking in relation to the conflict between ludic and narrative contract in the dystopian science-fiction game *Bioshock* (2K Games, 2007) and defined by writer Tom Bissell as "unscripted and gamer-determined—the 'fun' portions of the 'played' game. The difference between the framed narrative and the ludonarrative are what make story in games so unmanageable: One is fixed, the other is fluid, and yet they are intended, however notionally, to work together" (2010: 37). The importance of the ludonarrative experience can be explained by an example from *Shadows of the Empire*; whether or not the player–avatar Dash Rendar escapes from Coruscant in the last level or goes missing, the player is not affected in the same way as the spectator is moved by the fates of the doomed protagonists in *Rogue One*. The fate of a game character emotionally affecting the player depends on the game's genre as well as on the player's preferred style of playing. RPGs especially enable players to get emotionally attached to their avatars or non-player characters but those options are only vaguely hinted at in the early *Star Wars* video games and were not fully developed until the early 2000s.

The fascination of the scenario behind the Death Star plans involves ludic configuration and remixing of genre syntax. As different standard situations and narratives, such as the space dogfight or the lightsaber duel, move within an already familiar genre semantic from traditional war and adventure movies,[3] the signs, symbols, and icons of the *Star Wars* universe can be interpreted by further expansions in different media and used for several configurations. While *Dark Forces* employs the level architecture of mid-1990s first-person shooters, *Rogue One* uses the same story but puts it into the context of a different genre syntax inspired by men-on-a-mission movies such as *The Dirty Dozen* (Robert Aldrich, 1967). The semantic remains the same, but the effect on the audience is quite different. The initiation of the domino rally-like cause-and-effect chain that will result in the destruction of the first Death Star gives the player a stronger feeling of agency than absolving the Death Star trench run several dozen times in a row—even if it remains an imaginary agency resulting from the ludonarrative experience, given that the outcome to the Death Star battle has been well known since 1977. These ludic experiences could be compared to anecdotes that actors can invent to make their characters more rounded even if they do not affect the main plot of the drama in which they participate.

From *Dark Forces* and *Jedi Knight* to *X-Wing* and *TIE Fighter* on, quite a few *Star Wars* video games developed an alternative route to the rail tracks of *Rebel Assault*. In the first-person shooters and space-combat games, players were able to influence the rhythm of the game through their own choices. Later games even made it possible to choose from branching paths and individual character arcs. The ethical position of the avatar could be adjusted to the player's preferences. The most elaborated form of genre combinations arrived in 2003's RPG *Star Wars: Knights of the Old Republic* by developer Bioware; it featured several minigames, from playing cards at a bar to racing tournaments and even a reprise of the space combat from the *Star Wars* arcade game on a smaller scale. The structure of the game consisted of several worlds that could be explored non-linearly, with the prologue and the showdown allowing for several alternative endings. Depending on the player's choices of companions and character traits, the story of a fallen Jedi could result in outcomes favoring the light or the dark side. The cineludic forms in *Knights of the Old Republic* functioned as a performance option. Since players decide the camera control and fighting style, they can combine several combat moves from a tactical and aesthetic point of view. Character animations reflect player actions, with the cineludic building blocks enabling players to become performer and director at the same time. The overall structuring principle followed not a fixed series of challenges but a flexible quest structure. Game scholars, such as Espen Aarseth, have discussed the rather episodic quest structure as an alternative to fixed narratives in game analysis: "Although we certainly also 'analyze' (if that is the correct word) quests while doing them, the in medias res analysis is restricted by the lack of complete knowledge, so we need to distinguish between strategic and reflective analysis" (2004: 369).

Quest structures allow for the hybridization of game concepts and ludonarrative configurations by the player. The RPG in its analog and digital forms promises to be one of the most significant examples for the development of ludic montage techniques in the combining of cineludic forms, narrative segments, and different gameplay concepts. It would be a promising topic for further research to consider the reappearance of well-known traditional game concepts within other game structures. They can be included as an Easter egg or a tribute to an overall aesthetic concept but they can also be a part of their own cultural repertoire like the arcade references in *Knights of the Old Republic*, which presents the space battle that had been crucial to the 1983 Atari arcade game as an intersection on traveling between planets. That first level from Atari's game appears as a self-referential passage, just like the new *Star Wars* films feature the obstacle courses from *Rebel Assault*.

The *Star Wars* franchise has demonstrated the creative potential of cineludic forms in transmedia gaming contexts in several ways, from traditional board games to pen-and-paper RPGs, from arcade machines to consoles and personal computers, and from real Kenner toy sand-boxes of Tatooine desert landscapes in the late 1970s and early 1980s to virtual sandboxes. These ludic variations correspond to the transmedia character of games in general, as Juul discusses: "While there is no single medium or set of props that is the ideal game medium, games do exist, and do contain recognizable features, whether as card games, board games, video games, sports, or even mind games" (2005: 48).

The cineludic form serves as a bridge between cinematic set pieces on screen that articulate the template for ritualistic, as well as game-like, stock scenes. Those can be integrated as ludic mechanisms into analog and digital games, and on a mental level they can inspire different forms of play. In this regard, the further adventures in more open-ended *Star Wars* game worlds can be interpreted as a belated, digital continuation of the Kenner toys. Newman offers several readings of the *Star Wars* toys and their relation to later concepts such as the *LEGO Star Wars* series that focus on the playful aspect of constructing scenarios and characters instead of being immersed in a prefabricated experience. After a first playthrough of each story-oriented level, the *LEGO Star Wars* games allow players to enter a mode of free play offering a huge selection of playable characters, including player-created avatars. A very interesting interpretation by Newman, based upon ideas by Jenkins and Jesse Schell, considers the idea of playing with the film via the possibilities offered by the toys: "Once collected, the figures can be used in combinations that defy, amplify and extend the films' narrative, spatial and temporal logic" (2017: 150). The idea of using the figures to create new scenarios, which Jenkins also discusses in relation to animated fan films (2006), indicates an aspect that enables several modalities applicable to the cineludic form. Play turns into a repertoire of optional stock scenes, character traits, and configurative narratives. A *LEGO Star Wars* set can recreate and extend scenes from the films and the transmedia universe but it can also be turned into a metagame collision by integrating figures from other series. The idea of playing with *Star War* could also include playful comments on the rules and regulations of the galaxy far, far away. The obvious construction of the LEGO world represents the playful opposite to the rather immersive involvement pursued in a *Star Wars* pen-and-paper RPG campaign. Instead of creating a seamless in-character form of play, the LEGO approach encourages an anarchic cartoon mentality that also can be found in the LEGO variations exploring the lighter side of the Dark Knight, Batman. In classic cartoons by Tex Avery and Chuck Jones, the characters were always able to manipulate the film frame or change the fiction by drawing new lines and objects—an idea that the animated *LEGO Star Wars* films and television series (2005–present) continue, with Darth Vader appearing on set for a prequel-era scene and being gravely disappointed when told that his appearance is restricted to stories set during the first trilogy. In the *LEGO Star Wars* video games, the difference between ludic story game and playful free navigation is even included in the game design itself; after succeeding in a level's linear mode, players can return to the same level with the freedom to explore as their character of choice.

With promising larger-scale projects, like LucasArts's bounty-hunter game *Star Wars 1313* put on indefinite hiatus, it is quite significant that the remaining main areas of *Star Wars* gaming are not certain genres but open-ended representations of the galaxy far, far away, with gradual modalities not only between the light and the dark side but also between hyperrealism and cartoon. The online RPG *Star Wars: The Old Republic* (Bioware, 2011) continues the navigable world of *Knights of the Old Republic*, extended by a social component. Every guild can decide on what playing style to pursue and whether the players have to stay in character. The hyperrealistic tendency of the cineludic form that inspired *Rebel Assault* and *TIE Fighter* can be found in the detailed representation of the galaxy four thousand years before the events of the films. The hyperrealism affects not only the aesthetic representation but also the details of the rules as well.

The *Star Wars Battlefront* games (Electronic Arts, 2015–17) continue the arcade situation placed in the middle between hyperrealistic representation and cartoonish abstraction, including the *Battlefront II* loot box controversy that came under attack as a hidden form of gambling in some European countries (Hood, 2017). On the other side of the galactic playground are the cartoon modalities of the LEGO games that accentuate the playful and configurable characteristics of the constructed world. The games provide different portals and windows on the world of *Star Wars*, from the archival play of a long time ago on an arcade machine far, far away, up to the latest additions of the LEGO sets. In terms of the cineludic form, the history of the *Star Wars* games can be regarded not only as a productive detour from the conventional hero's journey but also as a pop-historic archive of the intertwined exchanges between films and games which we see in the transmedia patchworks of today.

Notes

1. For a comprehensive collection on the discussion, see Wardrip-Fruin and Harrigan (2004).
2. For comprehensive writings on the term "involvement," the research by game scholars Gordon Calleja and Britta Neitzel is highly recommended.
3. Film scholar Rick Altman established the terms "genre syntax" and "semantics" as essential parts of his pragmatic genre model developed in his seminal study *Film/Genre* (1999): "[W]e can as a whole distinguish between generic definitions that depend on a list of common traits, attitudes, characters, shots, locations, set, and the like—thus stressing the semantic elements that make up the genre—and definitions that play up certain constitutive relationships between undesignated and variable place-holders—relationships that might be called the genre's fundamental syntax" (219).

References

Aarseth, Espen (2004), "Quest Games as Post-Narrative Discourse," in Marie-Laure Ryan (ed.), *Narrative Across Media. The Languages of Storytelling*, Lincoln: University of Nebraska Press, pp. 361–76.

Altman, Rick (1999), *Film/Genre*, London: BFI Publishing.

Backe, Hans-Joachim (2015), "Between 'One for All' and 'Free for All'—Transmedia Storytelling," in Deutsches Filminstitut, Eva Lenhardt and Andreas Rauscher (eds), *Film & Games: Interactions*, Berlin: Bertz, pp. 90–5.

Bissell, Tom (2010), *Extra Lives: Why Video Games Matter*, New York: Pantheon Books.

Bolter, Jay David and Richard Grusin (2000), *Remediation: Understanding New Media*, Cambridge, MA: The MIT Press.

Brooker, Will (2000), *BFI Film Classics: Star Wars*, Basingstoke: Palgrave Macmillan.

Brooker, Will (2002), *Using the Force: Creativity, Community, and Star Wars Fans*, New York: Continuum Books.

Chapman, James (2000), *Licence to Thrill: A Cultural History of the James Bond Films*, New York: Columbia University Press.

Deutsches Filminstitut, Eva Lenhardt, Andreas Rauscher (eds) (2015), *Film & Games: Interactions*, Berlin: Bertz.

Egenfeldt-Nielsen, Simon, Jonas Heide Smith, and Susana Tosca (2008), *Understanding Video Games*, New York: Routledge.

Fullerton, Tracey (2008), *Game Design Workshop*, Burlington: Elsevier.

Gunning, Tom (2000): "The Cinema of Attractions. Early Film, Its Spectator and the Avant-Garde," in Joanne Hollows (ed.), *The Film Studies Reader*, London: Arnold, pp. 161–5.

Guynes, Sean, and Dan Hassler-Forest (eds) (2018), *Star Wars and the History of Transmedia Storytelling*, Amsterdam: Amsterdam University Press.

Hall, Stefan (2018): "Franchising Empire: Parker Brothers, Atari, and the Rise of LucasArts," in Sean Guynes, and Dan Hassler-Forest (eds), *Star Wars and the History of Transmedia Storytelling,* Amsterdam: Amsterdam University Press. pp. 87–101.

Hood, Vic (2017), "Star Wars Battlefront 2 Loot Boxes Investigated by Belgian Gaming Commission," <https://www.eurogamer.net/articles/2017-11-16-star-wars-battlefront-2-loot-boxes-investigated-by-belgian-gaming-commission> (last accessed 23 July 2018).

Hutcheon, Linda (2007), *A Theory of Adaptation*, New York: Routledge.

Jenkins, Henry (2004), "Game Design as Narrative Architecture," in Noah Wardrip-Fruin and Pat Harrigan (eds), *First Person: New Media as Story, Performance, and Game*, Cambridge, MA: The MIT Press, pp. 118–30.

Jenkins, Henry (2006), *Convergence Culture: Where Old and New Media Collide*, New York: New York University Press.

Juul, Jesper (2001), "Games Telling Stories?," in *Game Studies* Vol. 1 Issue 1, July, <http://www.gamestudies.org/0101/juul> (last accessed 1 July 2018).

Juul, Jesper (2005), *Half-Real: Video Games between Real Rules and Fictional Worlds*, Cambridge, MA: The MIT Press.

Ryan, Marie-Laure (2012), "Possible Worlds," in Jan Christoph Meister and Wilhelm Schernus (eds), *The Living Handbook of Narratology*, <http://www.lhn.uni-hamburg.de/article/possible-worlds> (last accessed 23 July 2018).

Wardrip-Fruin, Noah, and Pat Harrigan (eds) (2004), *First Person: New Media as Story, Performance, and Game*, Cambridge, MA: The MIT Press 2004.

PART III

ANIMATION: ADAPTATION ACROSS NATIONS, INDUSTRIES, AND PLATFORMS

7. GHOSTLY BOUNDARIES: TRANSNATIONAL TENSIONS AND ADAPTING ANIMATION IN THE *GHOST IN THE SHELL* FRANCHISE

Brian Ruh

In the runup to the March 2017 theatrical release of *Ghost in the Shell* (Rupert Sanders, 2017)—a live-action Hollywood adaptation of the popular Japanese manga and anime series—among the usual commercials and trailers was a promotional website at iammajor.me. This promotional website featured the Major, the film's main character, saying lines like, "I am hunted. I am the hunter. I am fearless. I am coming for them. I am Major." The website prompted users to identify with the Major's strength and to "upload who [they] are along with [their] picture to post on [their] social media accounts" (Busch, 2017). One Twitter user, Valerie Complex, took this opportunity to advance a criticism of how the film engaged with the issue of race. In particular, the main character of the original manga and anime series was named Major Motoko Kusanagi, but in the live-action adaptation was played by American actress Scarlett Johansson, who is of Dutch and Jewish descent. This caused controversy in many circles, with some critics arguing that such a prominent role should go to someone of Japanese or Asian background.[1]

The tools at iammajor.me would allow individual users to upload an image and run it through a filter to give it a "glitch" aesthetic, a color palette associated with the look of the film, and the words "I am major" and the *Ghost in the Shell* film logo superimposed at the bottom. Complex used the website to create three new images that criticized the *Ghost in the Shell* film itself. One featured an image of Japanese actress Rinko Kikuchi with the words, "I am the woman that should've been cast," while the other two had images of Johansson from the film with the phrases, "I am not Japanese" and "I am in love with white feminism" (Complex, 2017). Her tweet was widely retweeted, and others created their own variations in a similar vein, criticizing the film's casting choices (Green, 2017). One could certainly argue that Paramount Pictures should have taken well-publicized critiques of the casting choice into consideration before launching such a marketing campaign. Or, as Sanderson et al. conclude in another context, "[O]rganizational personnel operating at the intersection of social media and public relations must be attuned to the 'pulse' of the social media audience and carefully contemplate how specific public relations strategies will be interpreted and acted upon" (2016). Of course, the suggestion to try to envision possible ways consumers may interact with social-media campaigns, particularly in ways that may work against the product or idea being promoted, does not address core issues relating to why someone might feel inclined to hijack a social-media campaign.

The example above highlights the tensions involved in adapting a title like *Ghost in the Shell* from a comic- and animation-based franchise into a live-action film featuring a major Hollywood star and a budget, including promotion, of around $250 million (Shepherd, 2017). These tensions include expectations of fans of the original versus what the creators think will play best to a mass audience; the differences in storytelling techniques between comics, animation, and live-action film; and the specifics of a story rooted in the bubble economy of late 1980s Japan, as well as the cultural and technological shifts that have happened since. In this chapter, I shall demonstrate the interrelation of these concerns through a discussion of the "stateless" nature of anime, and how the *Ghost in the Shell* franchise and its live-action adaptation brought these elements into dialogue with one another.

Japanese Anime and Stateless Specificity

In "Conceptualizing Anime and the Database Fantasyscape," I have previously discussed the various discursive elements that have come together to shape what people commonly think of as "anime" (2014). In it, I interrogate the Japanese nature of anime, particularly the concept of *mukokuseki.* Often translated as "stateless," the word has come to describe something that is either unmarked by national elements or is an amalgam of many such elements in a non-hierarchical melange. This concept is important in the discussion of Japanese film history because cultural specificity had a significant impact on how film and, later, television and other audiovisual products were viewed and received outside of their countries of origin. The concept also influenced whether and how particular products became part of the global conversation around film and, it could be said, for animation, as well. Though designs and techniques commonly associated with anime are often seen as "Japanese" now, they arrived at this point only through a circuit of transnational adaptations and appropriations of various styles. Indeed, while one may now identify the style of a show like *Astro Boy* (*Tetsuwan Atom*, 1963–66) with Japan, that was not the case when it began airing on American television in 1963, the same year it debuted in Japan. As I have asserted, "In order to create a space for Japanese animation to thrive, it needed to secure its own marketplace niche by first becoming less Japanese and more 'global.' It was only after that that the form could be recognized as Japanese" (Ruh, 2014: 169). In other words, to travel well outside its country of origin, Japanese film and animation first had to adopt a certain visual language that would allow them to participate in a global conversation.

The seemingly "stateless" nature of anime can create a tension with regard to whether or not bodies in anime are viewed as Japanese. There has been a number of arguments on both sides of the issue. In *Anime from Akira to Princess Mononoke: Experiencing Contemporary Japanese Animation*, Susan J. Napier discusses how the representation of people in anime may be an attempt to participate "in a transnational, stateless culture" (2001: 24). She sees the worlds of anime as their own idiom, related to, but distinct from, actual physical bodies within the real world. This is part of the attraction to many viewers both within Japan and outside of it, although it serves different purposes for the different populations. Napier quotes *Ghost in the Shell* anime film director Mamoru Oshii, saying he suggests that "offering an alternative world to its Japanese audience [. . . is] part of a deliberate effort by modern Japanese to 'evade the fact that they are Japanese,' quoting a provocative statement by the director Miyazaki Hayao to the effect that 'the Japanese hate their own faces'" (2001: 25). Napier seems to play down the cultural specificity of this statement, asserting, "It is not just Japanese audiences who search for more varied forms of electronic entertainment . . . or who are tired of their own faces" (2001: 26). In fact, the popularity of anime outside of Japan seems to rest on the fact that "within the stateless fantasy space that anime provides, both Japanese and non-Japanese can participate in trying on a variety of what might be called 'postethnic' identities" (Napier 2001: 26).

Other critics have noticed similar tendencies, but have arrived at different conclusions about their meanings. Kenji Sato, for example, has described what he sees in anime as "ethnic bleaching," stating that the "characters of anime show the Japanese—who so aspire to Western traits—as they would like to see themselves" (1999: 24–5). He continues with the assertion that manga and anime "are the only two media capable of portraying reality the way Japanese feel it should be" (1999: 25). We can see that Sato is approaching the *mukokuseki* nature of anime from a different direction than Napier, who sees the "stateless" function of anime as potentially freeing, giving viewers the opportunity to identify (at least temporarily) with the characters on screen. Sato, on the other hand, is concerned that "increasingly, anime embodies the

Japanese consciousness of reality," and that in so doing, "the Japanese people . . . have engineered their own extinction" (1999: 27). Although "extinction" may be a rather exaggerated way of stating the case, it is certain that, in a lot of media, Japanese cities and aesthetics serve as playgrounds for (frequently white) foreigners while Japanese and Asian bodies fade into the background or are erased entirely. As Wong Kin Yuen puts it, "Following the success of *Blade Runner*, such cyber-thrillers as *Johnny Mnemonic* (1995), *Hackers* (1995), *Lawnmower Man 2* (1996), and *Strange Days* (1995) have also selectively used motifs of 'Asian' design environments, together with their visual icons, to portray cultural difference and to create visual pleasure from postmodern pastiche" (Wong, 2000), yet none of these films features characters of Asian descent. The relative absence of Asian characters was a criticism picked up even more recently in response to *Blade Runner 2049* (Denis Villeneuve, 2017), against which it was argued "there's no excuse for imagining a world that's so regressively homogenous [*sic*]" (Emerson, 2017).

The tension in these two interpretations of *mukokuseki* take on an additional dimension when dealing with cyberpunk works that problematize technology and the body. Jay Scott Chipman, for example, analyzes the usefulness of the *Ghost in the Shell* franchise to open "new possibilities for imagining cyborg myths for the new millennia [*sic*]" (2010: 171). Generally, the stories within the various *Ghost in the Shell* permutations focus on the character of Major Motoko Kusanagi, a cyborg operative in a secretive government agency called Section 9. Chipman concludes that the stories told through the *Ghost in the Shell* manga and anime "help contemporary individuals address, confront, and accept or reject the possibilities for increased realization of hybrid beings in the developed world" (2010: 189). One way in which the franchise accomplishes this is by focusing on how Kusanagi reacts and adapts to the technological, social, and political spaces around her. Chipman says, "It is difficult to not identify, as contemporary (literal, virtual and/or social) cyborgs, with Kusanagi" (2010: 189). In the process of becoming a global guide for how to negotiate a complex identity within technological worlds, Kusanagi somehow becomes less Japanese, though, and any kind of cultural specificity evaporates. This is intriguing because, as Steven T. Brown points out, the name "Kusanagi" is "rich with cultural connotations in Japanese history and mythology," notably as the name of a legendary sword "comparable in importance to the sword Excalibur in the history of Britain" (2010: 51). At the beginning of the *Ghost in the Shell* manga, writer and illustrator Masamune Shirow notes in the narration that the name "Major Motoko Kusanagi" is "obviously an alias" (2017: 8), because it is not a name that anyone would have been naturally given. Even so, the name imbues the character with a sense of power, but one that may be missed outside of a Japanese cultural and linguistic context. Additionally, at one point in the 1995 animated film, Kusanagi refers to an American character by using the word *hakujin* (literally "white person"), which would seem to set him apart from her and from most of the other characters. To focus solely on the elements of Kusanagi's character that are *mukokuseki* and flow easily across borders is to erase a part of the character's identity.

Science fiction (SF), and the cyberpunk subgenre in particular, has often had issues with race and technology, particularly with regard to the use of Asia as a signifier of futurity. Toshiya Ueno has developed the key criticism of the association between Japan and the future; in his essay "Japanimation and Techno-Orientalism," he states, "If the Orient was invented by the West then the Techno-Orient was invented by the world of information capitalism. In Techno-Orientalism, Japan is not only located geographically . . . as a satellite in orbit, but also projected chronologically by being located in the future of technology" (1996: 3). He goes on to liken techno-orientalism to a semi-reflective two-way mirror situated between Japan and the West (Ueno, 1996: 4). We can look through it and see what is on the other side, but it also reflects the image of ourselves in the act of looking. Consequently, what we see is what is "out there," as well as what we bring to the process ourselves. This is the case not only for the West looking at Japan but for Japan looking at the West. It is easy to make mistakes and misunderstandings in this manner because sometimes it is difficult to distinguish our own reflection from the object on which we are attempting to focus. Even if we ask a compatriot on the other side of the mirror for clarification, their view may be as muddled as ours. As Takayuki Tatsumi put it, "What cyberpunks seem to consume is not merely Japan, but their own science fiction projected in the future called Japan, while what the Japanese audience seems to consume is not merely American SF of the 1980s, but their own image synchronic with cyberpunk" (2006: 111).

This is not to say, though, that we should not strive for clarity as much as possible. Part of an investigation into a film such as *Ghost in the Shell* requires the acknowledgment of the historical forces that shaped the mirror through which we view the world. These forces have an impact not only on what we see but also on our way of seeing. Therefore, it is important that we attempt to assume different cultural perspectives to try to ferret out our potential lacunae. If perfect vision through Ueno's mirror is impossible, then multiple vantage points may bring us at least a little closer to mutual comprehension. In discussing techno-orientalism and cyberpunk, Baryon Tensor Posadas writes that it is important to consider Japanese SF and not just English-language SF that deals with Japanese themes. He states,

> While SF texts often employ colonial or Orientalist imagery with an eye towards its subversion or reversal, attending to Japanese SF texts reveals the limitations of this gesture by bringing to attention the structural pervasiveness of the gendered and racialized infrastructure that sets the

> terms of the genre's attempts to imagine other worlds and futurities. (Posadas, 2017: 146)

In other words, it is important to pay attention not only to how the West represents Japan, but also how Japan represents the West and itself.

We must keep in mind, though, that the things we talk about as being "Japanese" or "Western" are themselves products of a cycle of discourse; what may seem like a concept rooted in one culture may depend on how spatiotemporal boundary lines are drawn. What may seem culturally specific may turn out to be an import from somewhere else if we just look back a little further. Contemporarily, *mukokuseki* imagery contributes to this cycle by obscuring the transcultural circulation of visual elements. For example, Akira Kurosawa's samurai films inspired George Lucas on *Star Wars* (1977), noticeable in elements such as lightsabers and the tunics worn by some characters. The popularity of *Star Wars* in turn sparked a SF boom in the 1970s and 1980s in the United States and in Japan. This created fertile ground for films such as *Alien* (Ridley Scott, 1979) and *Blade Runner*, both of which inspired Mamoru Oshii when he was working on adapting *Ghost in the Shell* into an animated film. The 1980s boom in SF also led to the importation and adaptation of many Japanese anime series on American television, such as *Voltron* (1984–85) and *Robotech* (1985), which created an audience outside Japan ready for a film like *Ghost in the Shell* ten years later.

The Worlds of *Ghost in the Shell*

It is fitting that Masamune Shirow's *Ghost in the Shell* manga introduces its world with the following words: "It is the near future. The world is highly information-intensive, with a vast corporate network covering the planet, electrons and light pulsing through it. But the nation-state and ethnic groups still survive" (2017: 1). With such a setup, our present discussion of *Ghost in the Shell*—with mentions of online social media, race, and transnational adaptations—reflects the ideas Shirow initially set out in his cyberpunk future.

Ghost in the Shell began as a comic by Shirow, serialized in the manga magazine *Young Magazine Kaizokuban* beginning in 1989. It centers on Major Motoko Kusanagi, a special agent working for an elite, secretive squad of the Japanese government fighting terrorism and cybercrime. Kusanagi is a cyborg with an artificial body and an organic brain. In addition to depicting a world of high technology, the manga questions what it is like to live in a world that de-emphasizes the organic. In places, it asks if we are human just because people around us treat us as human. With its combination of military machinery and philosophical inquiry into the nature of humanity, *Ghost in the Shell* quickly became accepted as a key cyberpunk work although, as Shirow wrote of his efforts, "It's rather light cyberpunk ripping off the aesthetic, and much of it is just 'monkey see, monkey do' imitation" (Shirow, 2017: 350).

A few years later, in 1995, Mamoru Oshii's landmark anime adaptation of Shirow's work was released into theaters. Though not a hit in Japan, it became popular around the world owing to the involvement of the British company Manga Entertainment, whose Andy Frain and Laurence Guinness served as producers. "It wasn't very well received when it first hit theaters in Japan, so I was taken aback by its reception in the United States," said Oshii in one retrospective interview (Solomon, 2017). Though there has never been any indication that the content of Oshii's 1995 film was specifically tailored for a non-Japanese audience, the producers at least had an eye on non-Japanese markets, and its success outside of its country of origin was far from accidental.

Because Shirow's *Ghost in the Shell* consisted of many different chapters and stories, Oshii's adaptation had to pare down the manga into a single narrative thread. To do this, the film shifted focus to certain elements of the manga involving the saga of a mysterious hacker called the Puppet Master. Additionally, the film moves away from the Japan-centricity of Shirow's manga to "a place that, while clearly Asian, is deliberately impossible to identify as any one country" (Anonymous, 2017: 34). It is a "city overflowing with data" that, while strongly inspired by Hong Kong, is supposed to be "more than a near-future Hong Kong" (Anonymous, 2017: 35). This is closely related to the ideas of *mukokuseki*, and I would argue that the shift from Japan to a plausibly pan-Asian environment in the adaptation also signals a shift in theme and outlook. As Roh, Huang, and Niu write in the introduction to *Techno-Orientalism: Imagining Asia in Speculative Fiction, History, and Media*,

> differences in the technological signification of Japan and China manifest themselves in the fictive forecasts of the Asian-tinged future. If Japan is a screen on which the West has projected its technological fantasies, then China is a screen on which the West projects its fears of being colonized, mechanized, and instrumentalized in its own pursuit of technological dominance. (2015: 4)

These projections, however, are performed by not just those in the West, but also in some respects by Japan as well, which has its own complicated history with mainland Asia. In terms of *Ghost in the Shell*, it is certainly true that, while Shirow's characterization of Kusanagi is a powerful, outgoing fantasy, Oshii's heroine is far more questioning, pondering the nature of her own self, and this is reflected in the change of setting, as she is often dwarfed by the sheer amount of symbolic information that surrounds her.

In addition to the manga and Mamoru Oshii's filmic adaptations (including a second film, *Innocence* [2004], which interprets a different chapter of

Shirow's original manga), *Ghost in the Shell* was adapted into multiple media forms, such as video games for multiple platforms, novels, and a television series called *Stand Alone Complex* (2002–05). This series was directed by Kenji Kamiyama, who was one of Mamoru Oshii's protogés. The world of *Stand Alone Complex* was, in turn, expanded into its own series of novels, as well as a manga adaptation distinct from Shirow's original. Another direct-to-video series called *Ghost in the Shell: Arise* (2013–15) was directed by Kazuchika Kise, who had served as an animator on both of Oshii's films, as well as on *Stand Alone Complex*.

A live-action Hollywood adaptation of *Ghost in the Shell* had been in the works for a number of years before British director Rupert Sanders came on board. As a work of adaptation, Sanders's film is intriguing because it draws elements from many of the *Ghost in the Shell* anime versions which, in turn, have selectively incorporated elements of the manga. While both of Oshii's *Ghost in the Shell* films adapt stories from Shirow's first volume of the manga, they also incorporate other elements from different *Ghost in the Shell* chapters, as do Kamiyama's *Stand Alone Complex* and Kise's *Arise* series (Ruh, 2013). Therefore, taking characters and scenes from one part of the series and refashioning them in a different context in a new adaptation has been standard practice for *Ghost in the Shell*. In this regard, the live-action film was no different; Sanders has said that, when he was trying to come up with his own approach to the franchise, he compared himself to "a DJ sampling records," and said, "I was down in the basement sampling *Ghost in the Shell: Stand Alone Complex*, pulling out pieces and trying to knit them together" (Cohen, 2017: 16). Additionally, "Sanders and his writers dipped extensively into the story material of the previous iterations of *Ghost in the Shell*, especially *Innocence*" (Cohen, 2017: 16). On top of this, they felt their adaptation needed to include the original film's "must-have moments," specifically the major in her apartment, the chase with a man in thermoptic camouflage, and a climactic fight with a large tank (Cohen, 2017: 16). Kusanagi waking up in her apartment with the stunning skyline of a modern city visible through the window behind her was especially important. The official artbook of the film notes that "[a]lthough Sanders didn't want to be too slavish to the 1995 anime, the bedroom shot was one exception where he did try to match the source material' (Cohen, 2017: 62). Sanders compares the prospective absence of this bedroom scene to what would happen "if you go to *Star Wars* and nobody gets a lightsaber out" (Cohen, 2017: 62).

This high-tech city is notable for another reason: it has gone through a number of permutations throughout the franchise, from being obviously set in Japan in the original manga, to an unidentified city obviously modeled on Hong Kong in Oshii's 1995 film, back to being set in slightly different Japans in *Stand Alone Complex* and *Arise*. Sanders kept many of the same signifiers of a modern Asian metropolis, such as a proliferation of skyscrapers and glowing signs in both English and Japanese characters, but the remixing of old elements and additional designs led to a new city that is both familiar and uncanny to viewers of previous *Ghost in the Shell* works.

Adaptation, *Ghost in the Shell*, and *Mukokuseki*

In Oshii's first *Ghost in the Shell* film, Major Motoko Kusanagi is introduced as she carries out a targeted assassination of a corrupt official. The opening credits sequence presents a montage of the assembly of Kusanagi's cyborg body (or one that looks just like her). The first half of the movie features Kusanagi and the rest of Section 9 tracking down a man who hacked the brain of the foreign minister's interpreter ahead of an important summit meeting, and they chase him throughout the city. In the second half, Section 9 is contacted by a mysterious entity called the Puppet Master which turns out to be an evolved government hacking program. Seeing a way out of the constraints of Section 9 and a government that owns her body, Kusanagi merges with the Puppet Master and achieves freedom as a new form of net-based life. Sanders's *Ghost in the Shell* similarly begins with the creation of the Major; now she is named Mira Killian, however, and she receives an experimental cyborgization procedure from Hanka Robotics after her body was severely damaged in a terrorist attack that killed her family and made her forget most of her past. She adapts to the procedure and joins Section 9 as a counterterrorist operative where she pursues a mysterious antagonist named Kuze. Eventually she discovers that both she and Kuze had been antitechnology activists who were attacked by Hanka Robotics, had their memories altered, and had been forcibly experimented upon. In the film's climax, the Major has to fight a robot tank deployed by the Hanka CEO, and Kuze is killed in the process. Unlike Oshii's film, the Major returns to Section 9 in the end, in spite of the fact that her entire identity as the Major had been artificially constructed. Rather than escaping into freedom, the live-action Major ends up in a body she had been forced into, still in the service of the military government–corporate complex. Additionally, it turns out that Major Mira Killian used to be a young Japanese woman named Motoko Kusanagi. Not only did the main character of *Ghost in the Shell* become white in Sanders's adaptation but, in order to become the powerful Major, she had to become white within the diegesis of the story as well.

As Roh, Huang, and Niu argue, a work like *Ghost in the Shell* can allow "the insertion of Japan-as-signifier of futurity based on the constructed image of the West's Othering of technology and Asia. In other words, a Japan manufactured by the West can further ethnocentric or nationalist projects on both fronts" (2015: 8). The fact that such projects have connections to comics and animation should come as no surprise, at least according to manga author and critic Eiji Otsuka. In his introduction to the edited collection *Debating Otaku in Contemporary Japan*, he says that

> we cannot avoid facing the fact that the "aesthetics" of fascism are the root of "otaku" [geek] culture. This is connected to questions such as why is Miyazaki Hayao, who is politically an antiwar pacifist, at the same time a military fanatic who cannot deny the attraction and beauty of war machines? (2015: xxii)

This question comes up repeatedly in a lot of manga and anime: why the emphasis on weapons of war (such as tanks, guns, giant robots) even when the lesson is invariably that fighting is bad and occurs only when the protagonists are left with no other option? Otsuka does not say that the aesthetic origins of manga and anime necessarily make their current forms fascist per se, but "in the category of things that war and fascism have created we must at the very least include Japanese 'otaku' culture" (2015: xxiii).

This means that questions about nationalism, aesthetics, and fascism should not be separated from the Hollywood adaptation of *Ghost in the Shell*. It would not be a stretch to call Shirow's manga highly conservative in a political sense. As an example, in one of the early stories, the Major and the rest of her team raid a "welfare center" that is using child slave labor. Yet, when one of the kids asks Kusanagi if she has come to free them, she responds, "Do you just want to eat and contribute nothing, to be brainwashed by media trash? To sacrifice the nation's future for your own selfishness? Listen kid—you've got a ghost, and a brain . . . and you can access a cyberbrain. Create your own future . . ." (Shirow, 2017: 41). Here, she tells an exploited, young kid not to expect any help from the system that had been exploiting him and instead to pull himself up by his technological bootstraps. In the 1995 film, Oshii culled many of these elements, focusing more on Kusanagi's internal struggles. Yet, in a way similar to Miyazaki, Oshii certainly has his own attraction to military weaponry, and the *Ghost in the Shell* film includes such fetishism. In some ways, this fetishism of technology stands in for the fetishism of Kusanagi's body which is readily on display. We see her nude as she is being assembled in the beginning and as she is torn apart in the end. She is always shown in a rather detached fashion, however, that seems more clinical than arousing. This differs from Shirow's Kusanagi who is shown in a more conventionally attractive and sexualized manner.[2]

Arguments in favor of Johansson's casting can in some ways be traced back to the idea of *mukokuseki*. Though the Major is an unwilling subject of a government experiment, the technology demonstrates a kind of freedom and flexibility. Oshii has mildly critiqued the live-action film for hewing too closely to his work, saying, "I thought maybe they didn't have to make it *quite* so faithful," although he has supported their choice for the Major, noting in a special roundtable, "I think Scarlett Johansson is about the best person they could have cast for the part, and I'm happy to be able to say that I have no objections to that choice" (Anonymous, 2017: 154). One possible reason for this opinion is that in the role of the Major, Johansson embodies a phenotype that has been seen elsewhere in Oshii's films. Somewhat rare for a Japanese director, Oshii has not been adverse to shooting outside of Japan and working in languages other than Japanese. His film *Avalon* (2001), for example, was shot in Polish, and *Garm Wars* (2014) was shot in English. The films featured Malgorzata Foremniak and Melanie St-Pierre, respectively, in the lead roles and, while both characters embody the kind of cool beauty of Kusanagi in *Ghost in the Shell*, both are at least phenotypically white. In terms of the lead role, a lineage of resemblance can be drawn from Oshii's *Ghost in the Shell* to *Avalon*, *Garm Wars*, and Johansson in *Ghost in the Shell*. Therefore, it makes sense that the lead in *Ghost in the Shell* would appeal to Oshii.

Johansson's casting initially seemed to make economic sense as well. According to Box Office Mojo, Johansson ranks seventh by total gross of the films she has starred in, and she is the only woman in the top twenty (People Index 2018). Though a causal relationship cannot necessarily be inferred, there is at least a strong historical correlation between a film's gross and Johansson being in that film. Part of this is undeniably due to perceptions of her physical attractiveness; *Esquire* designated her "Sexiest Woman Alive" in both 2006 and 2013. Additionally, casting Johansson in the role of the Major in *Ghost in the Shell* is in keeping with the "cool" SF persona she cultivated in some of her previous films. Beginning with *Iron Man 2* (Jon Favreau, 2010), Johansson began playing the role of the intelligent and gymnastic superhero Black Widow in the Marvel Cinematic Universe. Subsequent performances included a role as a man-abducting alien in *Under the Skin* (Jonathan Glazer, 2013), the voice of an evolving artificial intelligence in *Her* (Spike Jonze, 2013), and a woman given extraordinary powers by an accidental overdose of a synthetic drug in *Lucy* (Luc Besson, 2014). Johansson's seeming bankability, coupled with her track record in thematically similar roles, would seem to support her starring role in *Ghost in the Shell*. Additionally, having a known white star in a prominent role tends to help a film receive the funding needed to get off the ground in the first place; as director Ridley Scott put it, discussing his film, *Exodus: Gods and Kings* (2014), "I can't mount a film of this budget . . . and say that my lead actor is Mohammad so-and-so from such-and-such . . . I'm just not going to get it financed. So the question doesn't even come up" (Foundas, 2014).

The casting of Johansson can be tied to ideas of *mukokuseki* and cyberpunk. As Oshii said in an interview,

> The Major is a cyborg and her physical form is an entirely assumed one. The name "Motoko Kusanagi" and her current body are not her original name and body, so there is no basis for saying that an Asian actress must portray her. Even if her original body (presuming such a thing existed) were a Japanese one, that would still apply. (Osborn 2017)

With such a defense of the casting choice, Oshii relies on the fact that the narrative of *Ghost in the Shell* destabilizes the coherence of the physical body and allows for the flexibility of play. The cyberpunk body is the *mukokuseki* body, released from the constraints of nation and race. This certainly seems potentially liberating, but it should not be accepted unquestioningly. As Sean McQueen argues in his analysis of cyberpunk, "Despite cyberpunk's confrontational veneer, its ideological weakness is its individualism, its often uncritical embrace of high technology, and its acceptance of, rather than resistance to, late capitalism" (2017: 5). The destabilization of the self in cyberpunk does allow for greater potential freedom, but it also allows other, perhaps more powerful, forces to intervene. Is it truly freedom when a body previously inscribed by culture and the nation-state is released only to be reinscribed by the forces of capital? To circle back to Oshii's justification, although it is certainly true that the narrative of *Ghost in the Shell* could support a non-Asian lead, it is also the case that it did not need to be a white woman, either.

In fact, the casting of the rest of the Major's teammates puts an intriguing spin on the case. The official *Art of Ghost in the Shell* book mentions that the actors formed a "multicultural ensemble that gives the film its own unique identity, differing from the manga and anime, which have a distinctly Japanese milieu." It goes on to explain that the film cast actors with roots in Japan, the United States, Canada, Zimbabwe, Singapore, Denmark, Australia, Britain, Turkey, and Fiji (Cohen, 2017: 35). Of particular note is the character of Ladriya, of which the book says, "[T]he introduction of this additional female agent is a first" for the franchise (Cohen, 2017: 47). This is certainly a good point; in almost all of the various *Ghost in the Shell* incarnations, Kusanagi is the only significant female character present and is certainly the only female character shown on her team. Unfortunately, Ladriya is still not given much to do, narratively speaking. The story of *Ghost in the Shell* is undoubtedly the Major's, and the casting of supporting characters does little to mitigate the main critique of her role.

Johansson's casting is problematic for two large reasons, one within the diegesis of the film and one outside of it. The film itself reveals that the Major used to be an activist named Motoko Kusanagi who was killed in a government raid and whose "ghost" was later put into the "shell" of a cyborg body. Setting aside the suspension of disbelief required for such a narrative (surely there were more suitable candidates for becoming an elite cyborg operative than an antitechnology activist with no prior military training), this act serves to double down on the erasure of Asian bodies in the world of *Ghost in the Shell*. This is far from a unique approach, though, as the concept of techno-orientalism itself is rooted in a view of "the Asian body as a form of expendable technology" (Roh, Huang, and Niu, 2015: 11). Though cyberpunk as a genre is more than thirty years old, in some ways contemporary examples are still rehashing some of the same ideas from the 1980s. As Posadas puts it, "[T]he persistence of techno-Orientalism in contemporary SF writings indicates that the colonial underpinnings of the SF genre is not merely an historical issue but one that pervades the genre's continued production of an imperial imagination in the present" (2017: 145). Within the film, then, we see a white woman taking center stage within a distinctly Asian environment. This is amplified by the fact that many of the other major characters are Caucasian as well. The film wants to have it both ways; it lays claim to the fantastic ideas and settings of Shirow, Oshii, and Kamiyama but it cannot deal with actual Asian bodies in prominent roles on screen. Therefore, the lack of opportunity for actors of Asian descent in leading Hollywood roles represents the other significant reason for objections to Johansson's casting. One could argue that such casting is not the fault of this singular film but is rather an indictment of a system offering little opportunity for people of color to gain the kinds of experiences and star in the kinds of films that would allow them to take a leading role in a film like *Ghost in the Shell*.

Conclusion

At first glance, it may seem like the issue of the Major's race within the world of *Ghost in the Shell* can be acceptably fluid both within the diegetic cyberpunk world of the franchise and the hyperconnected world in which we find ourselves (and that the genre presaged). After all, she seems to fit within what Takayuki Tatsumi called the "Japanoid"—"the post-eighties hyper-Creole subjectivity transgressing the boundary between the Japanese and the non-Japanese, and in so doing, naturalizing the very act of transgression" (2002: 16). In Shirow's manga, Oshii's anime, and Sanders's live-action adaptation, Kusanagi's "ghost" is able to occupy multiple "shells" thanks to the advent of advanced militarized consumer technologies. In some of the later episodes of the manga, and in Oshii's *Innocence*, Kusanagi even seems to be free from the very need for a specific shell to house her consciousness. Therefore, it might seem like we have grown beyond the concerns of the body. As Tatsumi puts it, we "are all enabled by high-technology to put on, take off, and hyperconsume any skin color just like any other brand new mode. Ethnically Japanese or not, we are all Japanoids" (2002: 16). In this way, Tatsumi's Japanoid is a kind of cyborg *mukokuseki* being. As the reaction to Sanders's *Ghost in the*

Shell adaptation demonstrates, the potential flattening of relations envisaged by such "hyperconsumption" may sound optimistic but it actually occludes an inherent inequality. Any form of identity based on consumption is based on the availability of capital which is far from evenly distributed among countries, races, and ethnicities. Therefore, while a white actress such as Scarlett Johansson is able to transgress "the boundary between the Japanese and the non-Japanese," this seldom happens in the other direction, at least on the global stage erected by Hollywood.

This is not to say, though, that Sanders's film did not have any positive effects. As with all adaptations, the release of the film prompted refocused attention on the original Japanese franchise. In early 2017, the lead-up to the film in North America saw the rerelease of Shirow's manga in hardcover, the release of the *Ghost in the Shell: Stand Alone Complex* series on Blu-ray for the first time, and the translation of a book of short stories set in the *Ghost in the Shell* universe. In this regard, the live-action film benefited the Japanese franchise by renewing focus on it, perhaps reminding audiences of something they had not seen in a number of years. Emphasizing the original, however, also drew inevitable comparisons with Oshii's film, given how much Sanders's film took from it as inspiration. The live-action version often came up short in such instances. For example, writing in *The Atlantic*, David Sims says that, while Oshii's film was a "genuinely revolutionary vision," the live-action version is a "flimsy copy of a copy, one that recreates some of the anime's set-pieces nearly shot for shot, but then pares away nearly everything else that made the original a classic" (2017).

In spite of such critiques, though, *Ghost in the Shell* distributor Paramount Pictures blamed the film's poor box-office performance on the controversy around Johansson's casting rather than on any of its structural flaws (Herreria, 2017). According to Box Office Mojo, the film made around $40 million in the United States and $129 million internationally on its $110 million production budget (Box Office Mojo 2017). These figures do not include the prints and advertising costs, however, which can run at double the production budget. With an estimated total budget of around $250 million, the film, even in a best-case scenario, still probably ran tens of millions into the red. In this way, the box-office failure of *Ghost in the Shell* illustrates the contradictory logic at work in Johansson's casting. In order for the film to be financially successful, the producers felt they needed to cast a white actress with name recognition who could bring in a significant audience but, because of the backlash against her casting, the film ended up losing money.[3]

An interesting contrast can be drawn between *Ghost in the Shell* and *Black Panther* (Ryan Coogler, 2018), a subsequently released Hollywood comic adaptation. Though it did not star a lead with the name draw of Johansson (but admittedly had a much larger budget and the power of the Marvel Studios brand behind it), the film featuring black actors from around the globe surpassed $1 billion in ticket sales worldwide. It is difficult to say that *Ghost in the Shell* was a failure while *Black Panther* was a success in these terms, though, because, on one level, the financial performance of *Black Panther* points to the ways large media corporations can sell the ideas of resistance back to an eager population (as Jennifer Gillan discusses in Chapter 9). At the same time, it's possible that the success of *Black Panther* could lead to genuine changes in ways *Ghost in the Shell* could not. In fact, the fictional nation of Wakanda in *Black Panther* could be seen as an example of *mukokuseki*—a stateless nation able to be transmitted through media to foment genuine change. Writing in *The New York Times Magazine*, Carvell Wallace made the case that *Black Panther* was a culturally important film: "[I]t doesn't matter that Wakanda was an idea from a comic book, created by two Jewish artists. No one knows colonization better than the colonized, and black folks wasted no time in recolonizing Wakanda. No genocide or takeover of land was required. We do with it as we please" (2018). Unfortunately, while the Hollywood adaptation of *Ghost in the Shell* had the inherent potential to be a significant film, it ended up rehashing a number of cyberpunk clichés regarding race and place. While the play of nationlessness and the fluidity of the SF imagination are potentially liberating, we need to be on our guard for what may try to step in and fill that gap.

Notes

1. One common suggestion, which began circulating long before the film was released, was Japanese actor Rinko Kikuch. See Barder, 2015 and Hamedy, 2016 for examples.
2. In some editions, the *manga* even includes four pages of a sex scene between Kusanagi and some of her girlfriends that takes place in a virtual reality.
3. Of course, as mentioned earlier, there were far more factors at work in the film's reception than the casting of the lead character. See Mendelson, 2017 for further details and discussion.

References

Anonymous (2017), *Ghost in the Shell, Readme: 1995–2017*, New York: Kodansha Comics.

Barder, Ollie (2015), "Rinko Kikuchi Would Be A Better Fit For The New 'Ghost In The Shell' Movie," *Forbes*, 24 August, <https://www.forbes.com/sites/olliebarder/2015/08/24/rinko-kikuchi-would-be-a-better-fit-for-the-new-ghost-in-the-shell-movie/> (last accessed 1 March 2018).

Box Office Mojo (2017), "Ghost in the Shell (2017)," *Box Office Mojo*, <http://www.boxofficemojo.com/movies/?id=ghostintheshell2017.htm> (last accessed 11 March 2018).

Brown, Steven T. (2010), *Tokyo Cyberpunk: Posthumanism in Japanese Visual Culture*, New York: Palgrave Macmillan.

Busch, Jenna (2017), "Tell Ghost in the Shell Who You Are at IAmMajor.me," *ComingSoon.net.*, 10 March, <http://www.comingsoon.net/movies/news/824533-ghost-in-the-shell-iammajorme> (last accessed 1 March 2018).

Chipman, Jay Scott (2010), "So Where Do I Go from Here? *Ghost in the Shell* and Imagining Cyborg Mythology for the New Millennium," in John Perlich and David Whitt (eds), *Millennial Mythmaking: Essays on the Power of Science Fiction and Fantasy Literature, Films and Games*, Jefferson, NC: McFarland, pp. 167–91.

Cohen, David S. (2017), *The Art of Ghost in the Shell*, San Rafael, CA: Insight Editions.

Complex, Valerie (2017), *Twitter*, 11 March, <https://twitter.com/ValerieComplex/status/840627652859961347> (last accessed 1 March 2018).

Emerson, Sarah (2017), "Cyberpunk Cities Fetishize Asian Culture But Have No Asians," *Vice*, 12 October, <https://www.vice.com/en_id/article/59dxmb/cyberpunk-cities-fetishize-asian-culture-but-have-no-asians> (last accessed 1 March 2018).

Foundas, Scott (2014), "'Exodus: Gods and Kings' Director Ridley Scott on Creating His Vision of Moses," *Variety*, 25 November, <https://variety.com/2014/film/news/ridley-scott-exodus-gods-and-kings-christian-bale-1201363668/> (last accessed 30 May 2018).

Green, Dylan (2017), "Ghost in the Shell Is Getting Called Out For Whitewashing In the Best Way," *Dorkly*, 13 March, <http://www.dorkly.com/post/82712/ghost-in-the-shell-is-getting-called-out-for-whitewashing-in-the-best-way> (last accessed 1 March 2018).

Hamedy, Saba (2016), "Fans Upset About 'Ghost in the Shell' Casting Recruit Rinko Kikuchi on Instagram," *Mashable*, 17 April, <https://mashable.com/2016/04/17/scarlett-johansson-ghost-in-the-shell-casting-instagram/> (last accessed 1 March 2018).

Herreria, Carla (2017), "Paramount Says 'Ghost In The Shell' Flopped Because Of Whitewashing Controversy," *HuffPost*, 5 April, <https://www.huffingtonpost.com/entry/paramount-admits-ghost-in-the-shell-did-poorly-because-casting_us_58e486cbe4b0d0b7e1660b61> (last accessed 1 March 2018).

McQueen, Sean (2017), *Deleuze and Baudrillard: From Cyberpunk to Biopunk*, Edinburgh: Edinburgh University Press.

Mendelson, Scott (2017), "Box Office: Five Reasons Scarlett Johansson's 'Ghost In The Shell' Bombed In America," *Forbes*, 3 April, <https://www.forbes.com/sites/scottmendelson/2017/04/03/box-office-five-reasons-ghost-in-the-shell-bombed-in-america/> (last accessed 1 March 2018).

Napier, Susan J. (2001), *Anime from Akira to Princess Mononoke: Experiencing Contemporary Japanese Animation*, New York: Palgrave.

Osborn, Alex (2017), "Original *Ghost in the Shell* Director Mamoru Oshii has no Problem with Live-Action Remake," *IGN*, 21 March, <http://www.ign.com/articles/2017/03/21/original-ghost-in-the-shell-director-mamoru-oshii-has-no-problem-with-live-action-remake> (last accessed 1 March 2018).

Otsuka Eiji (2015), "Foreword: Otaku Culture as 'Conversion Literature,'" in Patrick W. Galbraith, Thiam Huat Kam, and Bjorn-Ole Kamm (eds), *Debating Otaku in Contemporary Japan: Historical Perspectives and New Horizons*, London: Bloomsbury, pp. xiii–xxix.

Posadas, Baryon Tensor (2017), "Beyond Techno-Orientalism: Virtual Worlds and Identity Tourism in Japanese Cyberpunk," in Isiah Lavender III (ed.), *Dis-Orienting Planets: Racial Representations of Asia in Science Fiction*, Jackson, MS: University Press of Mississippi, pp. 144–59.

Roh, David S., Betsy Huang, and Greta A. Niu (2015), "Technologizing Orientalism: An Introduction," in David S. Roh, Betsy Huang, and Greta A. Niu (eds), *Techno-Orientalism: Imagining Asia in Speculative Fiction, History, and Media*, New Brunswick, NJ: Rutgers University Press, pp.1–19.

Ruh, Brian (2013), "Producing Transnational Cult Media: *Neon Genesis Evangelion* and *Ghost in the Shell* in Circulation," *Intensities: The Journal of Cult Media* 5, spring/summer 2013, <https://intensitiescultmedia.files.wordpress.com/2013/07/producing-transnational-cult-media-neon-genesis-evangelion-and-ghost-in-the-shell-in-circulation-brian-ruh.pdf> (last accessed 1 March 2018).

Ruh, Brian (2014), "Conceptualizing Anime and the Database Fantasyscape," *Mechademia* 9, pp. 164–75.

Sanderson, Jimmy, Katie Barnes, Christine Williamson, and Edward T. Kian (2016), "'How Could Anyone Have Predicted that #AskJameis Would Go Horribly Wrong?': Public Relations, Social Media, and Hashtag Hijacking," *Public Relations Review* 42, pp. 31–7.

Sato Kenji (1999), "More Animated than Life," *Kyoto Journal*, 41, pp. 22–7.

Shepherd, Jack (2017), "Ghost in the Shell Set to Lose $60 Million at the Box Office," *Independent*, 7 April, <http://www.independent.co.uk/arts-entertainment/films/news/ghost-in-the-shell-box-office-loss-60-million-a7671746.html> (last accessed 1 March 2018).

Sims, David (2017), "*Ghost in the Shell* Is an Incoherent Misfire of a Remake," *The Atlantic*, 30 March, <https://www.theatlantic.com/entertainment/archive/2017/03/ghost-in-the-shell-remake-review/521337/> (last accessed 1 March 2018).

Solomon, Charles (2017), "The Original 'Ghost in the Shell' Was a Watershed Film in Animation History," *Los Angeles Times*, 28 March, <http://www.latimes.com/entertainment/movies/la-et-mn-ghost-in-the-shell-history-20170328-story.html> (last accessed 1 March 2018).

Tatsumi Takayuki (2002), "The Japanoid Manifesto: Toward a New Poetics of Invisible Culture," *The Review of Contemporary Fiction*, 22: 2 pp. 12–18.

Tatsumi Takayuki (2006), *Full Metal Apache: Transactions Between Cyberpunk Japan and Avant-Pop America*, Durham, NC: Duke University Press.

Ueno Toshiya (1996), "Japanimation and Techno-Orientalism," *Documentary Box*, 9, pp. 1–5.

Wallace, Carvell (2018), "Why 'Black Panther' is a Defining Moment for Black America," *The New York Times Magazine*, 12 February, <https://www.nytimes.com/2018/02/12/magazine/why-black-panther-is-a-defining-moment-for-black-america.html> (last accessed 1 March 2018).

Wong, Kin Yuen (2000), "On the Edge of Spaces: Hong Kong Cityscape, Blade Runner, Ghost in the Shell" *Science Fiction Studies* 27: 1, pp. 1–21 <https://www.depauw.edu/sfs/backissues/80/wong80art.htm> (last accessed 1 March 2018).

8. HOW TO ANIMATE YOUR FRANCHISE: DREAMWORKS ANIMATION AND THE FRANCHISING OF *HOW TO TRAIN YOUR DRAGON*

Rayna Denison

Discussions of animation remain comparatively infrequent within the field of US media franchising. Animation appears at the fringes of the debates around character- and toy-based franchises such as *Transformers* (1984–present) and around comic-book adaptations like those for *Teenage Mutant Ninja Turtles* (1984–present), but overall there has been little focus on the roles that animation plays within franchising worlds (Brooker, 2012; Johnson, 2013; Kinder, 1991; Owczarski, 2015).[1] This is despite animation's long-standing presence within transmedia franchising in the United States and despite the fact that animated texts are often significant sites for franchise storytelling. (Johnson, 2013; Jenkins, 2006). The long history of media franchising in the United States provides us with plentiful examples of animation as a crucial factor in media franchising. Some animation-based franchises have been a part of culture for generations, such as *Scooby-Doo* (Hanna-Barbera, 1969–present) or *The Flintstones* (Hanna-Barbera, original series 1960–66). In other cases, animation has been used to keep a beloved franchise alive between live-action installments, as with *Star Trek: The Animated Series* (Filmation, 1973). More often, animation has become a significant element in mixed-media franchise worlds. Within those transmedia franchise worlds, a dominant medium can be difficult to discern, but animation is often a key part of a franchise's success and longevity. Such is the case for both the *Transformers* and *Teenage Mutant Ninja Turtles* franchises mentioned above. For both of those franchises, comic books, animation, live-action filmmaking, and toy production have all been the dominant franchise medium at one historical moment or another (Johnson,

2013; Kinder, 1991). Additionally, in recent years, animation has emerged as a synergistically inspired part of blockbuster franchise production, as when Marvel Television started its own in-house studio, Marvel Animation, in order to supply animated television series based on popular Marvel Cinematic Universe properties (Cavanaugh, 2014). Animation, therefore, plays many crucial roles within media franchising and needs to be considered as a flexible part of normal franchising strategy within US media culture.

Within these discussions of franchising and animation, the animated blockbuster presents the most obvious lacuna. Huge, transmedia worlds sprawl out from character- and brand-based intellectual properties (IP): from animated films such as *Shrek* (Andrew Adamson and Vicky Jenson, 2001), to *Ice Age* (Chris Wedge, 2002), to *Frozen* (Chris Buck and Jennifer Lee, 2013) and, of course, *Toy Story* (John Lasseter, 1995). To take *Toy Story* as an indicative example, not only does the franchise contain many famous toy brands within its narrative, it also includes three sequels to an original film, two spin-off sequels about protagonist Buzz Lightyear, a Buzz Lightyear television series, holiday specials, a series of short films, and a significant web presence. This in addition to *Toy Story*-inspired theme-park rides, video games, a theatrical ice-skating adaptation, and a live musical performed on Disney-owned cruise ships. The young target audience, scale, duration, variety, and transmedia qualities of this franchise may all suggest reasons why animation franchising has resisted investigation. And, while *Toy Story* is a remarkable and unusual example, it is by no means unique within US animation. Animation-based franchises therefore form the topic of this chapter. The aim is to think about how these franchises are shaped by the industries that create them; and about how those industries are shaped by them in turn.

Animated blockbusters have been some of the biggest hits in recent Hollywood history (Steadman, 2014) and yet, as Steve Neale has noted, their existence as blockbusters is rarely acknowledged within academic scholarship (Neale, 2003). Janet Wasko is one of the few to investigate the industrial machinery behind the hit-making process at Disney, mapping the release of *Hercules* (Ron Clements and John Musker, 1997) in order to think about how the different divisions within the Walt Disney Company worked to produce synergy (and thereby success) on the film's release (Wasko, 2001: 70–83). She therefore draws a direct connection between the synergies made possible by conglomeration, and the creation of animation-based franchises. Her instructive approach—a political economic consideration of how corporations produce film culture—is adopted in this chapter, tempered by Derek Johnson's assessment that we need to inquire "not just into the products of media *franchises*, but rather the process of media *franchising* constituted by complex social interactions within the industry structures supporting and driving cultural replication" (Johnson, 2013: 3, emphasis in original). Their collective emphasis on industrial practices and processes is the starting point for this chapter, in particular because the success of animated blockbusters, especially the discourses around Pixar, can make their hit status seem inevitable (McCulloch, 2015). By going back to the processes and practices of franchising, it becomes possible to reveal the challenges faced by industry when creating new animation-based franchises.

Despite the pun in the chapter title, then, this analysis is not intended as an instructional guide on how to create an animated blockbuster franchise. Rather, the aim here is to focus on a specific company, DreamWorks Animation SKG, and an example, *How to Train Your Dragon* (Chris Sanders and Dean DeBlois, 2010–present), which was not a guaranteed or inevitable hit. The aim is to reconstruct and reflect on the creation and management of this particular animation-based franchise through an examination of the company's annual reports, alongside over 150 articles in news, promotional, and trade journals from across the history of the franchise, in addition to reflecting upon aspects of the media texts produced during that process. The purpose of the investigation is threefold: first, to challenge the overdetermining arguments about transmedia franchising's links to conglomeration (see: Wasko, 2001 and Johnson, 2013); second, to argue for a foundational and reciprocal relationship between a studio and its franchises; and, finally, to suggest that the technologies used to create computer-generated animation (CG animation) provide a special case of franchising wherein shared technologies and imagery can be put to divergent uses.

On the surface, the selection of both DreamWorks Animation and *How to Train Your Dragon* might seem peripheral choices. Formed in 1994 as part of DreamWorks SKG, DreamWorks Animation has been a relative newcomer to the animation industry and, when it was devolved into a publicly traded company in 2004, it became one of Hollywood's smallest, non-conglomerated film studios.[2] It has been selected for precisely this reason: because its turn to franchising without conglomeration suggests that there is more nuance needed in our understanding of the relationship between franchising and corporate convergence (Jenkins, 2006). The efforts made by DreamWorks Animation to engage with franchising are reflected in clearly delineated partnerships and licensing arrangements that shifted and changed across the life of the studio, demonstrating the rising significance of franchising to the studio's bottom line. This study therefore begins shortly after DreamWorks Animation was taken public in 2004 and ends with its sale to Comcast–Universal in 2015 for $3.8 billion (Szalai, 2016). This roughly ten-year period coincides with the studio's launch of the *How to Train Your Dragon* franchise, offering useful parallels and developments for analysis. Though the first film was not released until 2010, its production at DreamWorks Animation was being discussed in the industry trade papers as early as September 2005. Across the subsequent ten-year period, *How to Train Your Dragon* went from tentative first release to emphatic franchise, and this trajectory mirrors DreamWorks Animation's industrial ambitions. As mutually reinforcing constructs, then, DreamWorks Animation was shaped by its franchising successes, just as the *How to Train Your Dragon* franchise became DreamWorks Animation's go-to example whenever launching new ventures or partnering in new kinds of production.

How to Create Your Franchise: Ancillary Advertising Production and the Uncertain Franchise Potential of *How to Train Your Dragon*

> *How to Train Your Dragon*, *Shrek Forever After* and *Megamind* ranked among the top 20 films of 2010. Together, they grossed approximately $1.6 billion worldwide—our biggest year ever at the box office. *How to Train Your Dragon* was not only among the year's best-reviewed movies, but it also had the distinction of being nominated for two Academy Awards. Thanks to its stellar performance, *How to Train Your Dragon* became our newest franchise. (Katzenberg, 2011)[3]

This excerpt, taken from the 2011 DreamWorks Animation Annual Review, makes it clear how contingent *How to Train Your Dragon*'s (Chris Sanders and Dean DeBlois, 2010) franchise status was on its release. In his assessment of the year, DreamWorks Animation's CEO Jeffrey Katzenberg prevaricates around *How to Train Your Dragon*'s financial success; for example, lumping the film together with two other hit films in order to inflate its box office. At the same time, Katzenberg talks up *How to Train Your Dragon*'s positive critical and industrial reception to justify the creation of DreamWorks Animation's latest animated franchise. As this suggests, the fate of *How to Train Your Dragon* was uncertain before the first film's release. This section investigates how that first film was promoted, focusing on some of the more unusual promotional tactics deployed by the studio. It will then examine the discourses around *How to Train Your Dragon*'s 2010 box-office results.

How to Train Your Dragon was not a smoothly orchestrated film. When first announced in media-industry trade paper *Variety* in 2005, *How to Train Your Dragon* was to be directed by "DreamWorks vets Lorna Cook ('Spirit: Stallion of the Cimarron') and David Soren" (LaPorte and Fritz, 2005: 1). A little under three years later, Peter Hastings was announced as their replacement, with Bonnie Arnold taking on the role of producer (a role she would retain, Siegel, 2008: 2). The directorial lineup changed a third time when Chris Sanders was asked to step in, taking him off his own project, *The Croods*. Sanders had joined DreamWorks Animation after Disney closed the animation studio in Florida at which he and Dean DeBlois had directed *Lilo and Stitch*

(2002) (Cohen, 2010: 4). In an interview published before the release of *How to Train Your Dragon*, DeBlois relates that DreamWorks Animation first approached Sanders in October 2008. Sanders then brought DeBlois on board specifically to help expand the action and adventure elements of *How to Train Your Dragon* (Anon., 2010). *Animation World Magazine* relates that "the pair was very rushed, with only a year to jumpstart a film that was already in production but deemed too kid-friendly and precious" (Desowitz, 2010). As these abrupt arrivals and departures suggest, unlike Disney or Pixar, DreamWorks Animation had yet to foster a stable pool of directing talent and, consequently, was still drawing upon directors from competitor studios or cannibalizing its own productions to complete projects.

Sanders and DeBlois were finally responsible for completing the adaptation of Cressida Cowell's *How to Train Your Dragon* book series into a CG animation film (Cowell, 2003).[4] The film version of *How to Train Your Dragon* is usually referred to as a loose adaptation of Cowell's books, retaining some key character names, such as that of the protagonist, Hiccup, who is the unlikely weakling heir to the Hooligan tribe of Vikings. In both the books and the films, the Hooligans are led by Hiccup's father, Stoick the Vast, and live in an age of dragons, some of whom are threats while others become their pets. The film, however, reinvents many of the situations and even some of the characters. Most notably the film begins with the dragons in conflict with the Vikings, and Bonnie Arnold pressured DreamWorks for the creation of a prominent young female character, Astrid (*Variety*, 2010e). The CG animation adaptation also removes the dragons' power of speech, making them more animalistic. This results in the starkest adaptation difference: Toothless the dragon. Reimagined from a small, ordinary dragon into a far larger "Night Fury," Toothless and Hiccup's relationship is much altered from the books (Figure 8.1) in which

Figure 8.1 Hiccup makes a bond with Toothless in the film version of *How to Train Your Dragon* (2010).

Toothless begins his partnership with Hiccup as a small common dragon, too small to be ridden.

With these changes in place, DreamWorks Animation reportedly spent more than the film's budget on trying to promote their new motion picture. Their production and marketing budgets were being "reined in" during the period, but films were still budgeted at a comparatively high rate: "around $140 million to $150 million to produce," while "P&A [prints and advertising] costs are also being kept at around $150 million to $175 million per pic" (*Variety*, 2011b). The pull of franchising, in which repetition can be used to offset high promotion and advertising costs, is clear in the light of these expenses. With a new film, such as *How to Train Your Dragon*, however, there were questions over how to generate interest in time for its release in late March 2010. It is here that the first examples of animation's distinctive contribution to franchise creation can be found.

The Winter Olympics in Vancouver, Canada in February 2010 provided an opportunity for promotional innovation. Using the preexisting human and dragon character models created for *How to Train Your Dragon*, a series of five "Dragon Viking 1010" television advertisements were produced by DreamWorks Animation. Narrated by Craig Ferguson (who plays the blacksmith and dragon-training teacher Gobber in the films), these advertisements told tales of the Viking origins of modern Winter Olympics sports, including snowboarding, bobsleigh, speed skating, and ski jump. These advertising "vignettes" thereby extracted the characters from their narrative and settings, repositioning them within the dominant Olympics frame, creating the first, albeit short, transmedia narratives within the nascent franchise. Jonathan Gray contends that these kinds of advertising texts are worth greater consideration in a franchising landscape because "many of these items are filters through which we must pass on way to the film or program, our first formative encounters with the text" (2010: 3). As a first formative encounter with *How to Train Your Dragon*, however, these advertisements were problematic on two fronts. Their connection to the Olympics limited the extent to which they could act as explanatory promotional texts. The key characters from the film remain unnamed within the vignettes although all of the protagonists are featured (Hiccup appearing on two occasions). The result is that the characters remain anonymous within these advertisements. The physics of the *How to Train Your Dragon* film is also altered to allow greater stunts without apparent consequences. In one, Stoick the Vast is shot into space while speed skating on a Gronkle dragon (Figure 8.2); in another, Hiccup is dumped off a snowboard by a Deadly Nadder dragon, falling thousands of feet into the snow below.[5] Instead of forging emotional connections between audiences and characters, then, these shorts exaggerate the film's comedy at the expense of making the characters memorable or meaningful to potential audiences.

Figure 8.2 Stoick the Vast, shot into space on a Gronkle for an Olympic tie-in advertisement titled *Speed Skating*.

Additionally, these vignette advertisements were not very well received. Tom Shales of *The Washington Post* reviewed US television channel NBC's coverage of the Vancouver Winter Olympics, writing that:

> It's quite possible that the nicey-niceness of the whole spectacle could get to be grating after a few more days, like being forced to watch "The Sound of Music" over and over again—from the first row. Commercials disguised as featurettes, like animation done to plug DreamWorks' [*sic*] forthcoming "How to Train Your Dragon," tax viewer patience. (Shales, 2010)

Shales's objection centers on the repetitive and oblique nature of the *How to Train Your Dragon* advertising vignettes. Later accounts suggest that the advertising strategy for *How to Train Your Dragon* failed overall. Sue Zeidler notes that this was part of a wider industry trend, arguing that, "The point is underscored by Dreamworks' [*sic*] latest animated tent-pole film. It spent $160 million to $175 million to market 'Dragon,' then had to revamp its promotional materials and TV commercials right before it opened" (Zeidler, 2010). With unstable stewardship and an expensive and failing advertising campaign, *How to Train Your Dragon*'s success would seem anything but assured when the film opened on 26 March 2010.

Offsetting some of the risk, DreamWorks Animation partnered with retailer Walmart in the United States to produce a merchandising deluge in the weeks before the release of *How to Train Your Dragon*. This partnership went beyond the sale of licensed goods: "For 'Dragon,' Wal-Mart had a hand in the creation of much of the merchandise it will sell and even its packaging, and it was kept in the loop during the production process" explains *Variety*'s Paul Bond (2010a). This partnership was reported as unusual for the time, although Derek Johnson notes that Bruce Stein had created a similar coproduction relationship between DreamWorks and toy company Hasbro back as early as 1995 (2013: 92–3). This new deal came post-independence, however, demonstrating DreamWorks Animation's attempts to foster relationships that had the potential to mimic the synergies produced through conglomeration. Together, DreamWorks Animation and Walmart created a life-sized Viking ship that was placed in New York's Times Square a few days before the film's release, and Walmart created more than a hundred *How to Train Your Dragon* products that were placed in specially designated and designed areas in more than 2,500 Walmart stores, which also included bespoke, scaled-down Viking ship displays and exclusive tie-in toys given away with in-store Happy Meals from McDonald's (Graser, 2010). Marc Graser, writing for *Variety*, contends that "The deal will put the DreamWorks Animation's toon in front of millions of consumers each day and guarantees the studio tens of millions of dollars [*sic*] worth of additional marketing support, bought by Walmart" (Graser, 2010). In addition to their own significant advertising spend, therefore, the Walmart deal helps to explain accusations about the "aggressive" marketing for *How to Train Your Dragon* (Stewart, 2010a). This deal between the relatively small DreamWorks Animation and Walmart also suggests that franchising is not the preserve of conglomerated studios. It reveals how DreamWorks Animation was attempting to forge cross-industrial links early on in the life of *How to Train Your Dragon* in order to generate synergies through industrial partnerships rather than utilizing internal conglomerated synergy.

With such high levels of advertising, expectations were also high when *How to Train Your Dragon* premiered (Zeidler, 2010). When the film failed to meet the top-end estimates of $70 million in box office in its opening weekend, DreamWorks Animation immediately suffered:

> First, the good news: "How to Train Your Dragon" is getting mostly positive reviews, and it set a worldwide Imax record for the biggest opening weekend of an animated film. Now, the bad news: Wall Street doesn't much care.
>
> Shares of DreamWorks Animation sunk 8% on Tuesday to $39.34 after "Dragon" posted a $43.7 million overall domestic box office opening and some analysts fired off negative reports to their clients. (Bond, 2010b)

As Paul Bond of *The Hollywood Reporter* explains above, *How to Train Your Dragon*'s failure to meet market expectations caused DreamWorks

Animation's stock to fall precipitously, albeit temporarily. These immediate correlations between box-office success and the valuation of the company's stock are indicative of DreamWorks Animation's reliance on hit filmmaking at this time and the impact that its films' success or failure had on the wider company. Again, this may help to explain why DreamWorks Animation was pushing towards franchising even when its films were less successful than initially predicted. Doing so allowed them to retain a more stable share price.

While noting the "soft" opening for the film, however, the trade press was also quick to point out that *How to Train Your Dragon* was not like its live-action blockbuster counterparts. Trades noted that "positive reviews will likely help 'How to Train Your Dragon' in the weeks ahead, when many children are out of school for spring break" (Barnes and Harris, 2010). *Variety*'s Andrew Stewart declares this to be a routine phenomenon, stating that, "films like 'Dragon' generally have a long life in theaters, particularly in advance of school holidays." He argues that *How to Train Your Dragon*'s opening weekend reflects an increasingly familiar pattern for animated blockbusters, perhaps especially those from DreamWorks Animation: "The studio saw similar results in 2005 after 'Madagascar' launched with a disappointing domestic take of $47 million, driving stock prices down 9%. But the toon played steadily for many weeks and went on to cume $556 million worldwide" (Stewart, 2010b). Stewart's analysis was to prove perspicacious as *How to Train Your Dragon* would regain the number-one box-office slot after five weeks on release, going on to take nearly $500 million at the global box office (McClintock, 2014). The larger implication of his analysis, however, is that animated blockbusters operate differently in the market when compared to their live-action film counterparts.

The animated blockbuster's success appears to rely on distinctive market factors, including: releases timed to capitalize on school holidays; the heightened importance of word of mouth and reception to slow-burn box-office accumulation; and, perhaps most importantly, worldwide translatability, with the world market more than doubling the box-office take for many films that become animated blockbusters. This means that animation franchising is reliant not just on transmedia production but also on the transnational replicability of those productions. Further, in terms of content, the failure of the initial marketing campaign for *How to Train Your Dragon* also shows how important it is for advertising to align with the content of an animated blockbuster film. Ann Globe, the head of worldwide marketing at DreamWorks Animation, was said to have demurred that "the toon posed a challenge for tub-thumpers given that it was less reliant on the comedic storytelling traditionally found in animated pics" (*Variety*, 2010c) while, a little later, arguing in interview that:

> "It's [success is] a testament to the film's playability," says Globe, noting wryly, "it's in no one's marketing plan to hit No. 1 in week 5. But there

was a real kind of discovery about it. The audience felt a sense of ownership about the movie." (*Variety*, 2010d)

Globe's comments imply that DreamWorks Animation was attempting to create an inhouse brand of animation blockbusters, one built upon broad comedy. The fact that *How to Train Your Dragon* did not accord to that strategy is further evidence of the uncertain status of the film at DreamWorks Animation on its release in 2010. What this first release in the *How to Train Your Dragon* franchise illustrates is that "transitioning out of our start-up phase" still found DreamWorks Animation experimenting in both production and advertising, allowing its new films to reshape and reimagine its corporate meanings (LaPorte and Fritz, 2005). This caused tensions on both sides: on the side of DreamWorks Animation, the company appears to have been continually on the back foot, struggling to manage market expectations for its works and juggling different styles of filmmaking in an attempt to ensure that the company had a breadth to its brand identity. For *How to Train Your Dragon*, there was clear resistance to producing a bespoke advertising strategy that departed from DreamWorks Animation's usual comedy-centered approach. The partnership with Walmart, however, also demonstrates the studio's commitment to *How to Train Your Dragon* and their desire to expand its world beyond film from the earliest moments of its existence.

How to Build Your Mythos: Television, Transmedia Storytelling, and the Maintenance of Franchising Logics

Having established *How to Train Your Dragon* as the newest franchise in its roster, DreamWorks Animation was quick to exploit the IP across film and television. A sequel was announced as early as 12 April 2010 (Barnes, 2010) and, by November, the company had announced that a trilogy of films would be produced along with a stadium show (*Variety*, 2010b). But, perhaps more instructively, DreamWorks Animation took this new franchise as an opportunity to expand its reach across cable television. Prior to the release of *How to Train Your Dragon*, DreamWorks Animation had made deals with Nickelodeon to produce and broadcast television shows based on its popular films, including a *Madagascar* (Eric Darnell and Tom McGrath, 2005) spin-off titled *The Penguins of Madagascar* (Nickelodeon, 2008–15) and *Kung Fu Panda: Legends of Awesomeness* (Nickelodeon, 2011–16) which acted as an extension of the main story of the *Kung Fu Panda* (John Stevenson and Mark Osborne, 2008) film. DreamWorks Animation, however, did not own the copyright to the television series produced for Nickelodeon. The distribution rights belong to Nickelodeon, giving DreamWorks Animation less opportunity to exploit or benefit from the IP. This appears to have caused

DreamWorks Animation to widen their televisual net in search of a more beneficial partnership.

With typical Katzenberg slickness, the DreamWorks Animation CEO announced a new arrangement for the *How to Train Your Dragon* television show, stating that "Cartoon Network was very excited and very aggressive in pursuing this with us . . . We love our partnership with Nickelodeon but we have this growing stable of franchises, and we need multiple homes" (*Variety*, 2011a). Katzenberg's sleight of hand is relatively obvious here: by suggesting that Cartoon Network was the company most keen on the new arrangement, he appears to have been attempting to mollify DreamWorks Animation's existing production partners at Nickelodeon. Further, he is able to suggest that DreamWorks Animation is growing and expanding to the point where its productions could not be contained within a single partnership any longer, demonstrating the strength of his studio. Beneath the marketing hyperbole, too, this deal suggests a shift in power, with DreamWorks Animation now able to retain greater levels of control over its animated films and television shows.

Despite the new relationship, however, much of the early franchising for *How to Train Your Dragon* followed a pattern of franchise production that was already well established at DreamWorks Animation. For example, the company was already regularly producing short animations based on its franchised films, notably holiday-themed shorts such as *Merry Madagascar* (David Soren, 2009) and *Shrek the Halls* (Gary Trousdale, 2005). Closely modeled on these patterns of production, a series of short animations was produced and released in the months following *How to Train Your Dragon*'s film release. The difference following the new Cartoon Network–DreamWorks Animation deal was that the television broadcast of the first of these *How to Train Your Dragon* extension texts, *Legend of the Boneknapper Dragon* (John Puglisi, 2010), aired on the Cartoon Network channel, rather than on Nickelodeon. It was broadcast in October 2010 as a way of promoting the release of the *How to Train Your Dragon* feature film on DVD (where it was also packaged as an extra). Like other examples from DreamWorks Animation, this initial series of *How to Train Your Dragon* shorts retained much of the feature film's voice cast. These shorts were also produced to a high animation specification, making extensive use of the feature film's settings and characters. These shorts allowed the studio to add depth to the story world of *How to Train Your Dragon*, fleshing out backstories of characters and the history of the Viking's interactions with dragons in much the same way as Henry Jenkins relates regarding *The Matrix* extension text, *The Animatrix* (2003) (Jenkins, 2006). *Legend of the Boneknapper Dragon* and *Book of Dragons* (Steve Hickner, 2011) were the most unusual in the series of three initial shorts (later followed by another after the release of *How to Train Your Dragon 2*), switching between the three-dimensional CG animation style of the feature film and a more simplified two-dimensional style of animation (Figures 8.3, 8.4, and 8.5), which was used to relate flashbacks and stories told by characters within the main narrative.

The deal with Cartoon Network to produce a new television series,

Figure 8.3 Gobber, the Boneknapper dragon and the hammerhead shark. Two-dimensional aesthetic in *Legend of the Boneknapper* (2011) flashbacks.

Figure 8.4 Gobber and the Boneknapper. Three-dimensional aesthetic in *Legend of the Boneknapper* (2011) contemporary segments.

Figure 8.5 Bork the Very, Very Unfortunate/Bold meets Whispering Death dragons in *Book of Dragons* (2011).

however, allowed for the greatest amount of transmedia storyworld extension (Smith and Pearson, 2015: 5–6). Two series were initially produced with Cartoon Network: *DreamWorks' Dragons: Riders of Berk* (2012–13) and *DreamWorks' Dragons: Defenders of Berk* (2013–14). Reviewing the first season for *Variety*, Brian Lowry states that "the program is pretty dazzling visually, and pretty effortlessly picks up where the narrative left off." In this observation, Lowry acknowledges the importance of forging connections between the film's narrative and that of the series while going on to note that:

> There's obviously a crass commercial element to all this—something has to keep the merchandising machine humming between movies—as well as the puzzling question of why DreamWorks' name belongs in the title, which probably doesn't mean much to the average tyke.
>
> Still, no one can accuse the studio of approaching these brand extensions of its show in slapdash fashion. (Lowry, 2012)

The two critical comments are particularly noteworthy in that Lowry is suggesting that the DreamWorks name has yet to coalesce into a meaningful brand for children. At the same time, he criticizes *Riders of Berk* for offering little more than a cynically calculated attempt to maintain existing audiences between films in the growing franchise. This negative interpretation of the television show comes within an otherwise positive review of the *Riders of Berk* series and seems to reflect a general weariness with television franchising practices for animated blockbuster films. The comments thereby speak to a wider cultural reflection on animation blockbuster franchising, wherein television texts are usually pushed into secondary, supportive roles, adjunct to feature films (see Gray, 2010 for more on the hierarchies between franchise texts). What is remarkable about the *How to Train Your Dragon* series is that the quality of its CG animation is compared favorably to that of the film, providing a further instance in which the CG animation technologies developed for animated film production could be exploited across media platforms and storytelling formats.

When Stuart Snyder Turner Broadcasting Animation (Cartoon Network's parent company) announced his company's new arrangement with DreamWorks Animation, there was, however, a degree of reframing: "We felt that the story and the characters would make a great television show . . . Now the work really begins, developing and fleshing out the show . . . I think it's a great story, a great premise for our audience, which loves comedy and action in combination." (*Variety*, 2010a) Perhaps unsurprisingly, Snyder attempts to reposition television as the franchise's new dominant medium. His emphatic focus on story in these comments, however, goes against the grain of arguments regarding medium specificity in franchising. For example, Anthony N. Smith and Roberta Pearson argue, "While technological convergence may appear to diminish media boundaries, distinct cultural, and more specifically, industrial conditions work to preserve them" (2015: 5).

Instead, for *How to Train Your Dragon*, showrunners Art Brown and Douglas Sloan spoke in interviews about their need to serve an overarching franchise narrative, using inherited character designs but balanced against considerable freedom to invent and innovate within the new *Riders of Berk* CG animated television series (Glennon, 2015) (see Figures 8.6 and 8.7 for changing character designs). This accords more with Derek Johnson's observations about media franchising, in which he states that the "products and content offered by media franchising . . . might be considered less in terms of unified brands and singular corporate interests, but instead as contested grounds of collaborative creativity where networked shareholders have negotiated the ongoing generation, exchange, and use of shared cultural resources" (2013: 7). The negotiations between DreamWorks Animation's feature production team and the ancillary production for television suggest that technological synergies are often held in tension with the needs of new media formats. Moreover, in the case of *How to Train Your Dragon*, it is clear from the instantiation of the television series that long-term franchise logics were more important than any one medium. Nevertheless, the driving medium for the franchise remains CG animated film, with the television narrative being sandwiched between the arcs of the CG animated feature films.

Snyder also argued, "We now have successful new shows that we can grow

Figure 8.6 Inherited character designs from *How to Train Your Dragon* in *DreamWorks Dragons: Riders of Berk* (2012–13).

Figure 8.7 Building towards *How to Train Your Dragon 2* (2014), retrofitting and aging character designs in *DreamWorks Dragons: Race to the Edge* (2015–18).

into the next big franchises. By partnering with premier entertainment companies like Warner Bros., DreamWorks Animation SKG and DC Entertainment, we are laying the groundwork for even further growth and expansion opportunities across the coming year" (Rose, 2011). In these comments Cartoon

Network's own franchising ambitions are laid bare. Despite its lack of ostensible conglomeration, then, DreamWorks Animation and Cartoon Network were beginning to form oligopolistic-styled brand and franchising networks that allowed each to access cognizant audiences and benefit synergistically from (at least some of) the other's textual production. DreamWorks Animation's extending network of deals was, in this period, starting to mimic the industrial forces of conglomeration.

How to Distribute Your Franchise: DreamWorks Animation, Distribution Experiments, and the Shaping of the *How to Train Your Dragon* Franchise

Television, be it as a launching pad for new franchise content or as a mechanism for retaining franchise consumers, has long been a key medium in DreamWorks Animation's franchising practices. For example, the television specials *Riders of Berk* and *Defenders of Berk* were used to fill in some of the temporal and story gaps between *How to Train Your Dragon* and its sequel which was set five years after the original film. When *How to Train Your Dragon 2* (Dean DeBlois, 2014) was put back to a 2014 release date, however, it overlapped with the creation of another new partnership deal. In 2013, *The New York Times* reported that "DreamWorks Animation, trying to lessen its dependence on the volatile movie business by aggressively expanding into TV programming, has decided to forgo cable television in favour of Netflix" (Barnes, 2014b). Partnering with the online streaming service Netflix opened up a high-profile online window for DreamWorks Animation, alongside its promotional DreamWorks television channel on YouTube and its ownership of Awesomeness TV and NOVA (Szalai, 2016).

In Jeffrey Katzenberg's eyes, the Netflix partnership made DreamWorks Animation "part of the television revolution" (Barnes, 2014a). This kind of hyperbole has been a common refrain across the history of the *How to Train Your Dragon* franchise, which has been variously utilized by DreamWorks Animation to sell industry and audiences on new production, distribution, and exhibition technologies. It was, for instance, *How to Train Your Dragon 2* that featured heavily in the promotional videos made by both Intel and DreamWorks Animation for their jointly produced Premo software. But, more commonly, it was distribution technologies that Katzenberg focused on and which he used *How to Train Your Dragon* to sell.

Before the release of the first *How to Train Your Dragon* film, Katzenberg did world tours promoting the benefits of 3-D exhibition to cinema owners. DreamWorks Animation had also struck a deal with the large-screen IMAX chain of venues to distribute their animated films in 3-D (Cieply and Barnes, 2010). As the time for *How to Train Your Dragon*'s release approached, however, James Cameron's *Avatar* (2009) had become such a huge hit worldwide that it was being held over on the most prestigious 3-D screens along with Tim Burton's *Alice in Wonderland* (2010). Despite being one of the major Hollywood proponents of the new digital 3-D technology, Katzenberg's own films were facing screen shortages. *How to Train Your Dragon*, for example, was due to go head to head on release with the retroactively stereoscoped *Clash of the Titans* (Louis Leterrier, 2010) from Warner Bros. The studios played chicken with the exhibitors. It was reported that DreamWorks Animation's distributor, Paramount Pictures, was "telling theaters that if they don't show the upcoming DreamWorks-produced 'Dragon' on a 3-D screen, then it will withhold from the theatre a 2-D version of the movie to play instead" (Verrier, 2010). There were also reports that Paramount had told exhibitors that they would withhold future films, including *Iron Man 2* (Jon Favreau, 2010) and *Shrek Forever After* (Mike Mitchell, 2010), should *How to Train Your Dragon* not receive the best 3-D exhibition, though the company denied the accusation (Kilday, 2010). DreamWorks Animation's investment in digital 3-D exhibition therefore reflected a significant investment on their part (in personnel and technology) and a high-stakes exhibition gamble in which their box office became dependent on achieving exhibition in the most expensive and high-profile cinema venues, from IMAX to other 3-D screen spaces.

Their gambling on new technologies of distribution is also clear from the Netflix deal which moved the production of *How to Train Your Dragon*'s television series from Cartoon Network to Netflix's online streaming service. Within those gambles, however, there is a high degree of consistency in production. For example, all of the television series have retained the voice cast from the feature films, including Jay Baruchel, America Ferrera, and T. J. Miller. Perhaps more importantly, the showrunners from *Riders of Berk* and *Defenders of Berk*, Brown and Sloan, were brought across to the new productions for Netflix. The Netflix series (six seasons at present), *Race to the Edge* (2015–18) has been able to slot into an ongoing production narrative. Its stories still take place between the two feature films, although they have aged the characters closer to the second film, and the new series has been tying up loose ends and introducing the second film's villain, albeit obliquely. The television shows, in some respects, are therefore about franchising stasis: maintaining and extending the fictional world created in the first film and bringing it into touching distance with the second film. In essence, the television media hold the franchise in stasis by resisting the changes wrought by *How to Train Your Dragon 2*, which introduced new characters, altered the relationships of some of the main characters, and even went as far as to kill off a major character. By placing the narrative of *Race to the Edge* before that of *How to Train Your Dragon 2*, but producing it after the release of the second film, the showrunners are forced to play a reflexive game with their audiences, trying to maintain their interest despite the fact that audiences know how the story will develop. The showrunners' answer to this quandary has been dragons. Sloan has said "What we found in doing the show is the audience loves new dragons. Any episodes about new dragons, they will love it [*sic*]" (Glennon 2015). New technologies of distribution are thereby balanced against a knowing audience and the heart of the franchising practice around *How to Train Your Dragon* at DreamWorks Animation: creating new dragon characters. While this suggests that characters drive the franchise, with their attendant potential for licensing and merchandizing, it is worth noting that this was the answer the showrunners found to the constraints placed upon them by the studio's franchising plans. In essence then, new dragons become not just a merchandising opportunity, they are also the answer to the need to limit character changes and development so as to maintain the overarching franchise narrative.

It is here that the discourses around DreamWorks Animation, *How to Train Your Dragon*, and franchising coalesce: in the balance struck between the studio's attempts to foster interindustrial relationships in technology and distribution, alongside the need to manage carefully the many onscreen representations of the characters that have emerged out of those processes. The relationship between *How to Train Your Dragon* and new digital 3-D is instructive here. Reviewing the film for *The New York Times*, A. O. Scott tells audiences:

> the real distinction of "How to Train Your Dragon"—the reason it deserves to be seen in a theatre with special glasses on, rather than slapped on the DVD player when the children are acting up—lies in those airborne sequences . . . When Hiccup first climbs on Toothless's back and urges the dragon to take wing, the hearts of the audience will soar with a primitive and durable delight. The techniques that enabled this feeling may be dauntingly complicated, but the feeling could not be simpler. (Scott, 2010)

In this review, the technological meets the textual and there is a joyous response to their commingling from the reviewer. The relationship between the studio and its franchise is equally well drawn in this response: from the early, faltering moments of uncertainty to the more recent uses of the franchise as a signpost for the studio's technological milestones, the *How to Train Your Dragon* franchise has become enmeshed with every new stage of its studio's development.

Moreover, that development has shaped the kinds of texts made around *How to Train Your Dragon*; from initial digital 3-D production through to the packaging of a whole series of television for release via Netflix, this particular franchise has been shaped by its distributive technologies and by the partnerships forged by its studio. Through such means we can see how franchising has moved to the center of industrial production in Hollywood. Franchising is no longer the province of conglomerates, but has become an everyday practice of media production. We can also see how important consistency has become within the production of franchises: the reuse of CG animation designs and technologies across film and television has allowed *How to Train Your Dragon* to excel between and across different media platforms from promotional advertising to feature production to television. While this franchise initially sparked from the genus of film, and while film might punctuate the franchise's production processes, it is now increasingly difficult to cite a single, defining media platform for *How to Train Your Dragon*. Instead, it is a mode of production—CG animation—that defines and unites its processes of *How to Train Your Dragon*'s franchise production. Animation now rests at the heart of Hollywood franchising and needs to be seen as the engine driving some of the most profitable and creatively experimental kinds of studio production.

Notes

1. It is worth noting that this lack is not mirrored elsewhere. For example, animation is a crucial part of the discussions of media franchising, normally called "media mix" in Japan. See, for example: Steinberg, 2012.
2. The decision to take the company public appears to have been led by cofounder of DreamWorks SKG, Jeffrey Katzenberg, who retained a controlling interest in the public stock of DreamWorks Animation, essentially ensuring that the company remained his. See: DreamWorks Animation SKG 2012 Annual Report: 30; Barnes, 2014: Section B, 1; DreamWorks Animation SKG Annual Report 2010, 2011: 1-2.
3. *How to Train Your Dragon* was nominated for Best Animated Feature and Best Score at the 2011 Academy Awards.
4. There is an earlier book specifically about Hiccup, the protagonist of *How to Train Your Dragon*, and a further eleven books in the *How to Train Your Dragon* series, but it officially began in 2003.
5. These shorts have now been archived on DreamWorksTV—a YouTube Channel, with a dedicated subsection for promoting the *How to Train Your Dragon* franchise: https://www.youtube.com/user/DreamworksAnimation.

References

Anon. (2011), *DreamWorks Animation SKG Annual Report 2010*, <http://www.dreamworksanimation.com/2010AR/docs/114527_008_DreamWorks_BMK.PDF> (last accessed 1 June 2018).

Anon. (2013), *DreamWorks Animation SKG 2012 Annual Report*, <http://www.dreamworksanimation.com/2012AR/ (last accessed 1 June 2018).

Bond, Paul (2010a), "'Dragon' Ship Charts New Course for Wal-Mart," *The Hollywood Reporter*, 20 March, <https://www.hollywoodreporter.com/news/dragon-ship-charts-new-course-21819> (last accessed 1 June 2018).

Bond, Paul (2010b), "DreamWorks Ani Shares Drop by 8%," *The Hollywood Reporter*, 29 March, LexisNexis Academic.

Barnes, Brooks (2010), "Late to the Game, DreamWorks is Bringing Kung Fu Panda World to the Web," *The New York Times*, 12 April, Section B, p. 9.

Barnes, Brooks (2013), "DreamWorks and Netflix in Deal for New TV Show," *The New York Times*, 18 June, Section B, p. 4.

Barnes, Brooks (2014a), "Netflix Teams with DreamWorks Animation to Create Cartoon Series," *The New York Times Blogs* (Media Decoder), 12 February, <https://www.nytimes.com/2013/06/18/business/media/dreamworks-and-netflix-in-deal-for-new-tv-programs.html> (last accessed 1 June 2018).

Barnes, Brookes (2014b), "DreamWorks Animation Aiming Beyond Film," *The New York Times*, 18 June, Section B, p. 1.

Barnes, Brooks and Rachel Lee Harris (2010), "'Dragon' Rules the Box Office," *The New York Times*, 29 March, Section C, p. 2.

Brooker, Will (2012), *Hunting the Dark Knight: Twenty-first Century Batman*, London: IB Tauris.

Cavanaugh, Patrick (2014), "Marvel's Guardians of the Galaxy Series Announced at NYCC," *Marvel.com*, 10 October, <https://news.marvel.com/tv/23444/marvels_guardians_of_the_galaxy_animated_series_announced_at_nycc/> (last accessed 1 June 2018).

Cieply, Michael and Brooks Barnes (2010), "'Avatar' Faces Traffic Jam at 3-D Screens," *The New York Times*, 30 January, Section C, p. 1.

Cohen, David S. (2010), "Disney to Shut Doors at Imagemovers," *Variety*, 15 March, p. 4.

Cowell, Cressida (2003), *How to Train Your Dragon*, London: Hodder and Stoughton.

Dela Rosa, Manny (2010), "'How to Train Your Dragon,' Interviews with Writers/Director Dean DeBlois and Chris Sanders" *YouTube*, 24 March, <https://www.youtube.com/watch?v=IMRylI5EBZ4> (last accessed 25 May 2018).

Desowitz, Bill (2010), "DreamWorks Unleashes the Dragons," *Animation World Magazine*, 26 March, <https://www.awn.com/animationworld/dreamworks-unleashes-dragons> (last accessed 1 June 2018).

Glennon, Christopher (2015), "'DreamWorks Dragons: Race to the Edge' Interview with Showrunners Art Brown and Douglas Sloan," *ToonZone*, 26 June, <https://www.toonzone.net/dreamworks-dragons-race-edge-interview-showrunners-art-brown-douglas-sloan/> (last accessed 1 June 2018).

Graser, Marc (2010), "Walmart Flies 'Dragon,'" *Variety*, 22 March, p. 3.

Gray, Jonathan (2010), *Show Sold Separately: Promos, Spoilers, and Other Media Paratexts*, New York: New York University Press.

Jenkins, Henry (2006), *Convergence Culture: When Old and New Media Collide*, eBook, Revised Edition, New York: New York University Press.

Johnson, Derek (2013), *Media Franchising: Creative License and Collaboration in the Culture Industries*, New York: New York University Press.

Katzenberg, Jeffrey (2011), "Letter to Shareholders," *DreamWorks Animation SKG Annual Report 2010*, 24 February, <http://www.dreamworksanimation.com/2010AR/docs/114527_008_DreamWorks_BMK.PDF> (last accessed 1 June 2018).

Kilday, Gregg (2010), "'Dragon' to Slay Box Office Competition," *The Hollywood Reporter*, 25 March, <https://www.hollywoodreporter.com/news/dragon-slay-box office-competition-22030> (last accessed 1 June 2018).

Kinder, Marsha (1991), *Playing with Power in Movies, Television and Video Games: From Muppet Babies to Teenage Mutant Ninja Turtles*, Berkeley: University of California Press.

LaPorte, Nicole and Ben Fritz (2005), "DreamWorks Grooming Toons," *Variety*, 21 September, p. 1.

Lowry, Brian (2012), "DreamWorks Dragons: Riders of Berk," *Variety*, 4 September, Review Section, p. 16.

McClintock, Pamela (2014), "Box Office: '22 Jump Street' Soars to $25M on Friday;

'Dragon 2' Hits $18.5M," *The Hollywood Reporter*, 14 June, <https://www.hollywoodreporter.com/news/box-office-22-jump-street-711897> (last accessed 31 March 2018).

McCulloch, Richard (2015), "Whistle While You Work: Branding, Critical Reception and Pixar's Production Culture," in Roberta Pearson and Anthony N. Smith (eds), *Storytelling in the Age of Media Convergence: Exploring Screen Narratives*, Basingstoke: Palgrave Macmillan, pp. 174–89.

Neale, Steve (2003), "Hollywood Blockbusters: Historical Dimensions," in Julian Stringer (ed.), *Movie Blockbusters*, London: Routledge, pp. 53–4.

Owczarski, Kimberly A. (2015), "'More Than Meets the Eye': Transformers and the Complexities of Franchise Film Production in Hollywood," *Quarterly Review of Film and Video* 32, pp. 675–94.

Rose, Lacey (2011), "Cartoon Network Unveils 13 New Series, DC Entertainment Partnership," *The Hollywood Reporter*, 23 March, <https://www.hollywoodreporter.com/news/cartoon-network-unveils-13-new-170487> (last accessed 1 June 2018).

Scott, A. O. (2010), "No Slaying Required: A Viking Aids an Enemy and Wins a Friend," *The New York Times*, 25 March.

Shales, Tom (2010), "Good as Gold: NBC Wraps itself in Well-deserved Olympics Glory," *The Washington Post*, 16 February, Style Section, p. C02.

Siegel, Tatiana (2008), "D'Works Fires Up 'Dragon' Voices," *Variety*, 18 March, p. 2.

Smith, Anthony N. and Roberta Pearson (eds) (2015), *Storytelling in the Media Convergence Age: Exploring Screen Narratives*, Basingstoke: Palgrave Macmillan.

Steadman, Alex (2014), "'Frozen' Becomes the Highest-Grossing Animated Film Ever," *Variety*, 30 March, <http://variety.com/2014/film/news/frozen-becomes-the-highest-grossing-animated-film-ever-1201150128/> (last accessed 31 March 2018).

Steinberg, Marc (2012), *Anime's Media Mix: Franchising Toys and Characters in Japan*, Minneapolis: University of Minnesota Press.

Stewart, Andrew (2010a), "'Dragon" Slayer?," *Variety*, 26 March, p. 1.

Stewart, Andrew (2010b), "'Dragon' Breath," *Variety*, 30 March, p. 1.

Szalai, Georg (2016), "NBCUniversal to Acquire DreamWorks Animation in $3.8B Deal," *The Hollywood Reporter*, 28 April, <https://www.hollywoodreporter.com/news/comcast-acquire-dreamworks-animation-888103> (last accessed 25 May 2018).

Variety (2010a), "'Dragons' Drawn to TV," *Variety*, 13 October, TV Section, p. 5.

Variety (2010b), "How to Brand Your Dragon; DreamWorks Animation Pumps Up Toon Icons on Land, Sea," *Variety*, 22 November, p. 4.

Variety (2010c), "Market Mix Masters," *Variety*, 7 October, p. 1.

Variety (2010d), "Pushing Past Doubt; How to Turn 'Problem' Chatter into Boffo B.O. and maybe even Awards Gold." *Variety*, 22 November, Features Section, p. A1.

Variety (2010e), "Toon Toppers Duel for Kudos," *Variety*, 17 November, Features Section, p. A3.

Variety (2011a), "Cartoon Net Sets Shows," *Variety*, 24 March, p. 3.

Variety (2011b), "DWA's Franchise Future," *Variety*, 25 February, p. 1.

Verrier, Richard (2010), "'Dragon,' 'Titans' Clash for 3-D Screens," *The Washington Post*, 21 March, Style Section, p. E09.

Wasko, Janet (2001), *Understanding Disney*, Cambridge: Polity.

Zeidler, Sue (2010), "Studios Cut Marketing as Landscape Changes," *The Hollywood Reporter*, 9 April, <https://www.hollywoodreporter.com/news/studios-cut-marketing-landscape-changes-22460> (last accessed 25 May 2018).

PART IV

TELEVISION: NEW DIRECTIONS FOR A LEGACY MEDIUM

9. TV BRAND-CASTING, SVOD, AND OTT AT COMCAST AND DISNEY

Jennifer Gillan

The media industry trade press and popular commentators have made a habit of speculating about the fate of scripted fare on US broadcast networks amid the rise of OTT platforms for delivering television through the Internet "over-the-top" of a cable box (Stelter, 2013; Littleton, 2017; James, 2015; Adalian, 2014; Hibbard, 2009; Quartararo, 2008; and Sims, 2015). As I have argued elsewhere, narratives about the displacement of one model by another often overlook the coexistence of models, such as broadcast television, cable on demand, and Over the Top (OTT) distribution (Gillan, 2015). The dramatic downturn in broadcast television ratings resulting in part from the fragmentation of the general audience into niches, the consequent channel spectrum expansion, and the popularity of television streamed through an Internet connection has been well documented (Friedlander, 2018; Littleton, 2017; Stelter, 2013; Gillan, 2010; Napoli, 2011; Lotz, 2014; Strangelove, 2015; and Spangler, 2014, 2016a, 2016b). Yet, little scholarly attention has been paid to the extended shelf life of much scripted fare and its robust afterlife on alternative distribution platforms. The ongoing life of already-aired episodes is possible only because of the expansion of web and streaming platforms through which to disseminate full episodes and seasons (Tryon, 2013; Jenner, 2016; and Strangelove, 2015). After-show, preshow, and overflow content, such as behind-the-scenes shorts, sneak peaks, and webisodes, now coexist with the television program, transforming it into a franchise and keeping the story world, institutions, and characters alive between episodes and seasons and after the series finale (Brooker, 2001; Jenkins, Ford, and Green, 2013;

Jenkins, 2006; Ross, 2008, Gray, 2010; Gillan, 2010, 2015; Evans, 2011; and Grainge and Johnson, 2015). Aimed at engaging existing viewers and enticing potential new ones, these strategies also register the more general shift in Hollywood toward franchise development and management, a topic that has been the subject of several scholarly studies (Stork, 2014; McDonald, 2014; McCulloch, 2014; Johnson, 2013; Wolf, 2012; Hills, 2012; Gray, 2010; Mann, 2009; Grainge, 2008; Skopal, 2007; and Jenkins, 2006). Contributing to this discussion and looking specifically at broadcast sitcoms, I contend that their continued viability is dependent not only on multiplatform distribution and transmedia expansion of franchise iterations on multiple platforms such as webisodes) but also on the secondary function of content promoting other content (for example, when a segment of Disney–ABC TV's *Black-ish* [2014–present] promoted Disney/Marvel's *Black Panther* [Ryan Coogler, 2018]).

In that which follows, I address the issue of transmedia marketing circuits in which franchise content is structured not only to enable viewers to start anywhere in a transmedia content circulation loop but also to facilitate

Figure 9.1 Anderson/ Johnson talks about the cultural implications of *Black Panther* and offers the interpretative frame of "fellowship" through which to read the film.

the circulation of messaging about the particular appeals of the interrelated content. As a means of focus, I contrast the content-as-promotion strategies connected to *Black-ish* and other sitcoms across the Disney–ABC Television Group with those connected to *Parks and Recreation* (NBC, 2009–15), an ensemble comedy of particular value to the on-demand library strategies of parent company Comcast, which needed content for Seeso, NBCUniveral's short-lived Subscription Video On Demand (SVOD) platform, and then for Netflix, Comcast's OTT partner. The aim of comparing *Parks and Recreation* and *Black-ish* is to highlight how each sitcom functions as part of the content-promotion circuits that organize the transmedia business model of each conglomerate.

By contrasting the way in which each sitcom functions as content that doubles as promotion for other content, I introduce the difference between precise clustering of content at Disney and loose clustering that characterizes the content library approach of Comcast and Netflix. ABC has a tightly connected lineup of comedies, with *Black-ish* airing on the same night as other diverse family sitcoms, including *Modern Family* (2009–present). *Black-ish* also exists in a precise cluster with diversity-centric and tween-inclusive series and films across the channels and studios controlled by The Walt Disney Company. In contrast, Comcast and Netflix value *Parks and Recreation* because it offers looser connections to a range of other comedy content such as *Party Down* (Starz, 2009–10) and *Master of None* (Netflix, 2015–present) available in their content libraries. *Parks* is ideally suited to a selection of a loose cluster of content, but it began as part of a precise cluster of brandcentric NBC comedy prior to Comcast acquiring the network, a point I return to in the case study. This changing position signals how broadcast content has adapted to the challenges and opportunities of the convergent media industries.

The case studies of *Parks and Recreation* and *Black-ish*, and their respective connections to the promotion of two *Guardians of the Galaxy* films (James Gunn, 2014, 2017) and the first *Black Panther* film, highlight the key differences between the function of broadcast television in the tightly integrated Walt Disney Company and in the looser structure within Comcast NBCUniversal. *Guardians of the Galaxy* and *Black Panther* are significant franchises for the examination of precise and loose clustering of content and promotion that follows. Yet, this chapter is not about those Disney/Marvel franchises per se but, rather, about how broadcast television and its characters/actors can play an important role in the circulation of promotion for film and television distributed (and often produced) by a single media conglomerate.

Comcast, NBCUniversal, Netflix, and the Loose Clustering Approach

Franchise filmmaking has been a topic of interest across academic commentary (for example, Herbert, 2017; Stork, 2014; Johnson, 2013; Wolf, 2012; McCulloch, 2015; Thompson, 2007; and Buckland, 2003) but the role of broadcast television in the promotion of those franchises has been overlooked, as have the different ways in which media conglomerates leverage content to act as promotion for other content. While that process is very easy to track across the precise cluster of Disney–ABC Television channels, let us look first at the loose-cluster model as exemplified by the positioning of *Parks and Recreation* by Netflix and its partner Comcast NBCUniversal.

Parks and Recreation, the workplace sitcom featuring an ensemble of misfit workers associated with the Parks and Recreation Department in Pawnee, Indiana, now appears on Netflix, but it is not the same as NBC's *Parks and Recreation*. The episode content has not changed since its original broadcast run (April 2009–February 2015) but the value and public perception of the ensemble comedy have undergone a transformation, especially once its seven seasons could be watched on Netflix at each subscriber's own pace. This change is connected to the one NBC underwent during the same period when the network came under control of Comcast, a telecommunications conglomerate and the largest broadcast and cable television company in the United States (Dean, 2018). Comcast gained a controlling stake in the broadcast network in 2011, integrated the NBC peacock into the Comcast logo in December 2012, and took full ownership of NBC and Universal in March 2013 (Arbel, 2016; Brown, 2016; and Chozick and Stelter, 2013). Comcast retooled the NBC business model, which tended to be brandcentric and protectionist (Anderson, 2007: 287), and entered into two significant deals with Netflix. The first dates to 2014 and the second to 2018 (Snider, 2016; Brown, 2016; and Spangler, 2018). *Parks and Recreation* became a valuable asset for Comcast's on-demand platform and for Netflix's streaming library, interfaces that aim to transform viewers of a single ensemble television program into viewers of an array of other programs and films clustered together in loose groupings related by genre, format, cast, and production team.

Parks boasts a cast of well-liked and broadly appealing comedians (for example, Amy Poehler, Nick Offerman, and Aziz Ansari) which enables the sitcom to work exceptionally well as a hub for creating custom content clusters. Viewers of *Parks* invest over the seven seasons in the characters and the actors who play them. Comcast and Netflix speculate that this affective attachment can translate into a desire to see the actors in the other content available in their content libraries. The library model banks on leveraging a subscriber's investment in one series to create interest in another series or a film if it is generated by a "because you watched" list of available titles.

Comcast has a vested interest in being the hub from which viewers can watch whatever content they want regardless of the source (for example, HBO, Showtime, Starz, AMC, NBC, Netflix, and YouTube). The Comcast interface, accessed through a Smart TV with Internet connectivity, allows streaming content to be available in what appears to be the Comcast content library alongside cable and network fare. Once Comcast controlled *Parks*, it allowed Netflix to carry its seasons after they completed their runs on NBC. The network still retained the rights to the first distribution window so that the series would have more value to advertisers footing the bill and to *Parks*'s future syndicators (such as FXX). In turn, the availability of *Parks* on Netflix brought new viewers to the series, allowed other viewers to rewatch it, and encouraged all viewers to sample other content with which the sitcom might be clustered. If the classic network deal is for a desirable primetime lineup to deliver audiences to advertisers, then desirable content clusters deliver subscribers to OTT and on-demand providers. Comcast and Netflix make for good partners, as Netflix is still many viewers' "go to" site for streaming content, and it has a library that complements the one accessible through Comcast's premium cable subscription (Adalian, 2017; Jarvey, 2017; Feldman, 2017, McAlone, 2017; Setoodeh, 2017; and Snider, 2016).

Netflix both acquires film and television content for its streaming on–demand library and produces originals, including *Master of None*, the auteurist comedy starring, cowritten, and produced by Aziz Ansari. The scripted comedy exists as part of a content cluster series because Ansari came to wider fame playing Tom Haverford on *Parks*, prior to which he had performed in an Internet comedy troupe, on a MTV sketch comedy show, and on stage as a successful stand-up act. Netflix invested in Ansari, starting with a stand-up special in 2013, followed by a second in 2015. The latter special doubled as promotion for the debut of *Master of None* in November 2015. The acclaimed slice-of-life comedy program is cocreated and inspired by the personal and professional experiences of Ansari, who plays aspiring Indian American actor Dev Shah, and Alan Yang, who was a writer on *Parks* (Framke 2015). Those Netflix subscribers who did not watch the 2015 season could catch up any time leading up to the May 2017 availability of season two. *Master of None*'s second season was even more critically acclaimed than the first, but each season won the Emmy for Outstanding Writing for a Comedy Series (VanDerWerff, 2017; Tallerico, 2017; and Gilbert, 2017).

While the episodes and seasons of *Master of None* do not necessarily have to be watched in order, such sequencing is still a marketable hook. After viewers consume a season (often by binge-watching it), they need more content to watch until the release of the new season. Within precise clusters, the drive is to watch in order, but in looser clusters like that of programming associated with Aziz Ansari, subscribers can start anywhere and move in any direction regardless of content chronology. *Master of None* fans awaiting the next season might sample or rewatch some *Parks and Recreation* which they can select from the content library and watch in any order. Alternatively, they can move between the sitcom and any other available content. Watching a content cluster, which includes *Master of None* and Ansari's stand-up specials, a subscriber might watch other stand-up specials, or catch Jerry Seinfeld or Judd Apatow in Netflix specials that combine stand-up with industry memoir from creators of former NBC comedy series. Comcast premium cable subscribers, who also have Netflix, could move from Ansari's comedy to the wide range of other stand-up specials on Netflix, HBO, Comedy Central, and other basic and premium channels carried by Comcast. They could watch Ansari in an episode of Seinfeld's *Comedians in Cars Getting Coffee* (originally on Crackle and now on Netflix, 2012–present) or move over to Showtime's *The Chi* (2018–present), written by Lena Waithe, one of Ansari's *Master of None* cowriters and Emmy winners, as well as a supporting actor in the series. In short, Netflix's and Comcast's partnership and combined content libraries make each subscription service a more attractive deal to existing and potential subscribers. To offer a wide range of compelling content clusters, Comcast, Netflix, and similar providers need to build up good libraries (or find partners) so there is enough appealing content not just to watch, but also to play a part in promoting a cluster of other content that subscribers might watch next.

Cast and crew connections are easy to leverage as we saw in the example of the bridge Ansari creates between *Master of None* and *Parks and Recreation*. Many other actors in the *Parks* ensemble can act as promotional hubs and help move viewers from the sitcom to clusters of loosely interrelated on-demand content on Comcast and Netflix (Leonard, 2013). As one commentator noted, Netflix could create a *Parks*-related content cluster through Adam Scott (who plays Ben Wyatt, the love interest of Poehler's Leslie Knope):

> *Parks and Recreation* is popular for Netflix and has good metrics (people watch the entire show, re-watch some episodes, and frequently rewind certain parts). One of the actors is Adam Scott (maybe some users rewind scenes with him in it), and they have the option to add a few cost-efficient movies with him. Do they do it? At the time of this writing, they do. There are 7 Adam Scott movies available to stream instantly [on Netflix], one an independent film where he is a main character. (Patel)

Scott, who guest starred in season two and became a *Parks* regular in season three, added a different comic style to the ensemble and had an established fanbase from his stint on Starz's "cater waiters" comedy series *Party Down* (2009–10).

Figure 9.2 *Parks and Recreation*'s Ben Wyatt (Adam Scott) reveals his live-action role-playing game to his wife Leslie Knope (Amy Poehler).

In 2017, Scott appeared along with Amy Poehler, Paul Rudd, and an ensemble of famous comedians in *Wet Hot American Summer: Ten Years Later* (Netflix, 2017). Scott has also had roles in Hollywood comedies such as *Step Brothers* (Adam McKay, 2008), starring Will Ferrell and John C. Reilly, and *Our Idiot Brother* (Jesse Peretz, 2011), starring Paul Rudd. Rudd connects to *Parks* both because he had a recurring role in season four and because he is linked in a content cluster with Rashida Jones (*Parks*'s Ann Perkins). She appeared in two films alongside some other *Parks*-affiliated comedians: Adam Scott (*Parks*) and Elizabeth Banks (*30 Rock* [NBC, 2006–13]) in *Our Idiot Brother*; Jason Segel (*Freaks & Geeks* [NBC, 1999–2000]), Aziz Ansari (*Parks*), Andy Samberg (*Saturday Night Live* [NBC, 1975–present); and Jaime Pressly (*My Name is Earl* [NBC, 2005–09]) in *I Love You, Man* (John Hamburg, 2009). These examples reaffirm the point that effective content clusters can be created using an ensemble comedy as a hub from which to customize a watchlist using a streaming library stocked with these different titles.

Back on *Parks*, the characters played by Jones and Scott are both closely associated with Chris Traeger, played by Rob Lowe. Scott and Lowe started as guest stars playing state auditors at the end of season two. After they were given jobs in Pawnee's City Hall and made regulars for season three, *Parks* became the ideal misfit ensemble and started ranking high on "best sitcoms" lists such as those posted online by *Ranker*, *Paste*, and digital magazine platforms like GQ.com. Lowe's association with *Parks* makes for another good content cluster as he was a young actor in a slate of 1980s films [for example, *The Outsiders* (Francis Ford Coppola, 1983), *St. Elmo's Fire* (Joel Schumacher, 1985), and *About Last Night* (Edward Zwick, 1986)] that have accrued enough nostalgia and mainstream cult status to make them some lower-cost, popular additions to a content library. Lowe also creates a bridge back to NBC because of his comic turn in the films of former *Saturday Night Live* comedian Mike Myers and his own stints as guest host of *SNL*. He is also best known for his role as Sam Seaborn on NBC's *The West Wing* (1999–2006). The Emmy-winning drama ran for seven seasons and is now available for streaming on demand via Comcast and Netflix.

NBC programs exist in a loose cluster when they are distributed on demand via the Comcast cable interface or via Netflix. *Parks* has this new role to play as part of their content libraries, but it began as part of a precise cluster of NBC comedy and was linked on the schedule with *The Office* (NBC, 2005–13) because Greg Daniels helped create both series. Originally, NBC wanted a direct spin-off, perhaps focusing on the office of another Dunder–Mifflin branch. Daniels and Mike Schur came up with a looser concept for the second series, calling it *Parks and Recreation* to signal the differentiation of the public-sector setting (Burns and Schildhause, 2015). *Parks* shares *The Office*'s small-town workplace conceit, its misfit ensemble structure, and its mockumentary format in which the characters sometimes directly address the camera. After some critics described the first season of *Parks* as too derivative—*The Office* only with a female boss (Poehler's Leslie Knope)—Daniels and Schur retooled the series to give it a more unique brand of misfit ensemble comedy (Richmond, 2011; Gilbert, 2013; and Sepinwall, 2009) while still aligning with NBC's self-proclaimed brand of "smart comedy" that combines silly antics with social commentary.

Parks also existed in a precise cluster with another smart comedy on the NBC schedule, Tina Fey's *30 Rock* (2006–13). The two comedies share the same initial dynamic between liberal female boss and her conservative male division head. Alec Baldwin's Jack Donaghy evolves from foil to friend of Fey's Liz Lemon and, on *Parks*, Nick Offerman's Ron Swanson has a similar relationship to Poehler's Leslie Knope. The two male characters became associated with popular sayings and, consequently, their faces and words are often featured on series-related merchandising, including mugs, posters, and shirts. While loose clusters of content could be created out of the library content associated with these well-liked actors, *30 Rock and Parks* were developed as star vehicles, respectively, for Fey and Poehler, who continue to be tightly connected in public perception and in content libraries. Fey and Poehler became linked with each other and with NBC's comedy brand through their respective *SNL* impressions of political candidates Sarah Palin and Hillary Clinton, their appearance in *SNL* primetime specials, their cohosting of the Golden Globe Awards, and their coanchoring (2004–06) of *SNL's* "Weekend Update." Tina Fey left *SNL* for *30 Rock*, taking costar Tracy Morgan with her and recruiting Alec Baldwin as her costar. Baldwin currently plays Donald Trump on *SNL*, and he guest hosted *SNL* twelve times between 1990 and 2005, and then four more times during the *30 Rock* years.

NBC has long treated *SNL* as a hub from which to launch its comedians in other fare but, between the mid-2000s and the mid-2010s, it experimented with using *SNL* comedians to headline its sitcoms and late-night lineup (Gillan, 2010). As noted, Fey left *SNL* to launch *30 Rock*, which offers a peek at what happens behind the scenes at a sketch comedy television program in the age of media conglomeration. For "Weekend Update" from 2006 to 2008, Poehler partnered with Seth Meyers, who then cohosted without her until he left *SNL* in 2014. He became the star of *Late Night with Seth Meyers* (NBC, 2014–present), a plum assignment he got when NBC moved up Jimmy Fallon to *The Tonight Show with Jimmy Fallon* (2014–present) at 11:30 p.m. Fallon also got his start on *SNL* and did a successful stint himself on "Weekend Update," coanchoring with Tina Fey from 2000 to 2004. All of these moves point to the brandcentric approach to television programming and casting in the mid-2000s at NBC (Gillan 2010).

While he has been a late-night host since 2009, Fallon left *SNL* in 2004 for feature films. This move was the standard one for the breakout performers that preceded him. The most successful *SNL* comedians graduated to starring roles in films unrelated to *SNL* and then established themselves as Hollywood's A-listers (for example, Bill Murray, Steve Martin, and Will Ferrell). The formation of NBC Universal in 2004 and the subsequent and gradual transfer of full ownership from General Electric to Comcast (between 2011 and 2013) led to a more brandcentric emphasis at the company now called NBCUniversal. A precise content cluster in a future content library could be created if Universal released the comedies starring the major breakout comics from *SNL* and other NBC programs. Universal Pictures also released Judd Apatow's *The 40-Year-Old Virgin* in 2005 (with *The Office*'s Steve Carell, who guest starred in a voiceover role on *SNL* from 1996 to 2011, and Jason Segel of *Freaks and Geeks*), Paul Feig's *Bridesmaids* (with *SNL*'s Kristen Wiig, 2008), and Nicholas Stoller's *Forgetting Sarah Marshall* (with Segel and *SNL*'s Bill Hader, 2008). In 1999/2000, Segel costarred in Apatow and Feig's brilliant but cancelled NBC dramedy, *Freaks & Geeks*.

Freaks brings us back to *Parks* because Rashida Jones was a teenage guest star on the high-school-set series, and she was the only actor that Greg Daniels brought with him from *The Office*. On *Parks*, Jones plays Ann Perkins, a Pawnee citizen frustrated at the well-meaning but ineffective Leslie Knope. While Ann starts out merely interacting with Leslie because she is the Deputy Director of the Parks and Recreation Department, she becomes her best friend and often her partner in wacky but effective Pawnee initiatives. By season two, Leslie changes from a more daffy character to a sunny striver who becomes a success story.

The changes made to *Parks* over the course of its run aimed to align better with the actors' personalities and their talents, including Chris Pratt's unexpected skill at pratfalls and improv, Aziz Ansari and Nick Offerman's comic asides, Aubrey Plaza's facial expressions, Adam Scott's fanboy behaviors, and Amy Poehler's infectious likeability, which made her better suited to play Leslie Knope as a workplace motivator rather than as a self-deluded and ineffectual boss (Burns and Schildhause, 2015). Over time, the producers also modified other characters and relationships (for example, Tom Haverford and Donna Meagle, or Andy and April the intern) to give all the members of the ensemble more screen time to showcase their particular skills. The producers also mixed things up by creating episodes with unexpected pairings of characters, thereby allowing the actors to display different comic interactions.

All the characters became better people through their interactions as pairs and as a group. The new characterizations made it clear that everyone has the capacity to see the good in one another and to find ways to help each other succeed. While many of the characters succeed in unexpected ways, Andy's change from Ann's comically shiftless boyfriend to Horatio Alger success story was most surprising. Through the help of Leslie and others at the Parks Department, Andy gets a job as the town hall "shoe shine boy," moves up to various small jobs in the department, hosts his own kids' television show, *The Johnny Karate Super Awesome Musical Explosion Show*, and ends up acting as a paid liaison for a charitable organization in which he gets to hang out all day with a British aristocrat who shares his childlike interests.

Though he started out playing this minor character, Pratt became one of the bigger stars through whom a content cluster might be created on Netflix. He has some classic scenes paired with Offerman's Ron and Scott's Ben, for example. Indeed, for many subscribers, *Parks* is only the workplace comedy in which to see Pratt, who was famously dismissed by James Gunn as "the chubby guy from *Parks and Recreation*" (Collis, 2014) before being cast as the lead in Marvel's *Guardians of the Galaxy* films. As a result of the deal Disney made to license its studios' post-2016 films to Netflix, a subscriber would only be able to cluster *Parks* with *Guardians of the Galaxy Vol. 2* (Littleton, 2017 and Graser, 2012) but a subscriber could move from watching Pratt in *Guardians of the Galaxy Vol. 2* to seeing him as Owen Grady, the lead in the

Figure 9.3 In *Parks and Recreation*'s "Soda Tax," Chris Traeger (Rob Lowe) and Tom Haverford (Aziz Ansari) help a reluctant Andy Dwyer (Chris Pratt) get in shape.

latest films in Universal's *Jurassic Park* (1993–present) franchise. Yet, Pratt's stint as Andy on *Parks* remains a key element of his star brand, as several popular press stories have focused on his career leap from *Parks*'s loveable doofus to Hollywood leading man (Collis, 2014; and Milliken, 2014). Pratt often becomes the pathway between *Guardians of the Galaxy* to *Parks*, which helps explain why he was seen as so pivotal to the series that production moved to London for the start of season six to accommodate his *Guardians Vol. 1* shooting schedule and keep him in a few more episodes.

Walt Disney Company and its Precise Clustering Approach

The Netflix and Comcast content loose-cluster library approach contrasts with the precise clustering that has typically characterized the organization and promotion of content associated with the Disney–ABC Television Group. This tightly integrated channel portfolio is demographically segmented with a channel for each of the following: toddlers, kids, tweens, teens, young adults, and adults. The progression of demographic targets and custom promotional content offered on each channel almost guarantees that a new film will reach the precise range of viewers that might be called to action to see the film in theaters upon its release. To help complete this circuit, Disney leverages its existing stars and transforms them into endorsers of its new content, and it does so in effective ways.

Starting in the 1950s, Disney coordinated theatrical and television production and made sure to create television programming, such as *The Mickey Mouse Club* (ABC, 1955–59), that doubled as promotion for Disney Studios output (Gillan, 2015). This model extends today to all the studios under Disney, including Lucasfilm, Pixar, Marvel, and Walt Disney Studios. A transmedia strategy favored by Disney is to turn a character from one piece of content into an endorser of another piece of content. While the original endorser was Walt Disney in his role as host of a television show in the 1950s and 1960s, the form has been revived in recent years.

This direct mode works best for Disney because of its position as a tightly integrated media and entertainment conglomerate which distinguishes it from the more loosely integrated Comcast NBCU. As a result, Disney–ABC has the advantage in creating and disseminating its content and promotion. Disney's precise cluster strategy is evident in the transmedia marketing related to Marvel's *Black Panther* in interstitial content airing during programs across the Disney–ABC Television Group. This content utilizes a format Disney pioneered: the hosted lead-in in which a television actor hosts a short program and gives framing commentary on a new feature film between snippets of related trailer footage.

Integrating trailers into television segments goes back to *The Mickey Mouse Club*, an afternoon program aimed at tweens of the 1950s. The lead-ins were typically hosted by one of the tween actors (called "Mouseketeers") offering interpretative frames for trailer footage for new theatrical releases, such as *Westward Ho the Wagons!* (William Beaudine, 1956), or upcoming television serials, such as *The Hardy Boys* (1956–57). The host would also interview an actor (or two) from the new film or serial. This example, one of many from Disney's two 1950s ABC television shows, indicates Disney's long history of coordinating theatrical and television production, and of creating television programming that doubles as promotion for the company's output (Gillan, 2015). The scale is different today because Disney now has a portfolio of television channels that can be leveraged to promote its full schedule of films not only from Walt Disney Studios, but also from Pixar, Lucasfilm, and Marvel.

The most overt hosted lead-ins today can be found in *Stars on the Set* (2012–present), a Disney Channel short series that combines the hosted lead-in footage with film star interviews. A hosted lead-in trailer commentary program, *Movie Surfers* (1997–present), plays most often on Disney XD, a cable channel that airs several animated series connected to the Star Wars and

Figure 9.4 Tom Holland (Peter Parker/Spider-Man) and Zendaya (Michelle "MJ" Jones) host an ABC special in which they promoted their upcoming movie, *Spider-Man: Homecoming*.

Marvel franchises. Trailers are common on both these channels, as well on ABC and Freeform, but these channels do not have any recurring interstitial programs. Their promotional shorts run infrequently as special programming just prior to the theatrical release of a film perceived to have a particular appeal to their target audiences. On rare occasions, ABC runs long-form programs, such as when Tom Holland (Peter Parker/Spider-Man) and Zendaya (Michelle "MJ" Jones) appeared as hosts of a special in which they promoted their upcoming movie, *Spider-Man: Homecoming* (Jon Watts, 2017) and talked about what it meant to have Spider-Man (controlled by Sony) "home" again with the other superheroes in the Marvel Universe. This special aired on June 15, 2017 across a two-hour block on ABC, but snippets could be easily edited to air on Disney Channel.

Zendaya was the ideal endorser because she had a supporting role in the new film, played the lead in her second Disney Channel Sitcom (*K.C. Undercover* [2015–18]), starred in two Disney Channel original movies, and was a contestant on ABC's *Dancing with the Stars* (2005–present). As is typical for Disney Channel stars, she also appeared in music videos, dance-related segments,

Radio Disney spots, and in many kinds of interstitials since the start of her first sitcom, *Shake It Up* (Disney Channel, 2010–13), in November 2010. She is emblematic of the way Disney utilizes its stars as endorsers of some new content or entertainment associated with the Disney conglomerate (Gillan 2015).

ABC sometimes leverages its sitcom characters/stars to participate in some form of integrated promotion (such as characters talking about a film or series) or in interstitial shorts for the bigger releases, especially for films that have a precise connection such as that between the Star Wars franchise releases and the fanboy main character in *The Goldbergs* (2013–present), or between *Black-ish*'s successful African American adman and Marvel's *Black Panther*. While promotion on Disney Channel had been running for months, the promotions on ABC and Freeform began immediately before *Black Panther*'s official February 16, 2018 release. Two very different styles of hosted lead-in specials aired: one during an episode of ABC's *Black-ish* ,and the other between segments of its college-set spin-off, *Grown-ish* (2018–present). The latter sitcom had just premiered in January 2018 on Freeform, Disney's young, female-skewing cable narrowcaster. With this precisely clustered pair of sitcoms, Disney–ABC Television had a unique opportunity and created interstitials in which the African American main character of each series promoted the extended President's Day weekend release of *Black Panther*, the first Marvel film to have an almost all-black cast and a range of female characters.

ABC's *Black Panther* interstitial, which first aired on February 6, 2018, utilizes Dre Johnson, the main character, and Anthony Anderson, the actor who plays him on *Black-ish*, as a bridge between television and film. This interstitial features an animated Dre Johnson/Anthony Anderson acting as host and commenting on the cultural significance of the film. Continuity is established between the sitcom and the short because Anderson contributes the voiceover in both, and the commentary sounds like something his character Dre would say in an episode. He shares what it means to him to have a black superhero. He talks about how he had seen many white superheroes as a kid and felt left out. He implicitly credits Disney with barrier breaking for creating a black superhero film and enabling black viewers to have a superhero of their own. Of course, the animated host in this short is some combination of a character and the actor/comedian who plays him, as well as the producer who created him, so the commentary could also reflect any of their personal feelings. This kind of interstitial blurs the lines between content and promotion, between actor and character, between character and creator, between PR and personal commentary, and it moves the output of the studio away from economic production and toward a mode of personal expression. This personalized dynamic makes the message about the film's cultural importance more effective as it

Figure 9.5 Anthony Anderson's *Black-ish* character Dre Johnson enacts this hosted lead-in's call to action: take a group to see *Black Panther*, a watershed film.

includes a call to affiliation (feel what I feel, look at the world through my eyes) and, thus, is more likely to translate into a call to action (buy a movie ticket and/or take a family or a group to see the film).

Black-ish's animated short is an original piece of content about content, but it has a similar function to the audio commentary used on Blu-ray and DVD special features. Robert Alan Brookey and Robert Westerfelhaus claim that audio commentaries "can direct the viewer toward preferred interpretations of the primary text while undermining unfavorable interpretations" (2002: 24). Dre's/Anderson's commentary has a rhetorical function: to assert the preferred interpretation that *Black Panther* is a culturally significant film for viewers committed to diversity in representation and a "must-see" film for black viewers. The interstitial that aired during the commercial breaks of a February 2018 episode of *Grown-ish* had a similar rhetorical function. It encouraged potential viewers to read *Black Panther* as a culturally significant film in terms of representation of black characters and of women as central characters and as warriors. Attending the film on opening weekend, both interstitials imply, should be viewed as a cultural event.

Figure 9.6 Shuri (Letitia Wright) shows her warrior side, a characteristic emphasized for the female-skewing demographic of the Freeform channel.

Both hosted lead-ins mix footage from the film with commentary about it. Similar to the audio commentaries about which Brookey and Westerfelhaus write, the hosted lead-in as a format "collapses exposure to promotional material into the experience of viewing" (2002: 23). It is composed of snippets from the trailer edited into commentary on it. This kind of intratextuality brings scenes from the film and commentary about it "in close proximity, temporally and spatially" (2002: 23), thereby making viewing and commentary feel like a continuous experience. As Brookey and Westerfelhaus contend, this kind of dynamic increases "the chances that promotional tactics will reach their target audience" (2002: 24).

The target teen and twenty-something audience of Freeform—a cable narrowcaster that skews female—aligns especially well with Disney's third phase of Marvel films which tend to be light-hearted action-adventure films known for humorous banter and an expanded range of characters beyond the usual white, male superheroes. *Black Panther* was an especially good fit for promotion through an association with *Grown-ish*, as several of the actors, including the lead, Yara Shahidi, are at least part African American by heritage. Disney–ABC audiences first got to know Shahidi when she started on ABC's *Black-ish*

Figure 9.7 The *Black Panther* animated hosted lead-in concludes with the title card for *Black-ish*, signaling Disney's transmedia marketing circuit connecting the Marvel Film and the ABC sitcom.

in 2014 as the fourteen-year-old Zoey Johnson, the oldest child of Dre and his mixed-race wife, Rainbow Johnson.

The writers then created and aired the college spin-off for Shahidi on Freeform. Zoey quickly establishes an eclectic and diverse social circle, and then talks directly to the camera about her personal interactions and about professional aspirations in the fashion world. This established rapport with viewers makes the character a ready-made endorser, and Freeform found the ideal spokesperson in Yara Shahidi. She developed a following not only as an actress on two sitcoms, but also as a political activist with a champion in Michelle Obama, a fashionista who has been featured in *Teen Vogue,* and a frequent user of Instagram, Twitter, and other social media utilities through which she promotes both sitcoms, as well as other Disney–ABC content. *Teen Vogue* contributed to the content-promotion circuit when it partnered with the series to give Zoey's character an internship at the magazine and a storyline about the uses and abuses of Instagram in the workplace. To create continuity, this internship storyline began in *Black-ish* before being picked up again in the middle of the first season of *Grown-ish*.

Figure 9.8 Yara Shahidi's interview of Chadwick Boseman (Black Panther) in a hosted lead-in airing during *Grown-ish*.

In the Freeform interstitial, Shahidi conducts a chat-style interview with three of the main actors from *Black Panther*, Chadwick Boseman (Black Panther), Michael B. Jordan (Killmonger), and Letitia Wright (Shuri). Within the interviews, which aired in segments over the course of the episode breaks, there are particular taste cluster appeals. For example, Shahidi and Wright discuss the use of Kendrick Lamar's music in the film, a topic presumed to appeal to the twelve- to thirty-four-year-old target audience for the cable channel, but also linked to the videos that surfaced of Wright demonstrating her ability to rap freestyle. In the shortened versions, which ran in a single commercial break in later episodes, the entire Freeform interstitial is cut into standalone segments so that Shahidi just appears to be doing one-on-one interviews. This editing changes the emphasis a bit. The standalone interview with Wright, for example, feels more like feminist commentary when presented without the interviews of the two male actors.

While Shahidi is identified in an onscreen caption by her full name, the hosted lead-in form relies on the blurred line between actor/character to tap into the affective connection viewers have with the character and actor. Established viewers already feel as if she is a trusted confidante because she

Figure 9.9 Yara Shahidi's interview of Letitia Wright (Shuri) in the *Grown-ish Black Panther* hosted lead-in.

often speaks directly and exclusively to the viewer in literal asides to the camera. Disney–ABC uses these kinds of interstitials to reinforce Disney's desired framing of the film and of itself as a conglomerate committed to diversity. They share similarities with the DVD features focused on the "making of" a film, a piece of docutainment through which a studio and/or production team mediates knowledge about itself (Caldwell, 2008: 2–3). John Caldwell claims that "making-ofs" are the equivalent of electronic press kits used in the official studio publicity distributed to the press (2008: 299). Hosted lead-ins are also creative ways to circulate PR without making it seem like PR. Those appearing in these interstitials act as if they are expressing personal opinions about the film, and the commentary may be aligned with their actual opinions, but it is a part of their jobs to promote the film and present it in these ways.

The Disney–ABC Television Group's use of these various kinds of hosted lead-ins is a mode of brand management, raising awareness about feature films released by the parent company and signaling the use value of broadcast sitcoms in an era of transmedia conglomerates.

Also relevant to this discussion is the hosted lead-in for *Black Panther* trailer footage that ran on Disney XD on February 3, 2018 as part of its recurring short-form series, *Movie Surfers*. In the segment, two hosts, Janelle and Jaden, split the lines equally. Janelle introduces the segment, saying, "*Black Panther* is almost here, but there are still so many secrets," which gives Jaden the chance to assure viewers, "Luckily, we've got you covered." Between trailer snippets, they tell viewers that they are giving them the "inside scoop." *Movie Surfers* is an example of a content–promotion hybrid that explicitly showcases the output of Disney's various studios while functioning as entertainment in its own right. On Disney XD, where the show often airs, there are plenty of Marvel's animated series, such as *Spider-Man* (2017–present) to keep viewers watching television while the shorts running between them encourage viewers to see the next Marvel movie in theaters (for example, *Spider-Man: Homecoming* hit theaters on July 7, 2017, with the next film arriving in 2019). The *Black Panther* installment of *Movie Surfers* ends with the two hosts reminding viewers to keep watching Disney XD. Janelle wraps up the segment, saying, "Now you know what we know," and then Jaden reminds viewers to "stick around. We're finding out more straight from the source."

Over on Disney Channel, the stars or directors of Marvel films are often the sources of insider information on the short-form series *Stars on the Set*. The hosts are typically affiliated with Disney Channel so they are familiar to viewers. They act as endorsers of the film by offering commentary between trailer and interview snippets. Typically, the hosted lead-in ends with an exclamation such as: "I can't wait to see this one on the big screen." Recurring host/endorser Morgan utters the line and shows the requisite enthusiasm for getting the inside scoop directly from Chris Pratt about an exciting chase scene and the "cool controllers" on the ship used in *Guardians of the Galaxy Vol. 2*. Pratt endorses the film as well, assuring viewers that the ship is very cool: "I felt like I was on a Disneyland ride that was built for me that I didn't even have to wait in line for." The statement is a double promotion as this hosted lead-in/interview is taking place at Disneyland's Hollywood Studios and the set up is that Morgan was taking a tour of the attraction and just happens to run into Pratt.

This transformation of television actor into film endorser is effective because of the familiarity of the television actor's everyday persona. This sense of ordinariness complements the aura of specialness associated with access to behind-the-scenes information about a Hollywood production and its stars (Dyer, 1986: 39, 49). The interviews provide the stars with an opportunity to shed some of that aura and become more relatable and appealing when they share personal details about themselves and the production. The value added is that these shorts purport to offer access to stars as they "really" are offscreen. In the *Guardians 2* segment mentioned above, Pratt is relaxed and silly while seeming a lot like Andy Dwyer. Both character and actor can be framed in terms of the "unlikely success story" narrative (Collis, 2014).

In contrast, the *Stars on the Set* short covering the first *Guardians of the*

Galaxy film positions Pratt as a relative unknown. Morgan has to introduce him and the character, telling viewers, "I am here with Chris Pratt, aka Peter Quill," and then turns to Pratt: "Who is this guy?" (a line that could apply to both Pratt and Quill). Pratt seems like a nervous newcomer. He tells viewers about his character ("He wants people to take him seriously") and then some trailer clips play that poke fun at the character's desire to be called "Star Lord," especially when an exchange between characters shows that it's a running joke in the film because the only one using the nickname is Quill himself. It recalls the character of Andy Dwyer creating alter egos for himself in *Parks and Recreation*, such as "Bert Macklin, FBI."

Stars on the Set and *Movie Surfers* are two examples of recurring short-form series that exemplify Disney's skill at finding creative ways to repackage repurposed footage from films released by a Disney-affiliated studio as part of new interstitials that are content in their own right. Disney Channel is uniquely suited to air this kind of program because it is the only cable brandcaster, meaning that everything that runs between episode segments of its sitcoms and films promotes Disney in some way. Through this structure, Disney turns 30-second spots into company voice-messaging from "Friends of Disney Channel," such as in spots focused on *LEGO Friends* sets that promote the power of friendship and align with the messages in the channel's tween-address sitcoms and movies (Gillan, 2015). Disney Channel's recurring interstitials rely on several key interpretive frames: immersion (in the universe of the film and its story world), insights (into studio and production history from "eyewitnesses"), and fellowship (among the characters; among the actors; and between the viewers and the actors who are positioned like friends in these interstitials).

The *Stars on the Set* segment on *Guardians of the Galaxy* highlights perceived appeals for tweens such as the best-friend dynamics between Rocket and Groot, characters spun off into a 2017 eponymous animated series for Disney XD and included in the ensemble of several of the channels' other animated programs within the *Guardians of the Galaxy* and *Avengers* (2012–present) franchises. This customized focus exemplifies how Disney's transmedia circulation of content and "content about content" is enabled by the demographic segmentation across the portfolio of Disney–ABC channels. While it uses a precise cluster dynamic, Disney–ABC has a similar call to action as Comcast NBCU: keep people watching entertainment content, watching out for upcoming content, and choosing what to watch next from their available content.

Looking Forward: Controlling the Content Library

Guardians of the Galaxy Vol. 2 became available in December 2017 on Netflix only because a licensing deal Disney made in 2012 had just gone into effect.

Netflix subscribers cannot watch *Guardians of the Galaxy Vol. 1* because the deal with Disney includes only the new theatrical releases starting in 2016 (Graser, 2012; Sarkar, 2018; Pepitone, 2017; and Littleton, 2017). This stipulation prevents the precise clustering of the complete franchise that would be possible on a Disney-owned streaming service (of course, that platform would not have *Parks*). *Guardians of the Galaxy* 2 may not be available for long on Netflix streaming because Disney has announced plans to launch its own platform for streaming television in 2019 (Kastrenakes, 2017 and Littleton, 2017). The successful slate of Marvel television programs coproduced for Netflix is supposedly secure, but that could change (Fleming, 2018).

Foreseeing future problems with the deals it was making with the major content owners, Netflix began investing heavily in producing originals in 2013 and has been successful in acculturating subscribers to become serial binge watchers of its original series and its licensed programming such as *Parks* (Moore, 2017; LaPorte, 2017; Roettgers 2017; Littleton 2017; McAlone, 2017; Spangler, 2013, 2016b; and Jenner 2016). Seeing the value in this approach, Comcast strengthened its existing partnership with Netflix in 2018 even as Disney reaffirmed its plans to launch a streaming service and to cease licensing much of its content to Netflix (Spangler, 2018 and Fleming, 2018). Then both conglomerates tried to outbid each other to attain Fox, with Disney prevailing over Comcast with a $71.3 billion bid and antitrust approval (Steinberg, 2018; Lang, 2018; and Shields and McAlone, 2018).

Comcast has already moved toward a content-library model, and acquiring Fox's library would have enabled it to create as many loose clusters for as many kinds of subscribers as possible. Disney has more interest in the precise clusters that the Fox library enables it to create or complete. Through the acquisition, Disney would bring over some more Marvel characters and the rights to the first *Star Wars* film (renamed *Episode IV: A New Hope*) while adding other franchises (such as *Avatar*, *X-Men*, *Fantastic Four*, and *Night at the Museum*). Looking forward, Disney also recognizes its need for control of streaming rights to content and, in the acquisition, it is likely to get the rights to stream: *Modern Family* (a Fox show broadcast on ABC); *This is Us* (2016–present) (a Fox show broadcast on NBC); *Empire* (2015–present) (a Fox show broadcast on Fox); and *Atlanta* (2016–present), *The Americans* (2013–18), *American Horror Story* (2011–present), and *Legion* (2017–present) (all on FX). In addition, Disney would acquire Fox-controlled back-catalogue shows, including *Buffy the Vampire Slayer* (1997–2003), *Firefly* (2002–03), and *The X-Files* (1993–2018), as well as animated series such as *The Simpsons* (1989–present), *Family Guy* (1999–present), and *Futurama* (1999–2013). The combination of this library and its existing ones would give Disney control not only of a variety of precise cluster franchises but also of a range of standalone series and films for its planned OTT platform, resulting in a competitive advantage in the marketplace (Shields and McAlone, 2018 and Bishop, 2017).

Disney might want to maintain its own precise content clusters, but content can align across output from different companies as indicated by the titles recommended together under a subscriber's "because you watched" tab on Netflix. *Parks* is currently available on Netflix alongside *Inside Out* (Pete Docter, 2015), a Pixar film starring Amy Poehler as the avatar for the emotion Joy who "saves the day" by rallying a misfit band of emotion avatars, and *Guardians of the Galaxy Vol. 2*, with its ragtag band of unlikely heroes joining together to save the Universe. The interpersonal relationships of the characters in both films have some humorous parallels to the misfit ensemble in *Parks*. Pratt's Andy and Poehler's Leslie are particularly aligned with the optimistic cultural outlook and a "can-do" spirit of these and other Disney-affiliated films and series. Pratt's upbeat worldview, combined with Leslie's (and by all accounts, Poehler's) ability to inspire others, become infectious, and the end of *Parks* feels like it was a series about what people can accomplish together.

Viewers ranking *Parks* on their top-sitcom lists indicate that they have a taste for series that celebrate optimism and the mutual support of a work-based friend group that we also see in *Guardians*. Of course, there could be many other elements of the film franchise or sitcom that appeal to different "taste communities," to use Netflix's term for how it divides viewers and connects them to potential clusters of content (Roettgers, 2017). The reasons why people enjoy the entertainment they do, and whether they want to consume precise or loose clusters of content, are obviously complex. One's taste in entertainment depends, to borrow language from Christine Gledhill, on "a range of determinations, potentially resistant or contradictory, arising from the differential social and cultural constitution of readers or viewers—by class, gender, race, age, personal history, and so on" (1990: 70).

Comcast and Netflix know people's tastes are elusive so they divide subscribers into taste communities (Adalian, 2018). Disney is still wedded to a more demographic approach probably because each channel in its portfolio attempts to match up to a demographic category. The ways that taste communities are a more effective organizational principle than demographics are subjects worthy of future analysis. After the Fox deal closes, and if Disney launches a successful streaming platform, we will have more data to consider in relation to how much content Disney controls, how committed it remains to demographics, and how much it begins to utilize a combination of taste communities and demographics for presenting content to its subscribers who might be enticed by both precise and loose content clusters never to leave the site. If that happens, Disney has a shot at unseating Netflix as the current go-to site for OTT, and thereby undercutting Comcast in the process.

References

Adalian, Josef (2014), "Why 1994 Was One of Network TV's Last Truly Great Seasons," *Vulture*, 2 September, <http://www.vulture.com/2014/08/1994-95-week-opening-essay-ratings-graphic.html> (last accessed 18 April 2018).

Adalian, Josef (2018), "Inside the Binge Factory," *New York Magazine*, 11 June, <http://www.vulture.com/2018/06/how-netflix-swallowed-tv-industry.html> (last accessed 25 June 2018).

Anderson, Christopher (2007), "Creating the Twenty-first-Century Network: NBC in the Age of Media Conglomerates," in Michele Hilmes (ed.), *NBC: America's Network*, Berkeley, CA: University of California Press, pp. 275–90.

Arbel, Tali (2016), "Comcast NBC Deal: Customers Are Starting to See the NBC Effect on Comcast," *AP News*, 26 October, <https://www.apnews.com/10aac36ac0b0436b85895736140474f6> (last accessed 25 June 2018).

Bishop, Bryan (2017), "Disney's Streaming Service Has Won, and It Hasn't Even Launched Yet," *The Verge*, 11 November, <https://www.theverge.com/2017/11/11/16637732/disney-star-wars-marvel-pixar-streaming-service-netflix> (last accessed 25 April 2018).

Brooker, Will (2001), "Living on *Dawson's Creek*: Teen Viewers, Cultural Convergence, and Television Overflow," *International Journal of Cultural Studies* 4 (4), pp. 456–72.

Brookey, Robert Alan and Robert Westerfelhaus (2002), "Hiding Homoeroticism in Plain View: The *Fight Club* DVD as Digital Closet," *Critical Studies in Media Communication* 19 (1), pp. 21–43.

Brown, Tina (2016), "Tina Brown Interviews Steve Burke, CEO of NBCUniversal at Royal Television Society London Conference," *Television Magazine* (October), *Royal Television Society*, https://www.rts.org.uk/article/steve-burke-%E2%80%98think-owner-not-renter%E2%80%99 (last accessed 26 May 2018).

Buckland, Warren (2003), "The Role of the Auteur in the Age of the Blockbuster: Steven Spielberg and Dreamworks," in Julian Stringer (ed.), *Movie Blockbusters*, New York: Routledge, pp. 84–98.

Burns, Ashley, and Chloe Schildhause (2015), "You Go Big Or You Go Home: An Oral History Of The Creation And Evolution Of *Parks And Recreation*," *Uproxx*, 23 February<http://uproxx.com/feature/parks-and-recreation-evolution-oral-history/> (last accessed 26 May 2018).

Caldwell, John Thornton (2008), *Production Culture: Industrial Reflexivity and Critical Practice in Film and Television*, Durham, NC: Duke University Press.

Chozick, Amy, and Brian Stelter (2013), "Comcast Buys Rest Of NBC In Early Sale," *The New York Times*, Late Edition, 13 February, p. B1.

Collis, Clark (2014), "How Chris Pratt Went from Zero to Hero," *Entertainment Weekly*, 18 July, pp. 26–31.

Dyer, Richard (1986), *Stars*, London: BFI.

Evans, Elizabeth (2011), *Transmedia Television: Audiences, New Media, and Daily Life*, New York: Routledge.

Feldman, Dana (2017), "Netflix Remains Ahead of Amazon and Hulu with 128m Viewers Expected This Year," *Forbes*, 13 April, <https://www.forbes.com/sites/danafeldman/2017/04/13/netflix-remains-ahead-of-amazon-and-hulu-with-128m-viewers-expected-this-year/> (last accessed 28 June 2018).

Fleming, Mike (2018), "Disney Unveils Inaugural Streaming Service Launch Slate to Town; No R-Rated Fare," *Deadline Hollywood*, 8 February, <https://deadline.com/2018/02/disney-streaming-service-launch-slate-star-wars-marvel-1202281846/> (last accessed 25 June 2018).

Framke, Caroline (2015), "*Master of None*, Netflix's Brilliant New Comedy, Is Immediately One of TV's Best Shows," *Vox*, 8 November, <https://www.vox.com/culture/2015/11/8/9690904/master-of-none-netflix-review> (last accessed 25 June 2018).

Friedlander, Jason (2018), "Three Predictions for OTT Video in 2018," *Broadcasting & Cable*, p. 32.

Gilbert, Matthew (2009), "Re-creation through *Recreation*," *The Boston Globe*, 9 April, <http://archive.boston.com/ae/tv/articles/2009/04/09/re_creation_through_recreation/> (last accessed 21 April 2018).

Gilbert, Matthew (2017), "In 2nd Season, 'Master of None' Avoids the Sophomore Slump," *The Boston Globe*, 15 May, <https://www.bostonglobe.com/arts/television/2017/05/10/more-ambitious-but-less-charming-master-none/xiQhMy84GVx99nW2XQSTOP/story.html> (last accessed 25 June 2018).

Gillan, Jennifer (2010), *Television and New Media: Must Click TV*, New York: Routledge.

Gillan, Jennifer (2015), *Television Brandcasting: The Return of the Content-Promotion Hybrid*, New York: Routledge.

Gledhill, Christine (1990), "Pleasurable Negotiation," in E. Deidre Pribram (ed.), *Female Spectators: Looking at Film and Television*, London: Verso, pp. 64–89.

Grainge Paul (2008), *Brand Hollywood: Selling Entertainment in a Global Media Age*, New York: Routledge.

Grainge, Paul and Catherine Johnson (2015), *Promotional Screen Industries*, New York: Routledge.

Graser, Marc (2012), "Disney Inks Exclusive Deal with Netflix," *Variety*, 4 December, 8 <https://variety.com/2012/digital/news/disney-inks-exclusive-deal-with-netflix-1118063070/> (last accessed 25 May 2018).

Gray, Jonathan (2010), *Show Sold Separately: Promos, Spoilers, and Other Media Paratexts*, New York: New York University Press.

Greenblatt, Bob (2017), "Don't Call the Morgue: Broadcast TV is Doing Fine," *The Hollywood Reporter*, 13 September, p. 70.

Herbert, Daniel (2017), *Film Remakes and Franchises*, New Brunswick, NJ: Rutgers University Press.

Hibbard James (2009), "It's a Horrible Thought: Demise of Broadcast TV is All Too Hot a Topic," *The Hollywood Reporter*, 21 September, pp. 7, 20.

Hills, Matt (2012), "*Torchwood*'s Trans-Transmedia: Media Tie-Ins and Brand 'Fanagement'", *Participations* 9 (2), pp. 409–28.

James, Dave (2015), "Broadcast TV Is Dying a Death of a Thousand Cuts," *TechRadar*, 4 June, <https://www.techradar.com/news/television/is-broadcast-tv-toast-1295801> (last accessed 25 June 2018).

Jarvey, Natalie (2017), "Netflix Content Budget May Grow to $8 Billion in 2018," *The Hollywood Reporter*, 16 October, <https://www.hollywoodreporter.com/news/netflix-content-budget-may-grow-8-billion-2018-1049308> (last accessed 25 June 2018).

Jenkins, Henry (2006), *Convergence Culture: Where Old and New Media Collide*, New York: New York University Press.

Jenkins, Henry, Sam Ford, and Joshua Green (2013), *Spreadable Media: Creating Value and Meaning in a Networked Culture*, New York: New York University Press.

Jenner, Mareike (2016), "Is this TVIV? On Netflix, TVIII and Binge-Watching," *New Media & Society*, 18 (2), pp. 257–73.

Johnson, Derek (2013), *Media Franchising: Creative License and Collaboration in the Culture Industries*, New York: New York University Press.

Kastrenakes, Jacob (2017), "Disney to End Netflix Deal and Launch Its Own Streaming

Service," *The Verge*, 8 August, <https://www.theverge.com/2017/8/8/16115254/disney-launching-streaming-service-ending-netflix-deal> (last accessed 25 June 2018).

Lang, Brent (2018), "Disney Ups Fox Bid to $71.3 Billion, Outflanking Comcast," *Variety*, 20 June, <https://variety.com/2018/biz/news/disney-21st-century-fox-counter-bigger-offer-comcast-1202851891/> (last accessed 25 June 2018).

LaPorte, Nicole (2017), "Netflix Offers A Rare Look Inside Its Strategy For Global Domination," *Fast Company*, 23 October, <https://www.fastcompany.com/40484686/netflix-offers-a-rare-look-inside-its-strategy-for-global-domination> (last accessed 25 June 2018).

Leonard, Andrew (2013), "How Netflix Is Turning Viewers into Puppets," *Salon*, 1 February, <https://www.salon.com/2013/02/01/how_netflix_is_turning_viewers_into_puppets/> (25 June 2018).

Lieberman, David (2011), "Comcast-NBC Deal Hits 'A Sweet Spot,'", *USA Today*, 19 January, p. 2b.

Littleton, Cynthia (2017), "Streaming Shake-Up: Networks Fight Back," *Variety*, 15 August, pp. 13–14.

Lotz, Amanda (2014), *The Television Will Be Revolutionized*, 2nd edition, New York: New York University Press.

McAlone, Nathan (2017), "This Chart Shows How Netflix Originals Keep Getting Better and Better," *Business Insider*, 12 April, <https://www.businessinsider.com/netflix-original-shows-vs-hbo-vs-amazon-2017-4> (last accessed 25 June 2018).

McCulloch, Richard (2014), "Whistle While You Work: Branding, Critical Reception and Pixar's Production Culture," in Roberta Pearson and Anthony N. Smith (eds), *Storytelling in the Media Convergence Age*, New York, Palgrave Macmillan, pp. 174–89.

McDonald, Kevin (2014), "The Cult of Comic-Con and the Spectacle of Superhero Marketing," in James N. Gilmore and Matthias Stork (eds), *Superhero Synergies: Comic Book Characters Go Digital*, Lanham, MD: Rowman and Littlefield, pp. 117–36.

Mann, Denise (2009), "It's Not TV, It's Brand Management TV: The Collective Author(s) of the *Lost* Franchise," in Vicki Mayer, Miranda Banks, and John Thornton Caldwell (eds), *Production Studies: Cultural Studies of Media Industries*, New York: Routledge, pp. 99–114.

Milliken, Mary (2014), "Chris Pratt On Going From Slob to Film Heartthrob," *Reuters*, 30 July, <https://www.reuters.com/article/us-film-chrispratt/chris-pratt-on-going-from-slob-to-film-heartthrob-idINKBN0FZ17E20140730> (last accessed 25 April 2018).

Moore, Frazier (2017), "How Does Netflix Know You Might Like 'Ozark'?", *Wisconsin Gazette*, 23 August, <http://www.wisconsingazette.com/entertainment/how-does-netflix-know-you-might-like-ozark/article_0667daf0-8819-11e7-b9d3-fbd7ffa76927.html> (last accessed 25 April 2018).

Napoli, Philip (2011), *Audience Evolution New Technologies and the Transformation of Media Audiences*, New York: Columbia.

Patel, Neil (n.d.), "How Netflix Uses Analytics to Select Movies, Create Content, and Make Multimillion Dollar Decisions," Neilpatel.com, <https://neilpatel.com/blog/how-netflix-uses-analytics/> (last accessed 25 April 2018).

Pepitone, Julianne (2012), "Netflix Inks Disney Deal, But No New Films 'til 2016," *CNN*, 4 December, <https://money.cnn.com/2012/12/04/technology/netflix-disney/index.html> (last accessed 23 April 2018).

Quartararo, Ron (2008), "Let's Praise (Not Bury) Linear TV," *Broadcasting & Cable*, 17 November, p. 24.

Richmond, Ray (2011), "TCA: NBC's 'Parks and Recreation' Basks in Pre-Comcast

Critical Glow," *Deadline Hollywood*, 12 January, <https://deadline.com/2011/01/tca-nbcs-parks-and-recreation-basks-in-pre-comcast-critical-glow-96197/> (last accessed 15 May 2018).

Roettgers, Janko (2017), "How Netflix Wants to Rule the World: A Behind the Scenes Look at a Global TV Network," *Variety*, 18 March, <https://variety.com/2017/digital/news/netflix-lab-day-behind-the-scenes-1202011105/> (last accessed 18 April 2018).

Ross, Sharon Marie (2008), *Beyond the Box: Television and the Internet*, Malden, MA: Wiley Blackwell.

Sarkar, Samit (2018), "What Disney's Streaming Service Means for Netflix Subscribers," *Polygon*, 7 February, <https://www.polygon.com/2018/2/7/16982030/disney-leaving-netflix-marvel-lucasfilm-pixar-streaming-rights> (last accessed 25 June 2018).

Sepinwall, Alan (2009), "One Angry Knope: *Parks and Recreation*, 'Beauty Pageant,'" *What's Alan Watching?*, 2 October, <http://sepinwall.blogspot.com/2009/10/parks-and-recreation-beauty-pageant-one.html> (last accessed 21 April 2018).

Setoodeh, Ramin (2017), "Has Netflix's Ted Sarandos Rescued (or Ruined) Hollywood?", *Variety*, 15 August, <https://variety.com/2017/digital/features/ted-sarandos-netflix-original-movies-shonda-rhimes-1202527321/> (last accessed 18 April 2018).

Shields, Mike and Nathan McAlone (2018), "Disney Has Won the $71.3 Billion Battle for Fox—Now it's Readying for a War with Netflix," *Business Insider*, 19 July, <https://www.businessinsider.com/disney-fox-takeover-deal-a-weapon-against-netflix-2017-12> (last accessed 21 July 2018).

Sims, David (2015), "Will *Parks and Recreation* Be the Last Great NBC Comedy?," *The Atlantic*, 25 February, <https://www.theatlantic.com/entertainment/archive/2015/02/will-parks-and-recreation-be-the-last-great-nbc-comedy/386057/> (last accessed 25 May 2018).

Skopal, Pavel (2007), "'The Adventure Continues on DVD': Franchise Movies as Home Video," *Convergence: The International Journal of Research into New Media Technologies* 13 (2), pp. 185–98.

Snider, Mike (2016), "Watch Netflix Over Cable—Yes, It's Happening as Comcast X1 Deal Goes Live," *USA Today*, 4 November, <http://www.usatoday.com/story/tech/news/2016/11/04/watch-netflix-over-cable-yes-s-happening-comcast-x1-deal-goes-live/93237570/> (last accessed 23 April 2018).

Spangler Todd (2013), "Netflix Survey: Binge-Watching is Not Weird or Unusual," *Variety*, 1 December, <https://variety.com/2013/digital/news/netflix-survey-binge-watching-is-not-weird-or-unusual-1200952292/> (last accessed 15 May 2018).

Spangler, Todd (2016a), "Binge Watching: 70% of Americans Engage in Marathon TV Viewing," *Variety*, 23 March, <https://variety.com/2016/digital/news/binge-watching-us-study-deloitte-1201737245/> (last accessed 15 May 2018).

Spangler, Todd (2016b), "Netflix Caused 50% of U.S. TV Viewing Drop in 2015," *Variety*, 3 March, <http://variety.com/2016/digital/news/netflix-tv-ratings-decline-2015-1201721672/> (last accessed 15 May 2018).

Spangler, Todd (2018), "Comcast Will Start Bundling Netflix Into Xfinity TV Packages," *Variety*, 13 April, <https://variety.com/2018/digital/news/comcast-netflix-bundle-xfinity-tv-packages-1202752477/> (last accessed 25 May 2018).

Steinberg, Brian (2018), "Comcast Makes $65 Billion All-Cash Bid for Fox," *Variety*, 13 June, <https://variety.com/2018/biz/news/comcast-bid-fox-disney-media-1202844736/> (last accessed 25 June 2018).

Stelter, Brian (2013), "As TV Ratings and Profits Fall, Networks Face A Cliffhanger," *The New York Times*, 13 May, p. A1.

Stork, Matthias (2014), "Assembling The Avengers: Reframing the Superhero Movie

Through Marvel's Cinematic Universe," in James N. Gilmore and Matthias Stork (eds), *Superhero Synergies: Comic Book Characters Go Digital*, Lanham, MD: Rowman and Littlefield, pp. 77–95.
Strangelove, Michael (2015), *Post-TV: Piracy, Cord-Cutting, and the Future of Television*, Toronto: University of Toronto Press.
Tallerico, Brian (2017), "Netflix's 'Master of None' Stakes Claim as One of TV's Best Shows," *Rogerebert.com*, 11 May, <https://www.rogerebert.com/demanders/netflixs-master-of-none-stakes-claim-as-one-of-tvs-best-shows> (last accessed 25 June 2018).
Thompson, Kristin (2007), *The Frodo Franchise: The Lord of the Rings and Modern Hollywood*, Los Angeles: University of California Press.
Tryon, Chuck (2013), *On-Demand Culture: Digital Delivery and the Future of Movies*, New Brunswick, NJ: Rutgers University Press.
VanDerWerff, Todd (2017), "Master of None Season Two Is So Warm and Welcoming That You Might Not Notice How Audacious It Is," *Vox*, 19 May, <https://www.vox.com/culture/2017/5/13/15628546/master-of-none-review-season-2-netflix> (last accessed 25 June 2018).
Wolf, Mark J. P. (2012), *Building Imaginary Worlds: The Theory and History of Subcreation*, London: Routledge.

10. NETWORK STREAMING: TV BROADCASTERS IN THE DIGITAL SPACE

Monica Sandler

In January 2015, NBC launched its SeeSo subscription video on demand (SVOD) platform, designed to deliver new programming along with a selection of the network's older television series. Branded as the ultimate source for "alternative comedy," the venture highlighted an established broadcaster's attempt at media franchising within the digital sphere (Brandman, 2017). Evan Shapiro, NBCUniversal Digital Enterprises executive vice-president, described the goals of the service in an interview following SeeSo's original launch: "We think the world needs more laughter, so we put together what we think is the best collection of comedy in one place" (Porch, 2016). The undertaking sought to build a platform based on the legacy of NBC comedy classics such as *Saturday Night Live* (*SNL*) (NBC, 1975–present) and more recently concluded series such as *The Office* (NBC, 2005–13), *30 Rock* (NBC, 2006–13), and *Parks and Recreation* (NBC, 2009–15). The service featured the exclusive rights to every *SNL* episode going back to the 1970s, along with other syndicated shows and a slate of twenty-three original series designed to complement the collection of classic NBC comedies. Less than two years later, the service shut down, becoming another failure in the ever-expanding streaming service marketplace. This chapter asks: in the world of big data-based entertainment companies such as Netflix and Amazon, how can old media incumbents like NBC compete?

The chapter will consider the trials and failures that old-school media have faced entering into the new media realm. Together with SeeSo, I shall focus on the programming and media-franchising strategies of CBS All Access and

CW Seed, two services which I argue reflect the emerging market of "broadcast streaming" services. But first we must consider a simple question: why are broadcasters creating these platforms? In 2010, when digital streaming markets first began to expand, an article in *Variety* noted, "The Hollywood studios, for their part, are finally accepting the notion that digital distribution is the future of the homevideo biz" (Graser, 2010). The article specifically addressed the expansion of licensing agreements between production companies and SVOD platforms such as Netflix and Transactional Video-on-Demand (TVOD) arenas such as iTunes. It continues:

> One thing is clear: Hollywood needs to start promoting new platforms for consumers to access entertainment and to quickly figure out a way to bring more revenues from those offerings. High profit-margins that came from sticking with the status quo are now a thing of the past, forcing business models to be reimagined on an almost weekly basis. (Graser, 2010)

At the time, DVD sales already had begun to decline as users searched for new and easier streaming options to watch content.

In the earliest period of digital streaming, between 2005 and 2010, television stations connected with streaming services to license out their series, ushering in a new era where content was now instantly available to consumers. Fox, NBC, and Disney each invested in Hulu, a platform that served the function of showcasing next-day network content. By 2013, as Netflix launched its original series initiative with *House of Cards* (Netflix, 2013–18) and the revival of *Arrested Development* (Fox/Netflix, 2003–present), those same SVOD services were suddenly broadcasters' biggest competition. Jamieson Cox, a pundit at *The Verge*, wrote at the time of SeeSo's launch: "If SeeSo proves viable, NBC has less reason to cough up its shows to streaming services that have become major competitors in recent years" (2015). We now seem to be living in the next stage of network streaming strategy where broadcasters are creating their own subscription services complete with new and original shows beyond what's on their network television stations, potentially liberating companies from their restrictive licensing contracts to outlets such as Netflix.

Exploring network-owned SVOD case studies, this chapter articulates the struggles and failures of old-school media to move into the digital sphere. While the SVOD strategies of media conglomerates will continue to evolve, this research specifically focuses on the period from 2010 to mid-2018, an era in which Comcast NBCU launched (and closed) SeeSo and also saw the first wave of self-created digital streaming strategies at conglomerates such as CBS and Warner Bros. Each of my case studies will examine differing strategies that broadcasting conglomerates have attempted to apply within their own streaming services. Launched in October 2014, CBS All Access predominantly functions as a location where viewers can stream CBS shows, with access to new episodes available the day after airing, as well as a large catalog of previous episodes of current and older CBS series. The platform presents an alternative approach for a network SVOD service in comparison to SeeSo's niche strategy; CBS All Access was designed to showcase the full gambit of CBS series both past and present, along with new web exclusives. CW Seed, the third service which I will examine, however, employs an entirely different approach; the platform is free to consumers, based on an advertising, rather than a subscription, model.

Within this chapter, I argue that examining these outlets can provide insight into not only the difficult path ahead for old-school media, but also the nature of how these applications are organized and structured. It is not my intention here to discuss contemporary network strategies as the "next big thing" in digital streaming but, rather, to point to the divide between old and new media and to examine a key historical shift away from the traditional television broadcasting models of the past century. Indeed, most of these platforms have existed in addition to their agreements with wider tech outlets. In this way, rather than thinking of each of the services as the be-all and end-all digital strategy, I should like to look at them instead as experiments that old-school media conglomerates are making as they attempt to weather viewers' continual transition away from broadcasting toward streaming. Does this strategy work? As Shapiro stated in a promotional interview (perhaps already foreshadowing the platform's demise): "Come, seeso for yourself!"

Do You Have a License for That? CBS All Access and Big Tech Licensing Agreements

At the launch of CBS All Access in October 2014, Leslie Moonves, President and CEO of CBS Corporation, proclaimed: "CBS All Access is another key step in the Company's long-standing strategy of monetizing our local and national content in the ways that viewers want it. This new subscription service will deliver the most of CBS to our biggest fans while being additive to the overall ecosystem" (Moraski, 2014). While Moonves's statements are obviously serving a promotional function, trying to sell consumers on the platform, his use of the word "ecosystem" highlights a significant aspect of CBS's growing digital strategy. Phil Simon states, "[I]n the age of the platform, every business stands to benefit from creating its own ecosystem—and engaging in others" (2011: 24). He explains that ecosystems are a seamless business model wherein companies build multiple platforms; outlets then work together across those branches to create an immersive experience, keeping users engaged with content across these many different realms (Simon, 2011: 24). All Access has

Figure 10.1 The user interface of CBS All Access, showcasing its original series, *The Good Fight* (2017–present).

become CBS Corporation's attempt to expand its broadcasting forum into the digital sphere. This was accomplished through its tight packaging and by holding together its streaming rights. *The New York Times* wrote at the time of All Access's launch: "CBS is an unlikely disrupter. It did not join its rivals ABC, Fox, and NBC when they started Hulu in 2008, for example. But it held on to the digital rights for its programming so that it could build its own web and mobile offerings, like CBS All Access" (Steel, 2014). This section will consider the role that content licensing has played within the network streaming apps, looking at how broadcasters have moved to create self-run distribution platforms as an alternative strategy to licensing content to their big tech competitors.

A *Deadline Hollywood* article from May 2018 notes a marked shift seen within that year's network upfronts. Nellie Andreeva notes: "It took a couple of years, but we arrived to a point this year where broadcast networks own or co-own every new scripted series on their schedule that is not a revival. The few freshman series that do not come from the networks' sister studios are co-productions" (2018). Nearly all shows discussed at upfronts that year were owned, or at least partially owned, by the unit production companies connected to the networks on which they would be airing. The article continues: "Because ownership comes with deficit financing series and steep losses on shows canceled early in their run, some observers expected that the current drive by the broadcast networks to own or co-own their content would subside. For now, there are no signs of that" (Andreeva, 2018). The article declares that under the old television model, broadcasters suffer a deficit at the launch of their own produced television series, one that, in previous generations, would probably not be made up with a show cancelled during its initial run. Instead, under the old model, networks were more incentivized to buy the licensing rights to a series for its first broadcast, rather than take the risk of losing money in their initial investments (Lotz, 2014: 83). In 2018, however, the new investment in distributor-owned shows and coproductions highlights a breakdown within the mechanism; maintaining a stake in a title means that the network, and the shows that become associated with it, will also have control over the licensing of the commodity in the future. In today's landscape, production companies are trying to control all realms of where a series can potentially generate revenue: creation, original run, and future syndication.

In addition, networks seem to be experimenting with determining their preferred online distribution mechanism. Lining up production rights indicates one of two things for digital streaming in the future. The first means that the network can control the licensing rights for the programming that becomes associated with it, and the second potentially could enable the building of a network application with their company-owned content. Within the streaming marketplace, licensing a series can mean lasting profits even to the most unsuccessful series. Cancelled series still often receive streaming deals, while some shows have been revived by tech platforms after being dropped by their original network. This was seen, for example, with *The Mindy Project* (Fox/Hulu, 2012–17), a television show produced by Universal Television that aired on Fox for three seasons before being cancelled; the series was soon after bought by Hulu, which not only retained streaming rights for the show's first run, but also aired three additional seasons thereafter (Andreeva, 2015). In addition, outlets such as Fox, NBC, and ABC profit from the next-day streaming rights Hulu contracts for self-produced series, and then subsequently can monetize a given series with updated deals on longer-term streaming availability. Indeed, production companies can gain very lucrative payouts by licensing their series to Netflix, Amazon, or Hulu,[1] or potentially they could control their commodities' future by owning their series outright and allowing them to stream to their own subscription platforms. Each of these entertainment conglomerates seems to be facing the same question: to license or not to license?

The question of which route would work best has become the subject of many conglomerates' experimentation as they have moved into the digital sphere. This perhaps is best embodied by CBS Corporation's many strategies that have emerged since 2015 across branches of the wider conglomerate. CBS was one of the earlier adopters of this pattern to own all of its series (Forbes, 2015). While the company had many of its series play on Netflix, Amazon, or Hulu prior to the launch of CBS All Access, it had been much more conservative about which shows were placed at streaming outlets (Fortune, 2017). CBS especially had been a major holdout on next-day streaming rights with Hulu; CBS All Access, however, in many ways functions as a Hulu-like application but exclusively for CBS series. The service features a livestreaming option where users can watch the broadcast of a series as it airs, next-day streaming for new programs, and finally a collection of older CBS productions such as

Cheers (NBC, 1982–93) and *I Love Lucy* (CBS, 1951–57) (Fortune, 2017).[2] CBS All Access created a situation, regardless of its success, that keeps intact all of the network's and its wider production company's commodities. Other network streaming services, however, have struggled because so many series connected to the brand are also available on other platforms. Part of the limitations of an application like SeeSo was that, when NBC went to create its service, it had already licensed out many of its series such as *30 Rock* and *The Office* to rival, and bigger, streaming platforms. This lack of exclusivity created less incentive for consumers to sign up.[3]

All Access, however, is not the only digital strategy that CBS Corporation is attempting to implement. While the conglomerate has fixated on controlling the back collection of CBS network series, it has also experimented with licensing through the premium cable subscription channel Showtime and its accompanying SVOD platform. CBS launched Showtime Anytime in 2015; the service was designed, similarly to HBO NOW, to provide access to the channel's content without a cable subscription. At the same time as this launch, Showtime also emerged as an add-on subscription option for Hulu Plus users, at a cost of $9 per month, and as a standalone option at $11 per month. Additionally, CW Seed is also a coproduction between CBS and Warner Bros., as an advertiser-based service. The launch of these models simultaneously gave users multiple options for digital access to this branch of CBS Corp. programming. In this way, the company is both dipping its toes into licensing and its own digital content distribution through the varied assets held within the conglomerate (Forbes, 2015).

CBS All Access represents the "go-it-alone" strategy for other broadcast companies waiting to see if CBS Corp. can continue to be a top content distributor in the digital sphere, just as it has been for decades over the airwaves (Schneider, 2017). Lining up these contracting rights in order to sell to tech distributors, however, notes another key element: the original television broadcast becomes an intermediary which may become irrelevant with continued cord cutting. Television conglomerates are trying to control the distribution patterns of their programming. Whether the future involves licensing agreements or their own distribution applications, they are preparing for a world without broadcasting.

How to Build a Streaming App . . . According to a Failed Streaming App: SeeSo Programming Strategies, *Star Trek: Discovery*, and the CW Seed Advertiser Model

Content has a significant role within broadcast streaming applications. Platforms such as CBS All Access feature not only their older series but also new, original content. Why are these apps making their own original programming? If these platforms are experiments, then, as television conglomerates consider their options within the digital sphere, original programs will have to be an additional inevitability. This section considers the relationship that the apps have had balancing a programming slate built around classic series with built-in followings alongside unknown, original content. Kelsey Balance, former Head of Programming at SeeSo, explains how her company's application sought to organize content: "Even though our original programming is actually driving 80–90% of our new subscriptions, our focus groups and continued research show that our audience supports the importance of the archival titles. It's a way of establishing our brand credibility that we can then leverage to show them why they should take a chance on our original shows" (Brandman, 2017). The selection of older titles represents a major strategy for luring users to explore the new original content. Each app has tried to sell itself around exclusive streaming rights to major cult series with huge followings (for example, SeeSo with *SNL*, CBS All Access with *Star Trek*, and CW Seed with *One Tree Hill*). Seeing these series requires paying for the app or, in the case of CW Seed, watching advertisements. Balance articulates these titles as essentially the building blocks to the applications:

> We started off creating a foundation of acquisitions, like all 41 years of *SNL*, *Monty Python*, *Kids in the Hall*. Those were the three acquisitions that we knew true comedy fans would love. Then, we got more things that true comedy fans would at least know about, like *Garth Marenghi's Darkplace*, *The Mighty Boosh*, and *Alan Partridge*. Then, there are the NBC comedies like *The Office* and *Parks and Rec*. From there, we ended up commissioning 23 original series. (Brandman, 2017)

Older programming was designed to feed back and forth into the new and original content. Indeed, this dynamic between old and new series was built into the app's infostructure. This goal of the platform was even rather embarrassingly ingrained into the platform's name. As Patricia Hadden, senior vice-president and head of marketing for NBCUniversal Digital Enterprises, explains, "The name plays off the curated experience and 'right brain comedy' programming filter. It's a reference to the mapping the comedy genome philosophy: the 'You came to SEE *The Office*, SO we'll show you other workplace comedies you'll love'" (Shields, 2015). While the name certainly notes one of the main goals of the application's programming model, this overt explanation highlights the cheesy and out-of-touch quality of the app.

As Hadden implies, much of SeeSo's original programming seemed most directly tied to creating shows similar to series held within the back collections. Programs such as *The UCB Show* (SeeSo, 2015–17), a sketch comedy series, featuring players from the Upright Citizens Brigade (UCB), emphasized established and NBC branded talent such as Amy Poehler; the comedian was one of the founders of UCB along with Matt Walsh (*Veep*) and Matt Besser (*The Hot Wives*), all of whom remained associated with the SeeSo show during its run. *The UCB Show* also seemed to complement SeeSo's older series collection, which included the Poehler-led *Parks and Recreation* and NBC's iconic sketch comedy series *SNL*. A similar example can be seen with Dan Harmon's *HarmonQuest* (SeeSo, 2016–17)—a seeming spin-off from his podcast *HarmonTown* (2012–present)—where he and guests played role-playing fantasy games before a live studio audience, with the final results presented in animated form. In this way, SeeSo's old and new series seemed intricately connected to one another as NBC continued to associate with talent associated with its wider network.

Balance described SeeSo's programming as "on par with lower-end cable" (Brandman. 2017). The platform featured a large array of stand-up comedy specials and low-cost live reality shows. Indeed, while Poehler was associated with *The UCB Show*, the series predominantly featured current lesser-known UCB emerging talent with much less involvement from the bigger-name stars associated with the company. In addition, many series featured more experimental series from comedians with established, smaller fan bases. These included *Bajillion Dollar Propertie$* (2016–present), from Paul F. Tompkins and Dan Ahdoot, both of whom had had success on VH1's *Best Week Ever* (2004–14) and their own podcasts. Many other series, such as *Hidden America with Jonah Ray* (SeeSo, 2016–17) and *Night Train With Wyatt Cenac* (SeeSo, 2016–17) centered around the popularity of the titular comedian's hosting.

While SeeSo did produce a number of cheaply made original series, none could be considered a "hit" that generated popular interest. Amanda Lotz, in *The Television Will be Revolutionized*, notes the role that "phenomenal television" plays within the ever-expanding digital programming landscape. She defines the term as "a particular category of programming that retains the social importance attributed to television's earlier operation as a cultural forum" (Lotz, 2014: 42). Prior to streaming, television broadcasters had produced massive hit shows, such as *All in the Family* (CBS, 1971–79) or *Friends* (NBC, 1994–2004), that dominated the cultural zeitgeist; it seemed as though everyone was watching. In the contemporary landscape, however, there are still crossover hits that maintain this form of cultural relevance despite the plethora of viewing options and probably smaller viewer numbers. Netflix's *Stranger Things* (2016–present) exemplifies this form of "phenomenal television"; the series dominates entertainment media outlets and has launched a diverse fan community, merchandising ventures, and a Universal Studios theme-park attraction (Haylock, 2018). This has been an area, however, where network streaming apps have struggled the most. Without a hit, SeeSo continued to rely on the appeal of their back collection and the cult following around licensed shows such as *Monty Python's Flying Circus* (BBC1, 1969–74).

I argue, however, that the programming strategy within CBS All Access has fixated more on the hope of creating one of these "phenomenal television" series to which Lotz alludes. Like SeeSo, CBS All Access has sought to take the interrelationship between their back catalog and new series to greater heights through its original programming strategies. Its first two series—*The Good Fight* (2017–present) and *Star Trek: Discovery* (2017–present)—were spin-offs of popular CBS brands. While *The Good Fight* has been successful enough to receive multiple season orders, I want instead to concentrate on the *Star Trek* series. Leslie Moonves describes the strategy behind *Star Trek: Discovery*:

> *Star Trek* is a family jewel; it's an important piece of business for us as we go forward.
>
> We're looking to do original content on All Access and build up that platform . . . All Access will put out original content and knowing the loyalty of *Star Trek* fans, this will boost it. There's about a billion channels out there and because of *Star Trek*, people will know what All Access is about. (Goldberg, 2015)

A case such as *Star Trek: Discovery* highlights the relationship that many of these back collections end up having to their original programming.[4] A *Hollywood Reporter* article also commented: "Moonves anticipated 'millions' signing up for All Access, which will allow diehards to watch the new series any time, any place and on any device they'd like—a move the exec said is consistent with how many viewers are watching CBS' content" (Goldberg, 2015). Placing *Star Trek: Discovery* on CBS All Access was a specific move designed to turn the cult franchise's extensive fan base into new subscribers.

New York Magazine's *Vulture.com* additionally commented on *Star Trek's* role in creating a lasting brand for CBS All Access:

> By adding a new Star Trek series into the mix, CBS is telegraphing that it has bigger ambitions for the service. After all, if CBS Television Studios were simply looking to maximize short-term revenue, it would have pursued a more traditional path for Trek and convinced any number of linear or digital networks . . . Instead, Moonves and his team decided to use the show to build up the value of All Access. Think of Trek as the *House of Cards* or *Orange Is the New Black* for All Access: a buzzworthy, exclusive series that, like Netflix's first big originals, reshapes consumer perceptions. (Adalian, 2015)

In this way, CBS All Access set itself apart in the beginning by launching with a franchise strategy, whereas Netflix took a few years to embrace a franchise model with its various Marvel TV series, *Fuller House* (Netflix, 2016–present) reboot, and blockbuster movies like *Bright* (David Ayer, 2017). All Access's original programming output in 2018, however, notes the limitations of the "prestige" programming strategy. Since launching in 2015, by July 2018 CBS All Access had added only four original series to the platform. With so many apps in the marketplace, platforms need to provide enough content to incentivize the monthly charge. While CBS has the solid back collection of CBS exclusives, it may lack the kind of original programming to retain users. At the same time, SeeSo seemed to take the opposite approach, producing a large range of original content with few series that gained wider public attention. It seems to be the age-old question of quality versus quantity. Both SeeSo and CBS All Access reflect some of the major limitations that can emerge from the creation of SVOD services: either having not enough content that consumers may care about or having a lot of content that no one cares about.

Broadcast conglomerates are also considering alternative strategies. CW Seed, a coventure with Warner Bros. and CBS, created its own slate of programming by working with advertisers, essentially taking the broadcasting formula and applying it to the digital sphere. Many series predominantly feature one particular promoter within programming, along with a number of cross-promotional ads for currently broadcasting CW series. In this way, the platform requires less advertiser financing than a normal broadcaster needs and is able to cross-promote within the company's wider ecosystem (Villarreal, 2015). Launched in 2006, The CW was designed as the CBS and Warner Bros. coventure that would hit a younger demographic than the other major broadcast networks. The network intended to target a demographic of twelve- to twenty-four-year-olds. Jennifer Gillan in *Television and New Media: Must-Click TV* notes the significance that the network was trying to accomplish, by syncing early on many of its shows to wider teen marketing arenas, citing examples such as *Gossip Girl* (The CW, 2007–12) and the availability of its musical soundtrack on iTunes (2010: 26–7).

The network's leaders were early adopters to linking their station to the digital outlets that their younger fans were using on a daily basis. A 2015 *Los Angeles Times* article, however, notes a study by digital-content company Defy Media that sought to identify the role that digital streaming has come to play within the youth marketplace, roughly a decade after the CW's founding. The study found that "those ages 13 to 34 watch 11.3 hours of free online video and 10.8 hours of subscription online video weekly—much more than the time reported for free online TV offerings from broadcast and cable networks (6.4 hours) or for regularly scheduled TV (8.3 hours)" (Villarreal, 2015). The younger viewers that The CW was trying to attract had moved online quicker than almost any other demographic. The *Times* article notes that, in 2015, The CW had seen its average demographic move to roughly age forty-two while the median age for CW Seed is twenty-three. In this way, Seed served the function of continuing The CW's wider demographic goals while also bringing viewers back to the broadcast network through on-platform advertisements for the station's programming. CW President Mark Pedowitz explains, "We recognized this was a good place to have the CW Seed develop its own identity and, at the same time, trade off uniquely in the digital space as CW's brand" (Villarreal, 2015). In addition to going after the teen market, CW Seed has worked to leverage the many already established brands from Warners corporate sibling DC. The service also features a number of animated programs such as *Vixen* (CW Seed, 2015–16) and *Freedom Fighters: The Ray* (CW Seed, 2017–present) based on DC commodities, expanding on The CW's lineup of superhero shows like *The Flash* (The CW, 2014–present) and *Supergirl* (CBS/The CW, 2015–present), a niche approach which I shall note at greater length in the next section.

In actuality, CW Seed has taken a similar technique to SeeSo in terms of building its series collections. The application showcases older CW shows and features exclusive streaming rights for "on-brand" teen series such as *The OC* (Fox, 2003–07), while its original programming additionally features lower-budget content. Most of these new Seed series contain shorter-form content, roughly ten minutes in length. These programs mimic content found on YouTube, a platform that attracts one of the largest teenaged audiences of any streaming or television network. Indeed, some Seed series directly emerged from YouTube web series; the web series *Husbands* (YouTube/CW Seed 2011–present) began as a modest self-produced YouTube series in 2011 and, through its popularity on the service, was picked up by CW Seed for its third season in 2013. The hope was that their viewership would soon after transfer to watching the program in roughly the same format on Seed (Ng, 2013). While the platform poached YouTube talent at its launch in 2013, many Seed series can now be streamed there as well.

Within this section, I have noted the role that content, both new and old, play within the broadcast streaming networks. Both SeeSo and CBS All Access face certain concerns that can emerge as they compete with bigger platforms for subscriptions. CW Seed represents an alternative approach for its content to reach potential viewers. It still remains unclear which, if any, of these strategies can work in the digital marketplace, though clearly SeeSo's strategy was unsuccessful.

Niche Work if You Can Get It: Niche Stream Platforms, Legacy Branding, and SeeSo

Right from its launch, NBC's SeeSo celebrated itself as a niche streaming service, designed to appeal to a more specific, smaller audience rather than compete in a mass marketplace. Shapiro points out that "by focusing on a specific, yet large niche, and providing a curated experience, we can help viewers find good stuff they might not or cannot find. SeeSo is your neighborhood comedy eatery, with daily specials from a chef you trust" (Morfoot, 2015). Lotz articulates that the streaming age has created a breakdown of the cable bundling system; cordcutters instead have been given more options to pick and choose what outlets they are specifically interested in. Within the digital space, "these alternative transaction or publishing models thrive on specialty, distinction, and niche taste" (Lotz, 2014: 39). She argues that digital streaming has caused everything to become more targeted and move away from the attempts to appeal to all audiences that had been the dominant strategy among broadcast networks before cable.

Writing in 2014 for her book's second edition, Lotz articulates tech platforms such as Netflix as one of these emerging "niche" subscription services (2014: 39). Moving forward to 2018, Netflix, along with outlets such as Amazon and Hulu, now concentrates on appealing to wider ranges of audiences, trying to bring in subscribers from every potential walk of life and background. Programmers on these smaller network-streaming services, however, have even gone so far as to argue that tech platforms are the new broadcasters. As Shapiro states:

> The three major aggregators in OTT (Over The Top) right now bear a striking resemblance to the three major broadcasters around 1978. They're trying to serve all people all the time. Those three services are terrific, but I think there's going to be a disruption in that status quo . . . We feel like we're providing the curated, niche experience that cable provided in the 1980s with the huge advantage of having *watched* the last 30 years take place. We can go right to original programming knowing that it's the ultimate destination for any successful new channel. We knew going into it that the market demands original, exclusive programming, so we are getting ahead of that. (Porch, 2016)

It is interesting that an NBC executive would linger on a comparison between SeeSo and a cable station. Lotz notes the significant role that the emergence of cable played in niche programming, describing a period in television history she terms "the multi-channel transition" that first began in the 1980s and greatly expanded going forward. This period saw the rise of massive cable bundle packages that gave consumers access to hundred of channels. She explains that "the emergence of so many new networks and channels changed the competitive dynamics of the industry and the type of programming likely to be produced. As the number of cable channels grew, this targeting became more and more narrow" (Lotz, 2014: 27). Lotz argues that there will be a continued expansion of this grouping of choice. We have now seen the growth

Figure 10.2 A sample of content available on SeeSo.

of mass-media tech platforms such as Netflix along with the smaller niche outlets, some of which surprisingly have been created by broadcasters. While previous generations in television programming attempted to appeal to mass audiences within the digital space, tech platforms have attempted to produce, as well as to buy, licensing rights for such a range of content that has appeal to almost any type of modern viewer.

Seemingly, there has been a role-reversal between old and new media. Balance notes:

> The fact that we don't have advertising has helped us have a lot more freedom. We don't need to think of our audience in terms of demographics, so we can just ask, "What's the point? What's the intention? What's the point-of-view? What's the tone? Does it fit the larger family of shows we've created?" (Brandman 2017)

In this way, the programming decisions seem to have been less about shows that individually brought in eyeballs and, instead, about building a unified collective brand across the site. The networks are using SVODs to brand themselves as a type of franchise. Indeed, SeeSo did not seem to target a specific demographic but, rather, a cluster of people across generations interested in the same topics. Shapiro articulates that they were looking to attract "comedy nerds—millennials, Gen X'ers, and boomers." He continues: "[Our] shows will pull an audience that defies age. We think the comedy nerd psychographic is much bigger than most people think it is" (Porch, 2016). In this way, much of the outlet's development and ultimate programming strategy indeed seemed to revolve around demographic research.

Even when describing these outlets as "niche," it is hard to think about services owned respectively by mass conglomerates Comcast and Time Warner as "small." Rather, the outlets that these major television companies have chosen to dip their toes into have been smaller and, in many cases, more experimental. This strategy represents a cautious attempt to sift out their best-possible digital options for the future. Time Warner (now called WarnerMedia under AT&T)

has been particularly aggressive within the realm of niche app production, having produced a broad range of specialized subscription services over the past several years. As well as the CW Seed coventure, there are outlets such as FilmStruck, "classic" film content site Warner Archive, Korean soap opera streaming site DramaFever, the gamer network Machinima, Uninterrupted (that is, a partnership with LeBron James and Maverick Carter), Ellen Digital Ventures (that is, a partnership with daytime talk-show host Ellen DeGeneres), and HBO's digital streaming service HBO NOW (lest we forget that WarnerMedia also owns HBO) (Flint, 2016). Services such as FilmStruck and Warner Archive, the former of which closed in late 2018 and the latter of which closed in late 2017, reflect the conglomerate's strategies for handling their massive collection of classic film titles which includes both Warner and MGM's Classical Hollywood productions. Should they license these titles out or keep them all together?

Beyond utilizing already established commodities, Warner Bros. has used CW Seed to expand its position in genre content. While the CW has become the primary home for a DC Comics television universe with series like *The Flash*, *Supergirl*, and *Black Lightning* (The CW, 2018–present), CW Seed has sought to add to the DC comics landscape with a number of animated series. This reflects Warner Bros.' wider commitment to DC content, which continues to be a top commodity for the company's cinematic ventures as well, with the 2017 releases of Patty Jenkins's *Wonder Woman* and Zack Snyder's *Justice League*. This being said, Warner additionally launched a DC service in 2018 predominantly focused on similar animated series to CW Seed, highlighting the disorganized nature of many of their streaming ventures (Little, 2018).

SeeSo was supposed to be the starting point for a similar multi-app strategy at NBC/Universal. Shapiro states, "With regard to individual brands and SVOD, our approach is to diversify the portfolio and create new opportunities to go to the consumer and not compete against ourselves on linear" (Porch, 2016). Shapiro also notes a plan for Comcast to create a number of niche applications in the future. Why? Because of market research. He continues, "When I got here, we embarked on a pretty substantial research project. We did 11,000 interviews of people who watch video online. We have nine different platforms in development right now. Based on the research, we're making decisions about what to do next" (Porch, 2016). The hope was that individuals interested in topics such as "alternative comedy" would add their application to their larger slate of monthly subscription services. SeeSo seems to have emerged because research indicated a public interest in a comedy streaming service. With series such as *HarmonQuest* and *The UCB Show*, SeeSo's launch centered on expanding NBC's long-existent comedy brand. SeeSo's concentration on alternative comedy drew directly from NBC's history of highly rated sitcoms. Their Thursday night lineup for decades has organized itself around a steady flow of thirty-minute comedy series, deemed at different points in their history as "Must See TV."

In this way, SeeSo seems to have emerged from NBC's legacy of comedic programming. As Gillan explains, "[A]ll programming elements on network and basic cable television in the U.S. are part of the promotional surround for advertising and company voice messaging," while audiences "address strategies in which calls-to-action double as calls to affiliation through their reliance on myths that appeal to viewers and serve industry needs" (2014: 8, 210). Gillan argues that, starting in the 1980s, NBC's "Must See TV" had equated the network with top comedy programming. Despite more modest hits going into the mid-2000s, NBC's Thursday slots, now branded "Comedy Night Done Right," featured a wave of series such as *The Office, Parks and Recreation,* and *30 Rock* that, Gillan claims, were designed as specialized, with appeal to smaller, more targeted demographics. She notes the significance of this strategy in 2015: "Whatever the outcome of its all-night comedy plan, NBCUniversal has certainly embraced brand-based TV for the immediate future" (Gillan. 2010: 210). Early ads for SeeSo focused less on NBC's brand in and of itself but, rather, on this legacy of NBC programming. Ads prominently celebrated SeeSo as the only location to stream all seasons of *SNL* and featured footage from the more recent "Comedy Night Done Right" series that, according to Gillan, had allowed NBC to embrace a more niche and targeted comedy audience roughly a decade earlier. In this way, the emergence of SeeSo seemed to serve as a continuation of NBC's broadcasting strategy of targeting smaller, more specialized communities of loyal comedy fans.

While SeeSo hosted many iconic NBC productions, such as *30 Rock* and *The Office*, the network also had licensed out these commodities to Hulu and Netflix, respectively. The app, in this way, reflects another possible route that networks are attempting outside of providing content to platforms that are now seemingly their competitors. NBC also took on the task of buying exclusive rights to a number of cult series like *Monty Python's Flying Circus* and *The Kids in the Hall* (CBC, 1988–95), rather than attempt exclusively to make use of NBC collections. This exposes one of the biggest limitations of the application because some of its most iconic comedy series, such as *Seinfeld* and *Friends*, were produced by other production companies and simply broadcast on the network. It can also highlight why there has been the more recent push among networks to privilege their in-shop-produced content (Andreeva, 2018). This notion of a unified network brand, however, seems to have become NBC's primary focus after the failure of SeeSo. An article in August 2018 announced a new potential streaming service entitled Watch Back. The platform will be intended less to create new content and more to function "as a marketing tool to pull viewers into the NBCU fold by offering samples of some full-length programming from across NBCU's networks as well as web-friendly clips"

(Jarvey, 2018). The article continues by noting that full seasons of actual NBC series would not be available. Instead, the platform seems as though it will be designed to entice viewers into watching NBC series across their vast array of networks. The conglomerate appears to be moving away from the niche strategy and toward a focus on building a unified and wider brand for its creative programming. SeeSo can then be seen as a passing strategy, highlighting the struggles that smaller and more targeted services face in the vast and constantly expanding digital streaming marketplace.

Back to the Bundle: the Paradox of Choice, VRV, and the Future of Network Streaming

On August 9, 2017, it was reported that SeeSo would be shutting down as a service. The statement announcing this decision proclaimed:

> Dear Seeso-ers,
> We're writing to let you know that later this year, Seeso will be shutting its comedy doors.
>
> Though we will be departing, much of our comedy will live on—and some of your favorite Seeso Originals have already found a new home.
>
> You can now find *HarmonQuest*, *My Brother, My Brother and Me*, *Hidden America with Jonah Ray* and *The Cyanide and Happiness Show* on VRV.
>
> We'll let you know as soon as we have further information, including where to find the exciting project *There's . . . Johnny!* and other Seeso Originals in the future (in the meantime, please note that *There's . . . Johnny!* will no longer premiere this month as previously announced. We love this show and can't wait for everyone to see it). (Otterson, 2017)

If the intention of SeeSo was to create an alternative strategy to licensing contract agreements, the outlet ended with a search for other platforms to buy their programming slate, confirming the yield that an already established outlet can find new viewers. Going niche helps to differentiate apps from a wide range of competing streaming platforms yet it can also be limiting. One pundit for *Engadget* wrote about SeeSo's failure:

> As a niche streaming service, [SeeSo] just didn't capture attention in the same way as all-encompassing alternatives like Amazon, Hulu, or Netflix . . . This doesn't mean that specialty streaming can't work, but it faces an uphill battle when you can spend just a few dollars more per month to access a much larger video catalog. (Fingas, 2017)

SeeSo's failure raises a wider question for all of these services: can each company stand by itself with its own subscription service exclusively tied to its own products?

The answer is a definite "No." Elissa Nelson notes in "Windows into the Digital World: Distributor Strategies and Consumer Choice in an Era of Connected Viewing" that "if using a product is too difficult to figure out or if enough titles aren't available, with the multiplicity of options, consumers can always look elsewhere" (2014: 69). She asserts that, with so many applications in a competitive landscape for everyone to grab top content, these marketplaces can very quickly become oversaturated, raising the question of how many subscription services a consumer may be willing to invest in regularly. Nelson continues, "Distributors have to find ways to effectively cut through the noise of the teeming digital distribution arena, while balancing the demands of different industry parties with the demands of the consumer" (2014: 69). Evan Shapiro echoed this sentiment in promotional interviews at the time of SeeSo's launch. He described some of the limitations that have grown from so much streaming content: "When people sit down to watch *something* on a Saturday night—they don't know what—that has become work" (Porch, 2016). He added that SeeSo's programmers looked to solve the consumer problem known as the "paradox of choice." Shapiro believes that the massive content selection on outlets such as Netflix and Amazon has made it difficult for consumers to make decisions on what to watch. He feels instead that smaller applications, such as the failed SeeSo, are much more maneuverable for users to find things to watch.

I argue, however, that this "paradox of choice" has instead emerged among consumers struggling to decide on which monthly subscription services they are willing to spend money. The latter part of this decade seems to be generating an all-out free for all within the streaming world, as outlets such as Netflix begin to dominate the pack. CBS All Access, Seeso, and CW Seed are three very different strategies, all signaling that traditional broadcasting has started to take seriously the need to move into the digital sphere. There are already so many other services and other film and television applications on the market, however. When does the bubble burst? When do consumers reach their breaking point of how many subscriptions they are willing to pay for? Industry figures such as Shapiro have compared these broadcast streaming services to cable channels, yet these are traditionally the kinds of networks that cable-bundling packages bolster. Indeed, on the cost front, one of cordcutters' biggest goals is to pay less than they would for a typical cable subscription. With so many apps out there, it can be easy for streaming costs to add up quickly to, or even exceed, the same amount as cable.

The announcement that SeeSo would close in 2017 noted that many of its original programs would find a new home on a service called VRV. Launched in mid-2016, VRV has brought together a large group of niche video content, making individual content more cost effective as a result. For $9.99 per month, subscribers get access to anime streaming services such as Crunchyroll and Funimation, and geek news services such as *Nerdist*. Essentially, VRV is a platform bundle service, modeled on tried and tested cable methodologies. Seeso first joined in early 2017 just as the platform began to show signs that it was struggling. Balance explained the decision:

> I think our goal is to try to be everywhere our audience could potentially be, and we love the strategy of what VRV is doing. We also had a very successful partnership with Amazon Channels, being a part of their larger service. So, with VRV and their audience demographic they were going for, it felt like a great fit for us, and we felt that there were definitely people there who were already Seeso consumers or who could be Seeso consumers, and our programming fits well in the larger, curated library they're building. (Brandman, 2017)

SeeSo's move to VRV both highlights its role within the niche programming community and aligns a major network with these smaller services.

Balance also notes SeeSo's relationship with Amazon Channels, a service that allows Amazon users to add additional streaming subscriptions directly to their Prime account, essentially using Amazon's Video streaming platform to access other services' content, similar to the subscription partnerships between Hulu and Showtime Anytime. A 2018 study, published in *Business Insider*, noted the effectiveness of outlets such as Amazon Channels: "[T]he sign-up rate is even higher for premium-cable TV apps: Amazon Channels accounts for 53% of HBO streaming sign-ups, 72% of Showtime OTT, and 70% of Starz OTT sign-ups. That compares to 47% of sign-ups directly through HBO's app, 26% directly through Showtime, and 30% directly through Starz" (Schomer 2018). Subscribers seem more interested in adding content to their existing big-tech memberships, a process which allows for fewer signing requirements and the ability to view programming all together within one connected ecosystem, rather than subscribing to a new service by itself. In this way, each of the platforms discussed within this chapter must reckon with "Big Five" data juggernauts (the FAANG group of Facebook, Apple, Amazon, Netflix, and Google) looming over them.

The question once again becomes: is it better for these broadcasters to continue their streaming ventures or simply to go down the route of licensing to the bigger tech companies? The answer to this question still remains to be seen; within this chapter, however, I have laid out the strategies and pitfalls that old-school media outlets continue to face in the digital landscape. These platforms highlight the role that an app's content plays as users pick and choose their monthly streaming service subscriptions and alternative approaches such as CW Seed that rely more heavily on advertisers. In the end, a statement from the 2010 *Variety* article noted at the beginning of this chapter continues to resonate: "High profit-margins that came from sticking with the status quo are now a thing of the past, forcing business models to be reimagined on an almost weekly basis" (Graser, 2010). While an outlet like SeeSo has already failed, it remains to be seen if a model like CBS All Access or CW Seed could prove itself as viable.

Notes

1. The payout potential from a licensing deal could be seen in Hulu's high-cost deal with Sony for the exclusive rights to *Seinfeld* (NBC, 1989–98) that was valued at roughly $875,000 an episode (Littleton 2015).
2. *Cheers* was a CBS Corporation production that aired on NBC during its original run; CBS All Access now houses the show because the original production company maintains the licensing rights to the original series.
3. These concerns about content licensing have also affected the strategies of other broadcasters such as Disney's ABC. By the end of 2019, Disney will launch its own streaming service, called Disney+. While it is still unknown what this platform will look like, one of Disney's emerging strategies in 2018 has involved letting its licensing agreements with wider streaming platforms expire. Disney has worked with both Netflix and Hulu on original Marvel series but its film and television content has gradually started to leave all other streaming services, setting up the company to have a wider slate of its own content when its service launches (Jackson, 2018). See Gillan, 2011.
4. While *Star Trek* had not necessarily been associated with the CBS brand since its original run in the 1960s, later series of the show appeared on Viacom's UPN when it was still owned by CBS Corporation.

References

Adalian, Josef (2015), "Will *Star Trek* Be CBS's *House of Cards*?," *Vulture*, 3 November, <http://www.vulture.com/2015/11/star-trek-cbs-all-access-house-of-cards.html> (last accessed 22 July 2018).

Andreeva, Nellie (2015), "Hulu Picks Up 'The Mindy Project' with 26-Episode Order," *Deadline Hollywood*, 15 May, <https://deadline.com/2015/05/the-mindy-project-hulu-season-4-pickup-26-episodes-1201427617/> (last accessed 22 July 2018).

Andreeva, Nellie (2018), "Upfronts 2018: Network Ownership Peaks, Digital Rights Demands Increase As Volume & License Fees Slide," *Deadline Hollywood*, 17 May, <http://deadline.com/2018/05/upfronts-2018-analysis-tv-network-ownership-digital-rights-license-fees-1202393504/> (last accessed 22 July 2018).

Brandman, Jake (2017), "Behind the Screens: SeeSo's Head of Programming on Curating Comedy for Streaming," *Observer*, 30 January, <http://observer.com/2017/01/seeso-interview-comedy-programming/> (last accessed 22 July 2018).

Cox, Jamieson (2015), "NBC's Comedy Streaming Service SeeSo Is Launching January 7th: It's Opening Up in the Wake of a Month-Long Free Beta Period," *The Verge*, 9 December, <https://www.theverge.com/2015/12/9/9879440/seeso-nbc-comedy-streaming-service> (last accessed 22 July 2018).

Fingas, Jon (2017), "NBC Will Shut Down its SeeSo Comedy Streaming Service this Year: The Niche Online Service Just Didn't Catch On," *Engadget*, 9 August, <https://www.engadget.com/2017/08/09/nbc-shutting-down-seeso/> (last accessed 22 July 2018).

Flint, Joe (2016), "Warner Bros. Creates Unit for Digital Platforms: New Division Will Include DramaFever, Warner Archive Instant and Investment in Machinima," *The Wall Street Journal*, 2 June, <https://www.wsj.com/articles/warner-bros-creates-unit-for-digital-platforms-1464896641> (last accessed 22 July 2018).

Forbes (2015), "Content Licensing Will be a Key Driver For CBS's Future Growth Amid Changing Media Landscape," *Forbes*, 11 September, <https://www.forbes.com/sites/greatspeculations/2015/09/11/content-licensing-will-be-a-key-driver-for-cbss-future-growth-amid-changing-media-landscape/2/#454e084755d7> (last accessed 22 July 2018).

Fortune (2017), "CBS's Drive to Own Shows Like 'MacGyver' Is Paying Off Overseas," *Fortune*, 1 June, <http://fortune.com/2017/06/01/cbs-programs-sales-overseas/ (last accessed 22 July 2018).

Gillan, Jennifer (2011), *Television and New Media: Must-Click TV*, New York: Routledge.

Gillan, Jennifer (2014), *Television Brandcasting: The Return of the Content-Promotion Hybrid*, New York: Routledge.

Goldberg, Lesley (2015), "Leslie Moonves Explains Why 'Star Trek' Went to CBS All Access," *The Hollywood Reporter*, 3 November, <https://www.hollywoodreporter.com/live-feed/why-star-trek-went-cbs-836710> (last accessed 22 July 2018).

Graser, Marc (2010), "Is Hollywood on the Road to Recovery? Showbiz Tries to Shake Off Recession, but It's a Slow Road Back," *Variety*, 25 September, <http://variety.com/2010/biz/news/is-hollywood-on-the-road-to-recovery-1118024585/> (last accessed 22 July 2018).

Haylock, Zoe (2018), "'Stranger Things' to Be Featured in Universal Studios' Halloween Horror Nights," *The Hollywood Reporter*, 4 April, <https://www.hollywoodreporter.com/news/stranger-things-be-featured-universal-studios-halloween-horror-nights-1099617> (last accessed 22 July 2018).

Jackson, Dan (2018), "Everything We Know About Disney's Secret New Streaming Service," *Thrillist*, 29 January, <https://www.thrillist.com/entertainment/nation/disney-streaming-service-netflix-coming-soon> (last accessed 22 July 2018).

Jarvey, Natalie (2018), "NBCUniversal Eyeing Late Fall Launch for New Video App Called Watch Back," *The Hollywood Reporter*, 9 August, <*https://www.hollywoodreporter.com/news/nbcuniversal-eyeing-late-fall-launch-new-video-app-watch-back-1126443*> (last accessed 12 August 2018).

Little, Morgan (2018), "DC Universe: Everything to Know about the Superhero Streaming Service," *CNET*, 20 July, <https://www.cnet.com/news/5-things-we-learned-about-dcs-new-streaming-service-at-comic-con/> (last accessed 22 July 2018).

Littleton, Cynthia (2015), "Hulu Sets Mammoth 'Seinfeld' Licensing Deal," *Variety*, 28 April, <https://variety.com/2015/tv/news/hulu-seinfeld-licensing-160-million-1201483537/> (last accessed 22 July 2018).

Lotz, Amanda (2014), *The Television Will Be Revolutionized*, New York: New York University Press.

Moraski, Lauren (2014), "CBS launches expansive digital subscription service," *CBS News*, 16 October, <https://www.cbsnews.com/news/cbs-launches-digital-subscription-service-cbs-all-access/> (last accessed 22 July 2018).

Morfoot, Addie (2015), "Comcast's 'SeeSo' is Hedge Against Millennials Leaving Cable-TV," *The Street*, 16 October, <https://www.thestreet.com/story/13326218/1/

comcast-s-seeso-is-hedge-against-millennials-leaving-cable-tv.html> (last accessed 22 July 2018).

Nelson, Elissa (2014), "Windows into the Digital World: Distributor Strategies and Consumer Choice in an Era of Connected Viewing," in Jennifer Holt and Kevin Sanson (eds), *Connected Viewing: Selling, Streaming, & Sharing Media in the Digital Era*, New York: Routledge, pp. 62–78.

Ng, Philiana (2013), "'Husbands' Co-Creators on Moving to The CW Seed and Future Beyond Season 3," *The Hollywood Reporter*, 15 August, <https://www.hollywoodreporter.com/news/husbands-creators-moving-cw-seed-606662> (last accessed 22 July 2018).

Otterson, Joe (2017), "SeeSo to Shut Down 'Later This Year,'" *Variety*, 9 August, <http://variety.com/2017/tv/news/seeso-shut-down-1202521126/> (last access 22 July 2018).

Porch, Scott (2016), "NBCU's Evan Shapiro Says Seeso Is Just The Beginning—Upwards Of 6 New SVOD Services Are Coming Our Way," *Decider*, 6 January, <https://decider.com/2016/01/06/seeso-evan-shapiro-interview/> (last accessed 22 July 2018).

Simon, Phil, 2011, *The Age of the Platform: How Amazon, Apple, Facebook, and Google Have Redefined Business,* Henderson, NV: Motion Publishing.

Schneider, Michael (2017), "Most-Watched Television Networks: Ranking 2017's Winners and Losers," *IndieWire*, <https://www.indiewire.com/2017/12/highest-network-ratings-2017-most-watched-hbo-cbs-espn-fx-msnbc-fox-news-1201911363/> (last accessed 22 July 2018).

Schomer, Audrey (2018), "Amazon Channels Accounts for More than Half of All DTC Sign-Ups," *Business Insider*, 18 May, <http://www.businessinsider.com/amazon-channels-accounts-for-more-than-half-of-all-dtc-sign-ups-2018-5> (last accessed 22 July 2018).

Shields, Mike (2015) "Here's Why NBCUniversal Named its Streaming Service SeeSo," *The Wall Street Journal*, 15 October, <https://www.wsj.com/articles/heres-why-nbcuniversal-named-its-streaming-service-seeso-1444942166> (last accessed 22 July 2018).

Steel, Emily (2014), "Cord-Cutters Rejoice: CBS Joins Web Stream," *The New York Times*, 16 October, *<https://www.nytimes.com/2014/10/17/business/cbs-to-offer-web-subscription-service.html>* (last accessed 22 July 2018).

Villarreal, Yvonne (2015), "CW Beefs Up its Seed App Lineup as Online TV Viewing Booms," *Los Angeles Times*, 27 October, <http://www.latimes.com/entertainment/envelope/cotown/la-et-ct-cw-network-seed-app-20151028-story.html> (last accessed 22 July 2018).

PART V

EMERGENT PLATFORMS: POSSIBLE FUTURES FOR THE MEDIA FRANCHISE

11. TRANSMEDIA-TO-GO: LICENSED MOBILE GAMING IN JAPAN

Bryan Hikari Hartzheim

A growing number of top-grossing and popular mobile game titles are extensions of larger media franchises that also stem from other media properties. These mobile games are minor excursions for most fans of the original franchise, often either developed well after the conclusion of the media from which the franchise initially developed or as a supplementary experience to more traditional media experiences based in film or television. And yet, these same companies which traffic in traditional media are valuing the mobile game as a lucrative and flexible "paratext," to borrow Jonathan Gray's term to describe the increasingly important ancillary texts of multimedia franchises today that include licensed comic books, glossy websites, and various forms of digital games (2010: 6). While mobile games developed from these franchise licenses initially began as recycled derivative products, the highest-grossing mobile games now have bolder production values from development studios, increasing input and involvement from the parent franchises, and, most importantly, the ability to adapt to accommodate changing fan interests and trends over time.

This chapter will look at the mobile game as a key paratext within the transmedia franchise. Licensed video games have a long history of critical and fan disfavor, and largely for good reason: they are often rushed products that have little bearing on the larger franchise itself (Elkington, 2009). Mobile games are similarly ancillary to the core texts, operating on development budgets that are a fraction of those in console gaming. This chapter will argue, however, that, despite their more limited interfaces and simpler gameplay compared with console-based licensed games, these mobile licensed games ironically offer deeper, more sustainable experiences that more closely integrate with their host franchise. While ostensibly superfluous to experiencing the franchise's narrative, franchise mobile games have become adept at repackaging the core assets of the franchise—its characters and scenarios—and presenting them in increasingly clever and nefarious ways.

As objects of critical analysis, mobile games were initially compared with other social/casual games for their exploitative psychological effects (Evans, 2016), coercive monetization tactics (Shokrizade, 2013), and repetitive or unbalanced gameplay (Alexander, 2013; Margalit, 2015). Such analyses tended to focus on the first waves of mobile games to appear on cell phones and smartphones in 2009. Ian Bogost's game *Cow Clicker*, and the resulting discussion of the game's deliberately shameless grinding gameplay, exemplify the discourse from this period (Tanz, 2011). In recent years, more varied analyses have emerged that interrogate mobile games for their formal, theoretical, and sociocultural properties as games rather than as exploitative services (Wilson and Leaver, 2016). This chapter adds to this expanding range of analyses, calling particular attention to licensed mobile games that extend or renew the life of the franchise through game mechanics that "affectively amplify" (Ash, 2012) core elements of the larger franchise itself. The analysis focuses particularly on licensed mobile games from Japan because Japan has not only the largest mobile games market in the world, but also far more licensed titles coming from its comics and animation industries than any other country. Japan's mobile games act as the ideal ancillary products to the bottom line of larger media franchises, and their aims for media convergence, in that they not only advertise the franchise's core texts to various audiences, but also actively prod continuous player consumption in both the game and the larger franchise's core texts.

A Case for the Paratext

The practice of creating video games connected to franchises in other media formats, such as film and television, has a long and varied history. Synergy strategies, where games use licenses from other media, has the effect of reducing risk by accounting for a given number of ensured sales from a preestablished consumer base of fans, be they followers of a sports league (*NBA 2K*, *Madden*), film series (*Star Wars, The Matrix*), or television show (*The Simpsons*, *LOST*). As Stephen Kline, Nick Dyer-Rutherford, and Greig de Peuter put it, "By capitalizing on narratives drawn from popular culture, marketing costs are drastically reduced because the expensive marketing work—building awareness—is already done" (2003: 226). Initially, licensors and developers favored such deals because they helped them get publishing contracts and increased the chance of the game turning a profit in a crowded market through name recognition among more casual players (Kerr, 2005: 72).

As such games have increased in prominence in popular culture, so has the study of such games as prominent examples of ancillary media within the larger media franchise. Ancillary licensed media typically were regarded as little more than tools to generate revenue based on other media texts. Such strategies could involve, for example, merchandise using visual references from a "high concept" film, (Wyatt, 1994) such as *Jurassic Park* (Steven Spielberg, 1993), or television series, such as the Sunbow–Marvel–Hasbro collaboratively produced *My Little Pony* (1986–87) or the various animated television specials created to hawk Kenner toys for *Strawberry Shortcake* (1980–85) (Englehardt, 1986). In Japan, as well, television shows were subjected to scrutiny for combining compelling and violent narratives for the purposes of selling an annually changing lineup of toy merchandise (Hori, 1996: 122–4). Video games which relied on such interdependence of media industries were likewise viewed critically in no small part because they could "lead to a reduction in the overall diversity of texts and the scope for radical innovation to emerge" (Kerr, 2005: 72).

Within the last dozen years, scholars have increasingly questioned this critical approach by examining ancillary media as additive texts that contribute their own unique narratives and experiences to the franchise. These media "paratexts," as Gray has observed, not only add to texts; rather, they "create texts, they manage them, and they fill them with many of the meanings we associate with them" (2010: 6). The paratexts of a franchise often serve vital functions for the film or television shows from which they derive, whether filling in narrative gaps or creatively building upon the franchise's world. In western discourse, this multimedia and platform storytelling is often characterized as "transmedia," a phrase that Henry Jenkins popularized (2006: 97), whereas in Japan such media convergence has typically been referred to as the "media mix," a term that emerged through the marketing strategies of Kadokawa Entertainment (Kawasaki and Iikura, 2009: 18). As Marc Steinberg notes, the term "media mix" has been the most widely used industrial term since the 1980s to describe "the phenomenon of transmedia communication, specifically, the development of a particular media franchise across multiple media types, over a particular period of time" (2012: 135).[1] Media-mix franchises revolving around transforming superheroes, mechanical robots, and magical girls have been dissected for their abilities to use multiple media—films, television shows, special effects, animation, toys, figures—to involve audiences in large transmedia experiences (Allison, 2006; Yokohama, 2006; Hartzheim, 2016).

Video Games and Media Convergence

Increasingly, video games have become an integral aspect of this multimedia experience. In describing the narrative strategy of "transmedia storytelling" in *The Matrix* franchise (The Wachowskis, 1999–2003), Jenkins describes the video game *Enter the Matrix* (Atari, 2003) as a key component in more heavily investing audiences in the film franchise: "The consumer who has played the game or watched the [animated] shorts will get a different experience of the movies than one who has simply had the theatrical film experience. The whole is worth more than the sum of the parts" (2006: 104). Hollywood films have perhaps the most conspicuous imprint on licensed games, from the notoriously poorly designed *E.T.: The Extra-Terrestrial* (Atari, 1982) to the more recent high-profile offerings of fantasy and sci-fi franchises adapted from best-selling novels and comic books such as Warner Bros. Interactive Entertainment's *Middle-earth: Shadow of Mordor* (2014) and *LEGO Marvel's Avengers* (2016), based respectively on the *Lord of the Rings* and Marvel franchises. Such games simultaneously allow players interactively to explore the worlds and characters of the films while aping the audiovisual design of the films themselves. As Robert Alan Brookey argues, games that strongly attempt to connect the player to the films are "designed rhetorically to extend the experience of a film and augment the practices of the film franchise" (2010, 28). Brookey's examples of Tolkien fantasyscapes and Marvel superhero side-stories suggest that the most successful of these licensed games fluidly connect the content in the games to the films through synergistic intertextual strategies such as opening credits, cutscenes, trailers and making-of videos, character designs and voices, and other bonuses appealing to fans of the original franchises. Thus, video games can be key paratexts in shaping how we ultimately view and consume licensed franchises. In Gray's words, "a videogame adaptation—or at least a good one—is not merely an attempt to rehash or to copy; it moves the story, its world, and its audience to a different narrative mode, wherein the audience can step into (parts of) the storyworld" (2014: 192).

Gray also notes that such games are constrained in the scope of the storyworld they can construct precisely because of licensing obstacles, such as limited access to scripts or filmmakers, or release schedules forced to meet the timing of the film or television series' release. In his discursive analysis on the compiling site Metacritic of reviews of games licensed from television shows, Gray says that, while short development periods might sabotage these games, critics have attacked their lack of immersion and their "limits of temporality, most notably endless repetition" (2014: 64). In this estimation, the games failed to capture the pace, timing, and viewing schedule associated with the television shows on which they were based. Trevor Elkington echoes these conclusions in his own analysis of reviews of games based on film and television licenses, finding that licensed games routinely scored lower on Metacritic, with critics lambasting the games for shallow gameplay, mediocre graphics, limited emergent or creative gameplay options, and an overreliance on cutscenes (2009: 217–19). Such gripes are echoed in studies of games based on popular television series such as the *LOST* and *24* franchises. M. J. Clarke observes that "tentpole TV" games contain complex scripts cowritten by screenwriters from the shows, as well as a "sheer amount of visual fidelity between the television series and the games themselves" (2012: 108). But these games must also operate under severe constraints, from rushed development cycles to "dead-end continuity" (where games have little narrative bearing on the television series) that limit the narrative and gameplay elements to side-missions and minigames. These critical comments on the impotence of game narratives with respect to their licensed tentpoles is evident in the perception of prestigious film franchises as well, from the continuity issues that plagued the James Bond-based *007 Legends* (Activision, 2012) (Fleury, 2014) to the marginal stories and characters presented in the series of games based around the Marvel Cinematic Universe.

Video games are, in many ways, a more involved paratext within Japanese media franchises. Japan's arguably most popular intellectual property (IP), *Pokémon* (1996–present), derived from a video game to become an international transmedia juggernaut, and many of the country's most popular franchises are media mixes that involve video games, either as their "mothership" or as an ancillary experience that frequently supplements anime and manga texts. According to Tsugata Nobuyuki, the English-language term "anime business" began to appear in media discourse in the 1990s to describe the convergence between the anime, manga, and gaming industries (2004: 198). English-language scholars have also pointed out that Japan's gaming industry benefits from, and is inextricably tied to, its anime and manga industries (Kohler, 2004; Allison, 2006). As Jennifer deWinter puts it, "In its simplest form, this means that properties are being adapted from one medium to another. In other cases, they are used in a type of serial narrative that ends in one medium and picks up in another" (2015: 320). This means that games occupy a position within Japan's cultural industries—and within its transmedia franchises—that has grown increasingly as important as older and more prominent visual media industries, such as film, television, animation, and comics, and their specific importance within such franchises often depends on the context of the franchise itself. Thus, some games, such as *Pokémon* or the Square Enix-produced role-playing game series *Final Fantasy* (1987–present), can spawn mainstream media-mix franchises for a broad audience range, while visual novel games, such as Type-Moon's *Fate: Stay Night* (2004–present) or 07th Expansion's *Higurashi When They Cry* (2002–14), explore complex multimedia narratives for older niche fans. Within franchises from which games are adapted, spin-offs can explore segments of a manga or anime franchise in greater detail, such as the various games based around the adventure series *One Piece* (Toei Animation, 1997–present), while still others can be derivative games that depict alternative realities or dimensions of the original property and its characters, such as the visual novel series *Kanon* (Key, 1999–2007). The financial importance of gaming content has led to increased involvement of video-game developers and publishers among the committees that handle media-mix projects; by 1998, Tokyo TV ratings reported that 60 percent of their sixty anime broadcasts were sponsored in part by games companies and audiovisual manufacturers (Clements, 2013: 194).

Going Mobile

Nearly all licensed games around the world have historically been adapted to PC, console, or handheld devices that were mostly dedicated gaming hardware. Mobile social video games have a much shorter commercial lifespan of around a decade, though they really began to see exponential growth, particularly in Japan, in 2009 (CESA, 2012). This year also coincided with the iOS introduction of Rovio's *Angry Birds*, which went on to generate over one billion downloads in its first three years (Dredge, 2012). *Angry Birds* also fits firmly in the developing market of the "casual game," or what Jesper Juul has described as games made for a broad audience. Signaling a wider social shift of gamers towards the mobile device and away from PC and consoles like Nintendo Wii, *Angry Birds* was the prototypical casual mobile game for smartphone users: set in a pleasant fictional world; easy to understand, intuitive game design; frequently interruptible gameplay sessions; lenient punishments and loud/colorful rewards (Juul, 2009: 50). Such games are ideal for the between moments—waiting in a line or while riding a train—of an active lifestyle. Indeed, scholars have indicated in survey studies that players of casual mobile games frequently did so out of boredom, even during domestic tasks such as cooking (Hjorth and Richardson, 2014). This type of play is especially familiar to Japanese *keitai* cell phone use, which Kenichi Fujimoto has described as "*nagara* mobilism," where "*nagara* (while-doing-something-else) refers to the state of multitasking separately, in parallel, and asynchronously while walking, moving, or playing" (80). As these scholars would argue, casual games like *Angry Birds* are ideal for mobile devices and the type of players who play them.

Angry Birds, like *Tetris* and *Snake* before it, was heavily responsible for introducing gaming on a new mobile device, in this case Apple's iPhone and its App Store as a platform for cheap and simple games. A large part of *Angry Birds*'s appeal was its hub of media and social networking convergence; Apple connected players via its Game Center social network, getting players interested in each other's in-game achievements in ways similar to status updates on Facebook or photo updates on Instagram. Moreover, the game's lack of a complex narrative, appealing character designs, and relative simplicity allowed it to be "easily adapted, remixed, and inhabited by others" (Leaver, 2016: 219). Because of this convergence across multiple social networks and contexts, *Angry Birds* became a viral transmedia franchise in its own right, spawning multiplatform sequels, licensed tie-ins with film franchises, political internet memes, cartoon fan remixes, and even its own blockbuster CG-animated feature film.

Though *Angry Birds* has come to signify mobile gaming in popular culture, however, there are few other examples of such large transmedia successes. More importantly, it represents a mode of mobile gaming—the single-transaction download—far less common among the top downloaded and grossing mobile games today. Most games in the App Store or Google Play platforms now operate on a free-to-play (F2P) model where the game is free to download and play, but requires players to purchase in-game currency depending on a variety of factors that can extend or augment the playing experience. As Elizabeth Evans diagrams, early F2P mobile games exhibit several similar characteristics: they are often some form of strategy-based simulation; contain a never-ending game world; have short, frequent playing times; and incorporate aspects of sociability as core experiential aspects (2015: 4–7).

A variation of Evans's format applies to one of the highest-grossing mobile games in Japan, *Puzzle and Dragons* (Gung Ho, 2012). The game revolves around puzzles, rather than strategy simulation, where the play of tile-matching determines the player's strength when attacking various monsters. The game's continuous addition of puzzles, monsters, and stages, however, means that play is ostensibly never ending. Each round of play can be as short as a couple of minutes, though this can last longer on more advanced stages. Initially, there was not much sociality involved in single-player mode, though Gung Ho introduced a multiplayer mode in 2015 to increase its social aspect. In-game currency, however, has a large role within the game. Where most mobile games at the time scattered purchases around the game, *Puzzle and Dragons* was unique in that it required players to acquire "Magic Stones" for all transactions (Suzuya, 2014: 20–1). Thus, players had to purchase Magic Stones if they wanted to replenish their health or stamina without enduring a waiting period. By simplifying the transaction system, players knew that, if they wanted a premium experience of more opportunities for longer play and better rewards, they would have to pay for a single in-game currency. Like *Angry Birds*, *Puzzle and Dragons*'s popularity led to its own multimedia experience with adaptations on the Nintendo DS, an arcade version developed by Square Enix, merchandise based on licensed characters, and multiple collaborative cross-promotional tie-ins with other media franchises, including *Angry Birds* (Grubb, 2013). These efforts resulted in over sixty million downloads at the time of writing, more than any other Japanese mobile game, and over $60 million in revenue per month.[2]

The popularity of the franchise—and mobile gaming in general—has inspired many other IP holders to enter the industry in search of similar revenue-generating hits that can prolong or renew the life of transmedia brands. Evans calls such IP-based freemium games "brand components" that amplify and support the value of the IP's brand (2016: 8). The mobile games *Snoopy Street Fair* (Beeline Interactive, 2011) and *The Simpsons: Tapped Out* (Electronic Arts, 2012), for example, remediate the visual and auditory pleasures of the animated franchises on which they are based, incorporating the design and even sense of humor of the original shows into the world of the game. The games also tap back into the series' intermediality, getting players interested in the franchise's characters, animation, and comics. *Snoopy Street Fair* includes a mission where players find pieces of *Peanuts* comic strips, with in-game links to purchase the Charles M. Schulz-authored comic collections, while *The Simpsons: Tapped Out* encourages players to watch specific episodes of *The Simpsons* television series. The games are limited in their incorporation of their respective franchise's other transmedia aspects, such as narrative and worldbuilding, largely because those franchises are episodic in nature and are either no longer serialized (*Peanuts*) or have become so entrenched in society that they have little flexibility with their narratives and worlds (*The Simpsons*).

The Role of Mobile Gaming in Japanese Transmedia

Despite the limitations of the above licensed mobile games, developers have become aware of the potential for mobile games to fit into larger transmedia narratives and worlds. Large game studios moved away from licensed console video games in the late 2000s in response to decreasing return on their significant investment, largely because of audiences becoming aware of the insufficient quality behind these games, as well as the rise and name recognition behind genuine game-based franchises such as *Grand Theft Auto* (Rockstar Games, 1997–present) and *Call of Duty* (Activision, 2003–present). But, in recent years, developers have returned to the movie-game business largely through adapting licensed social and mobile games. Here, the smaller budgets and shortened development time do not work against game design in the same detrimental ways as the console business (Kohler, 2013).

Building on Evans's initial observations of licensed mobile games, this chapter looks at the role that Japanese mobile licensed games play within their transmedia franchises, particularly with regard to the type of experiences that these games provide for players. Japan's development of arcade and console gaming has given way, in terms of sheer labor and revenue, to the mobile sector. While the country has had a long history of developing for on-the-go platforms—particularly within the Nintendo chronology of Game & Watch, Game Boy, the 3DS, and the console/handheld hybrid Switch—most of these platforms used business and gameplay models based on their console brethren.[3] The mobile social gaming arena, itself a development of browser-based gaming that became popular on desktops and Japanese *keitai* cell-phone culture, has grown with the widespread adoption of smartphones and tablets. From 2009, mobile gaming has experienced massive annual growth to the point that, in 2013, more developers had worked on mobile social games than on consoles for the first time in the history of the Japanese games industry (CESA, 2014). In the same year, there were more smartphone mobile gamers (28.61 million) than console gamers (27.83 million) in Japan, with revenues more than triple that of console games (Shin, 2015). With every smartphone sold to a first-time user, the player pool increases exponentially, and this increase in user base also means an increase in games developed for the App Store and Google Play. By 2018, the App Store alone featured over 800,000 game apps (Anonymous, 2018a). The vast majority of these games never find a player, but a few single-handedly have accumulated fortunes for their companies, such as Colopl's *White Cat Project* (2013) and mixi's *Monster Strike* (2013), the latter of which brought in over $1.3 billion in sales in 2017 to rank as the third highest-grossing game in the world (Cowley, 2018). All of these titles are original IP controlled by the studios that developed them. They have cultivated user bases using deep and clever game engines while also incorporating many different monetization strategies.

As developers in Japan seek to hedge their bets in an increasingly saturated mobile games market, however, many have sought partnerships with other studios in an industry primed for consolidation (Toto, 2016). This has also led to many developers partnering with publishers to make games based on well-known franchises to attract an established user base of players. A growing number of these games, which make their way to the App Store's top-grossing titles, are spun off from licensed properties. A survey of the top 100 grossing games conducted on June 1, 2018 shows that forty-nine were based on licensed properties, with the majority of them derived from IP based on anime, manga, or video-game franchises (App Annie Intelligence, 2018). Most of these titles are oriented toward male players in their twenties and thirties because of financial incentives. Free-to-play games tend to rely on high-spending players to support the vast numbers of players who tend not to spend anything on the game (Shaul, 2016). Male social/mobile games players still outnumber female players (though this gap is shrinking) but, more importantly, male spending on these games greatly outpaces female spending, in a similar way to other creative industries in Japan (CESA, 2018: 137, 144). The job-recruitment site Rikunabi Next found that working men spend, on average, more than three times as much on in-game transactions as women (Tanimoto, 2016), particularly in the important "whale" categories of big spenders who shell out over $100 to $500 ARPPU (average revenue per paying user). Very few women spend this amount, let alone exceed it, which dedicated (see: obsessive) male players often do (Anonymous 2018b).[4]

This means that the licensed mobile games market is saturated with creative content geared primarily toward male working-age players, with games based on properties ranging from classic superhero sagas (such as *Kinnikuman*, *Saint Seiya*), modern *shônen* adventure manga (such as *One Piece*, *Naruto*), military robot series (such as *Mobile Suit Gundam*, *Kantai Collection*), late-night *otaku anime* (such as *Fate/Stay Night*, *Sword Art Online*), and mobile iterations/spin-offs of console-based game franchises (such as *Monster Hunter*, *Dragon Quest*). These licensed games also exhibit a range of genres, often within the same game, *à la Puzzle and Dragons*; common mobile game genres such as puzzle, role-playing, battle, strategy, rhythm action, and monster hunting are the most popular genres of licensed games, with many games featuring a hybrid of two or more genres.

Three such licensed mobile games—*Love Live! School Idol Festival* (Bushimo/KLab, 2013–present), *Final Fantasy Record Keeper* (Square-Enix/DeNA, 2013–present), and *Dragon Ball Z: Dokkan Battle* (Banda–Namco/Akatsuki, 2015–present)—were played for over a year and analyzed for their role as game paratexts in their respective transmedia franchises. *Dragon Ball* is a popular action and adventure manga series revolving around fighting warriors and was eventually adapted into multiple animated television series, feature films, console video games, and collectible card games. *Dokkan Battle* is a puzzle–action hybrid where players connect tiles to perform powerful attacks against their opponents. *Final Fantasy* is an internationally recognized role-playing game franchise, spanning multiple console games, spin-offs, animated series, and even two extravagant CGI feature films. *Record Keeper* is a role-playing game in the same mode as *Final Fantasy* games for the Super Nintendo, eliminating the rogue exploration and dungeon-crawling aspects to focus almost entirely on quick-moving, turn-based battles. *Love Live!*, the newest of the three, is an animated idol girl group that originated as a series of magazine profiles and music videos but gained popularity through a late-night anime television series, several music concerts, and a surprise animated feature-film hit. *School Idol Festival* is the least recognized or attractive mobile game outside its home country, though its game is available in the App Store in several different languages. The game itself is in the genre of musical rhythm where players must time their finger taps on the screen to targets that move to the beat of one of several dozen different pop songs.

All three licensed games were selected based on several important factors. First, all the games come from Japanese transmedia franchises that have brand recognition and a built-in audience. Second, all the games originate in different media (for example, manga, video games, animated music videos) providing a good sampling of the range of Japan's content industries as represented in mobile game form. Third, though the audiences for these games are not necessarily mutually exclusive, they do represent different periods and ages of fandom, from the broad generational reach of mainstream titles, such as *Dragon Ball* and *Final Fantasy*, to the more niche, fan-driven idol market of *Love Live!* Fourth, the studios which develop these games are similarly protean in their backgrounds and approaches to game development; both Bandai–Namco and Square Enix have a long history in console-game development, with the former specializing in adapting licensed IP and the latter successfully developing its own series of original IP, while mobile game developer KLab started in web development before applying a shotgun approach to several original and licensed mobile-game titles and striking the jackpot with *Love Live!* Finally, all three games were among the top fifty grossing games in the App Store for over a year, with millions of downloads, showing that the audience was significant and spending enough to keep these games operating at a consistent pace. Thus, this trio of licensed mobile games provides a good cross-section for licensed mobile game paratexts in Japan over the last five years.

The three games are analyzed here as paratexts that shape, or are shaped by, the transmedia franchises from which they are derived. Because of this, my analysis focuses less on their specific game design mechanics or narratives and more on how such franchise elements, such as plots, characters, designs, sound samples, and musical tracks, act as conduits that *amplify* the affective values of their respective transmedia franchises. Jason Ash uses the term "affective amplification" to refer to how games can "generate and modulate between affective states" (Ash, 2012: 12); this means both the ability of games to shift affective states within players (such as the negative feelings associated with death and the positive feelings associated with avoiding it) and the ability of games to maintain such affective states for a long period. In the context of licensed games, however, I exclusively refer to this second ability of "affective amplification" to the games' ability to construct a constant state of positive feedback that players derive from accessing or possessing digital assets of the franchise (for example, collectable characters, playable songs, or even "story sequences"). In contrast to an original game, which must provide reasons for players to value assets such as characters or items through their relationship to the game's story or mechanics, licensed mobile games contain assets whose desirability derives from their associations with their franchise's narratives and worlds. The games are not "additive" to franchises in the sense of providing any new information to those who have experienced it; rather, they are designed around advertising and augmenting the most pleasurable affective aspects of the franchise to those both new and old to it. This is transmedia designed for those on the move: a loop of constant "juicy" fan assets, generating what can be called affective *franchise* amplification.

Unlike console-based games with a defined limit of assets, mobile games can quickly add more juicy assets to their game worlds owing to the presence of operations staff actively managing the game. Mobile games frequently update, with each update incorporating new assets to collect or experience. Thus, licensed mobile games possess a continuously updating supply of assets with predetermined affective value, with no physical limit to the digital assets the dedicated fan/player can possess or experience. All three games examined here amplify the affective values of their franchises in at least five important ways that advertise the franchise to new players, augment key moments in the franchise for its fans, and connect new players and established fans in the same game space. In short, these mobile games turn prior positive memories of the franchise into game rewards that must be consumed and managed endlessly.

Gacha characters

The single most important amplification (and monetization) device for Japanese mobile games is the lottery known as the *gacha*.[5] Originating in vending machines called *gashapon*, which dispense randomly selected toys and other items for a small cost, *gacha* lottery mechanics have a long history of being incorporated into video games as bonus stages and levels. Mobile games have put this lottery mechanic front and center, tempting players to make low-cost purchases through *gacha* charges which typically range from $3 to $5. Players attempt to acquire limited-time sets of digital characters, weapons, or premium items in each *gacha* which is rotated out after a few days or weeks for a new set of collectables. As Kiyoshi Shin notes, Japanese players have also been conditioned to such random-chance mechanics primarily through NTT Docomo's i-mode mobile Internet service and *pachinko* slot machines, and thus rack up three times as much in-game charges as those in the United States (Shin, 2015). Japanese players are also particularly enthusiastic about collecting characters through the success of franchises such as *Pokémon*, *Digimon*, *Yu-Gi-Oh!* and *Monster Hunter*, and avid players fork out huge sums of money every month to collect their favorites. One player spent over $6,000 in a single evening attempting to acquire a character in the Cygames-produced *Granblue Fantasy* (Nakamura, 2016).

For licensed mobile games, the hard work of building value in attractive characters has already been done for them. Thus, these games use the *gacha* to pump out characters and items that fans will desire immediately. The games often make these characters difficult to acquire to encourage increased spending, but almost every game will include multiple versions of the character and/or item. This character-centric model allows newer characters constantly to have value over older characters (for example, newer characters have higher statistical parameters than older characters), but the model also continually encourages players to return to the game to acquire alternative versions of their favorite characters. In practice, this kind of character customization speaks to how fans of Japanese franchises tend to treat their favorite characters: as customizable elements that do not necessarily need a reference point in the narrative. *School Idol Festival*'s main idol protagonists appear in the game in multiple colors and rarity values, and each high-quality iteration of the character is drawn with a different style or theme and provided with a different pose, hairstyle, and wardrobe. While characters are used for attaining ever-higher scores in the game's musical challenges, they function in the same way as bromide postcards for movie or musical stars, donning different forms and outfits that feed into fan desire for a never-ending template of alternative personalities.

Dokkan Battle plays with character collectability as well, with valuable or popular characters in iconic poses from the manga and anime, and in both normal and "superpowered" forms. The game goes a step further, creating "fusions" of characters as if they were combined with one another (a superpower from the series). The game provides original characters that are, nevertheless, based on the mythology of the franchise world. *Record Keeper* characters are not drawn with the same skill and quality as those in the other two games here, opting for digital sprite models that hark back to the 8- and 16-bit generation of many of its earlier titles. The game, however, uses weapons in addition to characters, enabling players to upgrade characters with tools pulled from various games in the franchise. By allowing players to customize characters and weapons across games, *Record Keeper* allows players

Figure 11.1 Gacha mechanics in *Love Live: School Idol Festival* (Bushimo, 2013–present).

to tailor their favorite characters to their personal tastes with the assets of the entire *Final Fantasy* universe.

Narrative iconography

As Trevor Elkington has stated, licensed games will often take two approaches to narratives on which their games are based: either a direct adaptation, or a side-story-like chapter in a transmedia story. Licensed mobile games often take the former path but, while they will stick fairly closely to the original story of the franchise, they will also often take a pared-down approach because of limitations with the hardware. Owing to the limited playtime habits of mobile players, licensed mobile games will often selectively choose key moments or scenes from the franchise and spin limited-time "events" around them rather than attempt direct adaptations of lengthy source material. This is, in many respects, similar to strategies that early "pre crash" games, such as *Superman* (Atari, 1978) and *Star Wars: The Empire Strikes Back* (Parker Brothers, 1982), employed, visualizing isolated scenes and "well-known, iconographic characters" from their original sources because of hardware limitations (Aldred, 2012). For franchises that have been in existence for several years or decades, this means that there is an ample amount of material for developers to mine. In *Dokkan Battle*, entire story arcs are boiled down to nostalgic core elements such as iconic manga panels, character poses, and lines of dialogue. With *Dragon Ball*, these scenes are especially recognizable for fans because much of the original manga's panels provided layouts for the anime. *Record Keeper* takes a detailed approach to this style of piecemeal narrative. Based on a memorable scene or dungeon from one of the franchise's games, events even use the background art and music from the games to provide additional affective pleasure. Both games, moreover, market these audiovisual references to appeal to previous viewers or gamers of the original franchises by telling their players to access famous scenes. *Record Keeper* is especially sly because not only does its advertising tell players to "relive your favorite *Final Fantasy* moments," but the game narrative itself is built on retrieving "memories" of the *Final Fantasy* franchise that have been spirited away by mysterious villains.

In other cases, some mobile games will not completely repurpose famous scenes or art but will, instead, create original stories that mirror or complement a dominant narrative in the franchise. *School Idol Festival* does this well, creating side stories of the established idol characters that occur between episodes of the dominant anime narrative from which the characters derive. Fans consume such small narratives less to understand or experience a specific scene or event, but rather to enjoy their memories of the characters' relationships and the general atmosphere during which the scene took place. Such narratives might reveal or enhance some overlooked aspect of the character's

Figure 11.2 Reliving narratives in *Final Fantasy: Record Keeper* (Square–Enix, 2013–present).

personality or simply provide an opportunity for fans to hear the original voice actress speak new lines of original dialogue.

Serial media coordination

Ideally, licensed mobile games drive sales of other media within the media mix, in ways similar to that in which *Snoopy's Street Fair* promotes *Peanuts* comics within the game. Licensed mobile games usually have the added benefit of running alongside a concurrently broadcasting animated television series. *Dokkan Battle*, for instance, was developed and released to coincide with the release of *Dragon Ball Super* (Toei Animation, 2015–18), the first *Dragon Ball* original animated television series in eighteen years. Produced by Toei Animation, the show featured new characters and settings that were immediately advertised as collectables in the game. With tight coordination between the animation studio and game studio, characters introduced in the television program also soon became available in events for the game. *Record Keeper* created a similar flow of events for the release of *Final Fantasy XV* in 2016 as assets from the AAA game were smoothly incorporated by its own studio into the mobile game as events and rewards. This is a fairly predictable model for media coordination, but *School Idol Festival* took a more ambitious path as players who downloaded the game following the release of the television series were presented with a different version of the game that more closely matched the experience of the show. Musical songs from the television series were immediately made available as playable, as opposed to unlockable, stages while subsequent seasons of the show saw the implementation not only of special new characters in the game, but entire new sets of songs and a redesigned interface to mimic more closely the audiovisual aesthetic of the new television seasons.

Like-minded collaborations

Considering the fairly simple engines of most mobile games, developers employ a number of strategies to encourage continued play. One such strategy is the collaboration, a marketing and acquisition tool where two or more companies become partners to promote one another's games. Characters from one game will appear in another for a limited time, and vice versa, exposing both sets of players to different games. As Serkan Toto points out, the strategy is used to market and acquire "new users for *existing games*—and are not to be confused with the simpler concept of cross promotion" (2013). This can take a lot of coordination and preparation, though collaborations with licensed mobile games can skip this intermediary step by appealing to characters of similar franchises within their own network. For *Record Keeper*, this means running collaboration events for characters and items for classic games under the Square–Enix umbrella such as *Romancing Saga* or *Seiken Densetsu/Secret of Mana*, as the RPGs have followings similar to those of *Final Fantasy*. The events can also function as focus groups for future mobile iterations of such games if players are enthusiastic for the events. *Dragon Ball*'s collaboration comes from the Akira Toriyama umbrella of authored manga but, even more broadly, under publisher Shueisha's wing. This is why the game features collaborations with characters and stories from *Dr. Slump*, Toriyama's first manga, and *One Piece*, Shueisha's all-time best-selling manga. Moreover, as the developers have access to veteran artists at Toei Animation, the game features official artwork that mixes and matches the outfits of the different characters for unique collectable interpretations.

Outside social/promotional events

Licensed mobile games also promote, and are promoted by, events outside of the game world which, in turn, direct players back to the game. Magazines, such as *Famitsu App*, run monthly features that break down special codes for players to receive premiere items. Such codes are also available through in-game missions. Shueisha holds the annual Jump Festa, a convention for fans of *Shonen Jump*, the manga magazine in which *Dragon Ball* ran from 1984–95. The convention brings together manga artists and writers but also holds press conferences, demo sessions, and prize giveaways for new games. *Love Live!* has its own fan conventions and musical performances but these rarely make their way into the game directly. Instead, the game conducts special marketing deals with specific locations such as bookstores and cafés. The game told players, for instance, to visit a maid café to receive a code for a limited-time character if they ordered a similarly limited-time, themed item off the menu. Thus, both café and character were aligned to promote one another,

Figure 11.3 Collaboration characters in *Dragon Ball Z: Dokkan Battle* (Bandai-Namco, 2015–present).

providing fans with a fun (if pricey) side quest in the process. This sort of player activity is ideal for licensed franchises where the game directs players on a series of missions outside of the game world with rewards that inevitably prod players back into the game world in an insidious loop.

Gacha, media coordination/collaboration, and events both interior and exterior to licensed mobile game worlds thus create an endless stream of assets that feed off of player familiarity with the franchise. Able to update the asset database, redesign interfaces, and collaborate or coordinate with media across related industrial networks on a weekly or even daily basis, mobile games are the most flexible licensed media texts for franchises that exist today. While console games are similarly able to update their games through periodic downloadable content (DLC), mobile games in Japan operate according to an event-based design (rather than the loop-based design of most Western games) that requires constant updates. This makes them uniquely able to attune their games to changes in the franchise media, trends in related media, and player activity or interest towards specific characters or scenarios within the franchise.

Regarding this last point, Japanese mobile games amplify affective value for the franchise on two levels: at a premium for preexisting fans of the franchise; and at a free (but deep) sampling for casual players new to the franchise. Fans play the games to access and experience old memories of the franchise but are able to acquire high-quality or upgraded versions of specific elements of the franchise (such as characters, weapons, scenarios) to which they are particularly attached. That these elements are difficult to acquire makes players spend an inordinate amount of time or money on these games, though players may turn against the franchise if their favorite elements are *too* difficult to get. Casual players, on the other hand, are given what is basically a stripped-down version of the franchise for free. Players unfamiliar with the franchise are provided with portals into various attractive elements via the game's interface.

The game serves as a miniature art gallery for the franchise's character and background designs, as a sample playlist for the various songs and performances by voice actors, and as an abridged version of the narrative provided in story modes. Players who find any of these elements attractive will be encouraged to purchase more complete texts in the franchise itself, often directly through the game via banner links set up on the surface of the interface or in splash pages that appear when the game boots up. Casual players interested in experiencing a more coherent narrative, meaningful character development, or richer audiovisual pleasures of the franchise might do so outside of the game world, at a cost.

Conclusion

Japanese licensed mobile games have a varied set of methods in which they use collectable assets to their advantage, but this does not say much about the quality of the play in the games themselves. As mentioned earlier, many of these games are derivative of other popular mobile game genres and frequently have little to do with the experience of the franchise itself. The gameplay often simply serves as a way to facilitate or obstruct the collection of attractive assets; this can be seen, for example, in *Dokkan Battle*, where "battles" are little more than connecting tiles on a game board, with the results of the puzzle playing out in automated sequences of avatars fighting one another.[6] The gameplay, as with most casual games, is meant to be quick and simple, with added depth coming in mastery of the game engine, more intense play during event periods, and a large amount of customization and upgrading available via the game's character-building menus.

What this simple gameplay also means is that hurdles for asset accumulation are low in terms of player skill. If players have deep enough pockets, they can become "gamer" supporters of their favorite franchise. This unlimited access to the franchise's most expensive digital goods is important for emerging franchises where fans are looking to consume licensed material but have few avenues for doing so. This was probably one of the main reasons for the success of the *Love Live: School Idol Festival* mobile game; it created a digital outlet for the anime and idol group that, like Kenner's *Star Wars* toys between the *Star Wars* tentpole movies, allowed fans to continue the franchise's experience between tentpole releases of the films and television seasons. *School Idol Festival* was even more lucrative than material products such as toys or dolls, however, as the game was not limited by supply lines and could provide a constant stream of "product" in different skins, side stories, and songs. Even the idea of the tentpole should be questioned in the case of *Love Live!*, because fans' primary entertainment experience truly depended on the point at which they entered the franchise; if it was between seasonal broadcasts of the anime, fans were more likely to experience the mobile game before any other media. Crafting a transmedia empire without a true mother ship is a win for *Love Live!* media rights holders ASCII Media Works/Kadokawa, but also for ancillary mobile game developers such as KLab who can now hope for tentpoles becoming affective amplification tools for *their* media.

As franchises begin to understand how mobile games can benefit their bottom line, there will probably be increased coordination between the games and franchise's other media texts as they are serialized. This increased fidelity to the franchise makes licensed mobile games that much more "in tune" with the franchise, though in service of a business model that is more about pleasing the fan and hyping the franchise than with any original gameplay experience. Indeed, we are already seeing many of the same strategies employed in licensed mobile games find their way into licensed console movie games. *Star Wars: Battlefront II* (Electronic Arts, 2017) and *Middle-earth: Shadow of War* (Warner Bros. Interactive Entertainment, 2017), for instance, riled fans of the *Star Wars* and *Lord of the Rings* franchises through incorporating a large number of microtransactions into the core game in order for players to unlock all of the game's content and remain competitive in the game's online arenas (Fernandez, 2017). Even less controversial DLC, such as paying to drive a Batmobile from *Batman v Superman: Dawn of Justice* (Zack Snyder, 2016) or a branded jeep from *Jurassic World: Fall Kingdom* (J. A. Bayona, 2018) in *Rocket League* (Psyonix, 2015), shows how tie-in events, promotions, and a stream of additional content have become commonplace among even the biggest franchises. Licensed games are still coming along in terms of the quality of the gameplay but, with their endless variations on the same theme via the mobile platform, they are going down a path that might have been exactly what the franchise's most dedicated fans wanted all along.

Notes

1. Steinberg actually defines the media mix in marketing terms, that which he dubs the "marketing media mix," and in the circulation of its narratives and characters or the "anime media mix." These two definitions both refer to media synergy, but each is historically and conceptually distinct because the former revolves around using media with the end goal of selling a particular advertiser's message or product, while the latter views synergy as a goal in and of itself. With the anime media mix, in other words, the consumption of any media helps sustain the entire media franchise. For the purposes of clarity, this chapter's usage of "media mix" will refer to the definition of the "anime media mix," as it translates most closely to conceptions of media convergence that scholars in other national and industrial contexts have analyzed.
2. These numbers actually reflect a decline in revenues for *Puzzle and Dragons* though it is still continually ranked in the top five grossing apps on a daily and weekly basis despite this fall. To address this calamity, Gung Ho fittingly marketed the game as a "national game brand," with a multimedia marketing blitz of *manga*, *anime*, apps, and toy products to enhance the visibility of the core game. See Watanabe, 2018.

3. While, at the time of writing, the Nintendo Switch is still early in its life-cycle, early signs indicate that Nintendo will treat the system in a similar way to previous consoles and portable devices in the emphasis on revenues being generated by game cartridge and download sales, rather than from recurring in-game micropayments and downloadable content charges.
4. Though industry professionals are quick to cite these numbers, at least a couple of additional caveats can explain this gap. According to the Ministry of Health, Labor, and Welfare's 2018 Basic Survey on Wage Structure, in Japan women earn on average far less than men, resulting in lower average purchasing power. Also, because so few games actually cater to working-age women, few female players might feel justified in spending like whales on the male-oriented games that do exist. Finally, while men in their twenties and early thirties exhibit the highest spending rates (CESA, 2018: 144), women in their thirties and forties exhibit similar spending rates on games as men in the same age range. As more women enter the workforce and maintain careers later into their lives, the industry might attempt to acknowledge and capitalize on this growing field of consumers with licenses that fit their interests. We are already beginning to see this, with male idol group games such as *Uta no Prince-sama: Shining Live* (KLab, 2017) and *Idolish7* (Bandai–Namco, 2015), or sports titles traditionally popular with female players, such as *Captain Tsubasa: Dream Team* (KLab, 2017) and *Prince of Tennis: Rising Beat* (Akatsuki, 2017) regularly appearing among the top 100 grossing games on the App Store.
5. Though *gacha* is associated with, and found in, most Japanese mobile games, the practice has found its way into popular mobile games developed in the United States. *Hearthstone* (Blizzard, 2014) and *Dungeon Boss* (Big Fish Games, 2015) both incorporate *gacha* mechanics, and various licensed games based on franchises such as *Star Wars*, *South Park*, and *Marvel* now feature *gacha* as ways in which to acquire popular and powerful characters. See Kanerva, 2016.
6. It should be mentioned that many previous *Dragon Ball Z* games also use these sorts of puzzle gimmicks, such as playing cards and dice, so one can consider *Dokkan Battle* a continuation of this tradition.

References

Aldred, Jessica (2012), "A Question of Character: Transmediation, Abstraction, and Identification in Early Games Licensed from Movies," in Mark J .P. Wolf (ed.), *Before the Crash: Early Video Game History*, Detroit: Wayne State University Press, pp. 90–104.

Alexander, Leigh. (2013) "The Five Reasons Freemium Sucks (According to QWOP's Developer)," *Indie Games*, 30 March, <http://indiegames.com/2013/03/the_five_reasons_freemium_suck.html> (last accessed 13 September 2017).

Allison, Anne (2006), *Millennial Monsters: Japanese Toys and the Global Imagination*, Berkeley: University of California Press.

Anonymous (2018a), "Count of Active Applications in App Store," *PocketGamer*, <http://www.pocketgamer.biz/metrics/app-store/app-count/> (last accessed 15 June 2018).

Anonymous (2018b), "Planet, smartphone game ni kan suru ishiki chôsa sumahoyûza no 4nin ni hitori ga mainichi geemu de tanoshindeiru" (Planet, Opinion Poll on Smartphone Users Shows One in Every Four Smartphone Users Play Games Everyday), *Social Game Info*, 5 March, <http://gamebiz.jp/?p=205267> (last accessed 15 June 2018).

App Annie Intelligence (2018), App Annie Store Stats, iOS Top App Charts, Japan,

June 1 2018, iPhone, Overall Grossing Rank, <https://www.appannie.com/apps/ios/top-chart/?country=JP&category=36&device=iphone&date=2018-06-01&feed=Grossing&rank_sorting_type=rank&page_number=0&page_size=100&table_selections=&metrics=free_rank,paid_rank,price,category,all_avg,all_count,last_avg,last_count,first_release_date,last_updated_date,est_download,est_revenue,wau&order_type=desc&order_by=grossing_rank> (last accessed 30 June 2018).

Ash, Jason (2012), "Attention, Videogames, and the Retentional Economies of Affective Amplification," *Theory, Culture, & Society* 29: 6, pp. 3–26.

Brookey, Robert Alan (2010), *Hollywood Gamers: Digital Convergence*, Bloomington: Indiana University Press.

CESA (2012), *2012 CESA White Paper*, Tokyo: Computer Entertainment Supplier's Association.

CESA (2014), *2014 CESA White Paper*, Tokyo: Computer Entertainment Supplier's Association.

CESA (2018), *2018 CESA Ippan seikatsusha chôsa hôkokusho: nihon geemu yûzaa & hiyûzaa chôsa* (2018 CESA Report on Investigations into Everyday Consumers: Users and Non-Users of Japanese Games), Tokyo: Computer Entertainment Supplier's Association.

Cowley, Rip (2018), "Tencent's Megahit Mobile MOBA Honor of Kings Generated $1.9 Billion in Revenues in 2017," *PocketGamer*, 30 January, <http://www.pocketgamer.biz/news/67432/tencents-megahit-mobile-moba-honor-of-kings-generated-19-billion-in-revenues-in-2017/> (last accessed 15 June 2018).

Clarke, M. J. (2012), *Transmedia Television: New Trends in Network Serial Production*, New York: Bloomsbury.

Clements, Jonathan (2013), *Anime: A History*, London: BFI.

deWinter, Jennifer (2015), "Japan," in Mark J. P. Wolf (ed), *Video Games Around the World*, Cambridge, MA: The MIT Press, pp. 319–44.

Dredge, Stuart (2012), "Angry Birds Games Have 200m Monthly Active Players," *The Guardian*, 10 October, <https://www.theguardian.com/technology/appsblog/2012/oct/10/angry-birds-200m-monthly-users> (last accessed 22 November 2017).

Elkington, Trevor (2009), "Too Many Cooks: Media Convergence and Self-Defeating Adaptations," in Bernard Perron and Mark J. P. Wolf (eds), *The Videogame Theory Reader 2*, New York: Routledge, pp. 213–37.

Englehardt, Tom (1986), "The Strawberry Shortcake Strategy," in Todd Gitlin (ed.), *Watching Television*, New York: Pantheon Books, pp. 68–110.

Evans, Elizabeth (2016), "The Economics of Free: Freemium Games, Branding, and the Impatience Economy," *Convergence: The International Journal of Research into New Media Technologies* 22: 6, pp. 563–80.

Fernandez, Matt (2017), "'Star Wars: Battlefront II' Controversy Signals Disconnect in Industry," *Variety*, 23 November, <https://variety.com/2017/digital/news/star-wars-video-game-controversy-microtransaction-loot-box-1202621913/> (last accessed 15 May 2018).

Flanagan, Martin, Mike McKenny, and Andy Livingstone (eds) (2016), *The Marvel Studios Phenomenon: Inside a Transmedia Universe*, London: Bloomsbury.

Fleury, James (2014), "'Sometimes the Old Ways are the Best': Ret-Conning in James Bond Video Games," in Michele Brittany (ed.), *James Bond and Popular Culture: Essays on the Influence of the Fictional Superspy*, Jefferson, NC: McFarland, pp. 247–66.

Fujimoto Kenichi (2005), "The Third-Stage Paradigm: Territory Machines from the Girls' Pager Revolution to Mobile Aesthetics," in Mizuko Ito, Daisuke Okabe, and Misa Matsuda (eds), *Personal, Portable, Pedestrian: Mobile Phones in Japanese Life*, Cambridge, MA: The MIT Press, pp. 77–102.

Gray, Jonathan (2010), *Show Sold Separately: Promos, Spoilers, and Other Media Paratexts*, New York: New York University Press.

Gray, Jonathan (2014), "In the Game: The Creative and Textual Constraints of Licensed Video Games," in Denise Mann (ed.), *Wired TV: Laboring Over an Internet Future*, New Brunswick, NJ: Rutgers University Press, pp. 53–71.

Grubb, Jeff (2013), "Finnish Friends: Rovio's Angry Birds Are Teaming Up with Developer Gung Ho to Appear in Puzzle & Dragons," *VentureBeat*, 4 November, <https://venturebeat.com/2013/11/04/rovios-angry-birds-are-teaming-up-with-developer-gungho-to-appear-in-puzzle-dragons/> (last accessed 2 November 2017).

Hartzheim, Bryan Hikari (2016), "*Pretty Cure* and the Magical Girl Media Mix," *The Journal of Popular Culture* 49, pp. 1059–85.

Hjorth, Larissa and Ingrid Richardson (2014), *Gaming in Social, Locative, and Mobile Media*, London: Palgrave Macmillan.

Hori Takahiro (1996), *Nihon no omocha anime wa kore de iinoka* (Are Japan's Toy Anime Good Like This?), Tokyo: Chirekisha.

Jenkins, Henry (2006), *Convergence Culture: When Old and New Media Collide*, New York: New York University Press.

Juul, Jesper (2009), *A Casual Revolution: Reinventing Video Games and Their Players*, Cambridge, MA: The MIT Press.

Kanerva, Taija (2016), "Cultures Combined: Japanese Gachas are Sweeping F2P Mobile Games in the West," *Game Refinery*, 11 February, <http://www.gamerefinery.com/japanese-gachas-sweeping-f2p-games-west/> (last accessed 25 September 2017).

Kawasaki Takuto and Iikura Yoshiyuki (2009), "Ranobe kyara wa tajû sakuhin no sekai no yume wo miru ka" (Do Light Novel Characters Dream of Multi-Work Worlds), in Ichiyanagi Hirotaka and Kume Yoriko (eds), *Raito noberu kenkyû josetsu* (Intro to Light Novel Research), Tokyo: Seikyusha.

Kerr, Aphra (2005), *The Business and Culture of Digital Games: Gamework and Gameplay*, London: Sage.

Kline, Stephen, Nick Dyer-Witheford, and Greig de Peuter (2003), *Digital Play: The Interaction of Technology, Culture, and Marketing*, Montreal: McGill-Queen's University Press.

Kohler, Chris (2004), *Power-Up: How Japanese Video Games Gave the World an Extra Life*, Indianapolis: Brady Games.

Kohler, Chris (2013), "Why Games Based on Movies Disappeared (And Why They're Coming Back)," *Wired*, 15 February, <https://www.wired.com/2013/02/movie-games/> (last accessed 4 March 2018).

Leaver, Tara (2016), "Angry Birds as a Social Network Market," in Michele Willson and Tara Leaver, *Social, Casual and Mobile Games: The Changing Gaming Landscape*, New York: Bloomsbury, pp. 213–24.

Margaret, Liraz (2015), "Why Can't We Stop Playing Games On Our Phones?," *The Next Web*, 29 July, <https://thenextweb.com/insider/2015/07/28/why-cant-we-stop-playing-games-on-our-phones/> (last accessed 10 October 2017).

Nakamura Yuji (2016), "$6,065 Spent in One Night Shows Dark Side of Japan's Mobile Games," *Bloomberg*, 10 March, <https://www.bloomberg.com/news/articles/2016-03-09/-6-065-hunt-for-blonde-avatar-exposes-dark-side-of-japan-gaming> (last accessed 2 November 2017).

Shaul, Brandy (2016), "Infographic: 'Whales' Account for 70% of In-App Purchase Revenue," *Adweek*, 1 March, <https://www.adweek.com/digital/infographic-whales-account-for-70-of-in-app-purchase-revenue/> (last accessed 15 June 2018).

Shin Kiyoshi (2015), "Smartphones Bring New Golden Age for Japan's Games Industry," *Nippon.com*, 9 April, <https://www.nippon.com/en/in-depth/a04102/> (last accessed 12 September 2017).

Shokrizade, Ramin (2013), “The Top F2P Monetization Tricks,” *Gamasutra*, 23 June, <https://www.gamasutra.com/blogs/RaminShokrizade/20130626/194933/The_Top_F2P_Monetization_Tricks.php> (last accessed 13 September 2017).
Spence, Kevin (2013), “Candy Crush Represents Everything that’s Wrong with Mobile Gaming,” 4 April, <http://www.gameskinny.com/1b054/candy-crush-represents-everything-thats-wrong-with-mobile-gaming> (last accessed 10 October 2017).
Steinberg, Marc (2012), *Anime’s Media Mix: Franchising Toys and Characters in Japan*, Minneapolis: University of Minnesota Press.
Suzuya Nidaime (2014), *Anata wa naze Pazudora ni hamatta no ka* (Why Did You Get Sucked into Puzzle and Dragons?), Tokyo: Futabasha.
Tanimoto (2016), “Dansei wa josei no x-bai mo geemu ni kakin!? Bijinesupaason no sumahogeemu riyô jittai chosa” (Men Spend X-times More on Games Than Women!?, Factual Investigation into Smartphone Game Use Among Businesspeople), 24 August, <https://next.rikunabi.com/journal/20160824/> (last accessed 12 April 2018).
Tanz, Jason (2011), “The Curse of *Cow Clicker*: How a Cheeky Satire Became a Videogame Hit,” *Wired*, 11 December, <https://www.wired.com/2011/12/ff_cow clicker/> (last accessed 10 October 2017).
Toto, Serkan (2013), “Examining a Unique Marketing Tool for Japanese Mobile Games: ‘Collaborations,’” *Kantan Games*, 2 May, <https://www.serkantoto.com/2013/05/02/marketing-user-acquisition-japanese-mobile-games-collaborations/> (last accessed 8 December 2017).
Toto, Serkan (2016), “The Major Trend in Japan’s Mobile Game Industry Now? Consolidation,” *Kantan Games,* 28 January, <https://www.serkantoto.com/2016/01/28/consolidation-japan-mobile-game-industry/> (last accessed 8 December 2017).
Tsugata Nobuyuki (2004), *Nihon animation no chikara: Hachijûgo-nen no rekishi wo tsuranuku futatsu no jiku* (The Power of Japan’s Animatoin: Two Axes Running Through 85 Years of History), Tokyo: NTT Shuppan.
Watanabe Takumi (2018), “‘Pazudora’ yuuza hangen! Gungho no shounenba” (“Puzzle Dragon” Users Halved! Gung Ho’s Crucial Point), *Toyo Keizai*, 3 February, <https://toyokeizai.net/articles/-/207320> (last accessed 3 February 2018)
Wilson, Michele and Tama Leaver (eds) (2016), *Social, Casual and Mobile Games*, London: Bloomsbury.
Wyatt, Justin (1994), *High Concept: High Concept: Movies and Marketing in Hollywood*, Austin: University of Texas Press.
Yokohama Yuji (2006), “‘Shinseiki Evangelion’ ni okeru monogatari sekai no kôsei: Media Mikkusu sakuhinron no kanôsei” (The narrative world and composition of *Neon Genesis Evangelion*: The potentiality for a media mix production theory), in *Nihon Bungaku* (Japan Literature), January, pp. 51–61.

12. LOCATING ESPORTS SPECTATORSHIP—STUDIO AUDIENCE(ING) AND SITES OF SPECULATION

Alexander Champlin

The year 2017 marked a landmark for esports (that is, electronic sport, organized video game competitions) in general and for the Los Angeles-based heart of the North American esports industry in particular. Most significantly, it saw the launch of LA's second dedicated, proprietary esports studio, the Blizzard Arena in Burbank, which would host the newly minted *Overwatch League* (*OWL*), as well as tournaments for *Hearthstone* (Blizzard Entertainment, 2014), *Heroes of the Storm* (Blizzard Entertainment, 2015), and *StarCraft II: Wings of Liberty* (Blizzard Entertainment, 2010). The most novel of these, Blizzard's *OWL*, is a tournament modeled on sports franchises and tied to particular cities around the world.[1] The same year (2017) also saw the reorganization of one of the most successful esports leagues, the North American and European *League of Legends Championship Series*, which adopted a new model of sports franchising in preparation for its 2018 season. Also based in Los Angeles, the *North American League of Legends Championship Series* (*NA LCS*) has been the perennial esports franchise for half a decade.[2] Its reorganization moved away from a model that rotated teams in and out of competition (a relegation system) in favor of a permanent structure whereby sponsor–owners buy franchise spots and grow these teams over subsequent seasons—a move with very visible buy-in from major, and crucially, traditional sports franchises (Garren, 2017).

In recent years esports competitions have grown in scale and audience, spawning sophisticated media franchises that operate on the scale of television production or organized sport. Esports first developed at the turn of the twenty-first century in Korea where they grew out of a juncture of cable television and Internet café culture. Between 2008 and 2012, emergent media platforms, such as Twitch and YouTube, as well as improvements to broadband infrastructures, gave rise to a global esports scene that has been growing steadily since. Both the launch of Blizzard's *OWL* and the restructuring of Riot's *LCS* portend a great deal about the future of competitive gaming in North American and global markets. In short, they signal an interest in concertizing, formalizing, and soliciting investment for permanent and durable esports organizations, infrastructures, and markets. These franchises are big; Riot sold permanent positions for $10 million to existing teams and $13 million to new teams, and Blizzard was rumored to have charged $20 million for its franchise spots (Kishner, 2017). What differentiates these franchises from traditional sports teams, though, is the management of the league as a whole falls to Riot and Blizzard, respectively, rather than franchise owners forming a commission. The economy of the league is driven primarily by in-game microtransactions and bolstering player bases (Volk, 2016).[3] This means that game developers and the franchise owners buying into these leagues are forming relationships around the perceived potential of this media phenomenon rather than fully established revenue streams.

These moves by Blizzard and Riot Games are timely. Riot has grown its *League of Legends* (*LoL*) franchise as a proprietary esport since 2011, turning *LoL* into a kind of miraculous dinosaur in the world of online gaming. Released at retail in 2009, its continued success has been the product of a Sisyphean process of patching, updating, and remaking the same game to keep it fresh and current. Blizzard's *Overwatch* has had a different trajectory. The game launched in 2016 to outstanding reviews and seemed poised out of the gate to evolve the MOBA[4]/Shooter/esport landscape but has struggled to find an audience as an esport (Grayson, 2017; Stubbs, 2017). Blizzard's *OWL* has attempted to develop an esports franchise that can revise and make good on the game's speculative potential. In both cases, Riot and Blizzard's moves to implement systems of franchising, to develop esports localization, and ground this in the studios where these competitions take place signal a desire to reinvest in and transform their proprietary titles into lasting and bankable media franchises. The current moment belies an attempt to cordon off, stake out, and preserve markets that have been identified. The future for both of these endeavors is uncertain but, together, they embody a kind of politics of their moment. They attract audiences who resemble sports fans but engage differently. They stake out new areas of content production and consumption, making use of post-broadcast Internet streaming platforms to reach their viewers. And they skirt commercial frameworks based on advertising in favor of new (and crucially) internal-facing economies of synergy[5] and anticipation.

In this chapter, I examine the history of the *NA LCS* studio in Los Angeles—

North America's first dedicated, proprietary studio for esports production—and, to a more minor extent, the Blizzard Arena. My research is based on five years of regular site visits to the *NA LCS* studio as well as site-specific research at other permanent esports studios—specifically three KeSPA (Korean e-Sports Association) studios, and the Blizzard Arena in Los Angeles. I use the history of the *NA LCS* studio's development, the broader growth of the LCS, and of esports (conceived generally) to contextualize developments in "audiencing"[6] and marketing around esports in North America. By "audiencing," I mean infrastructures that hail and instruct audiences. The history of this half decade in esports might best be understood through a political economy of esports audiences and the practices that hail and house these audiences. The esports studio, a bastion of this political economy, is where esports producers distill and reflect their viewer base in a carefully constructed media space with the aim of informing and directing audience behavior and loyalty.

My work complements and elaborates on the findings of other scholars who have taken up questions of audiencing around esports. Nicholas Thiel Taylor's study of two *Major League Gaming* (*MLG*)[7] events, one in 2008 and another in 2012, concludes that the studio audience for these events increasingly serves as a referent for the online audience who see themselves reflected in the crowd (Taylor, 2016). In the development of esports studios in the United States, this is palpable. In their promotion of the *LCS* studio, for example, Riot foregrounds this function: "[W]e knew how much the energy of a passionate crowd enhanced the experience for pros and fans watching at home. We created the opportunity for fans to gather and watch games live from the studio, bringing new energy to the weekly broadcasts" (Schmiedicker and Stewart, 2017). While I am inspired by this, I think the question of audiencing goes deeper. Benjamin Burrows and Paul Ramma argue that esport and livestreaming platforms serve as testing grounds for new patterns of production and new technological affordances that permit deeper interactive engagements, and I believe this position clarifies the significance of audiencing further (Burrows and Ramma, 2015). My position follows from an observation about Riot's profit model: The *NA/EU LCS*, Riot's wildly successful and widely emulated esport production, operates at a loss (Van Allen, 2017; Sinclair, 2016). What this suggests for audiences and production is not simply the preeminence of an online viewer but that imperatives for production are about influencing these viewers' behaviors in a synergistic way. It is a move away from marketability towards a kind of audiencing more managerial in scope.

Esport Foundations

Video games is a medium that has existed for over half a century,[8] and even the earliest iterations had agonistic elements.[9] The capacity to compete with other players in a game lends itself directly to more formalized kinds of competition. Julia Hiltscher's "A Short History of eSports" compiles a helpful gloss of early gaming competitions. In 1972, Stanford University held the first formalized video gaming competition, an "Intergalactic *Spacewar* Olympics" (Hiltscher, 2015). Iterations of video-gaming competition would follow as games became a household media form. Atari organized a *Space Invaders* Championship in 1980. Between 1982 and 1984, the television show *Starcade* (1982–84) featured players competing for high scores in popular arcade games. One of the most valuable pieces of video-game history is the custom cartridge used in the 1990 *Nintendo World Championships*, which features competition levels of several Nintendo Entertainment System games designed specifically for this event (Hiltscher, 2015). In the 1990s, following the home console boom of the late 1980s, several television shows, such as *Nick Arcade* (Nickelodeon, 1992), featured video-game competition.

Geologically, these tournaments and game shows demonstrate a long history of gaming competition. In terms of scope, scale, and presentation, however, they reflect an era markedly distinct from contemporary esports. What is salient about the modern wave of esports is an attendant media industry that markets this competition to an audience of spectators. After 1999, esports came to represent a more sophisticated and coordinated effort to turn video-game play into a spectator commodity deeply indebted to the competitive gaming boom in Korea. We could imagine roughly two eras in the history of video-game competition. This is not definitive but it is a useful way of bracketing competition. The first would be the period up to 1999, when players competed in video games, but sporadically, often informally, and when it was about spectatorship, it had more in common with game shows than spectator sports.

The second era would begin roughly with the Korean esports phenomenon of the late 1990s and early 2000s. Dal Yong Jin traces its genesis to a confluence of factors (2010). In the 1990s, Korea began a rapid development of its broadband infrastructures. This coincided with the late-1990s Asian financial crash which found a generation of young people out of work and occupying Internet cafés as spaces to search for jobs. The Internet café, called "PC Bangs" in Korea, served as a social space where people gathered to work but also to play games in their downtime. It was in these PC cafés that *StarCraft: Brood War* (Blizzard Entertainment, 1998) became an esport urtext and spawned formalized competitions, esport studios, and cable channels dedicated exclusively to video-game competition. The production model that modern esports studios seek to emulate was born in Korea just at the turn of the century.[10]

Outside Korea, where fan bases were more diffuse and broadcast was not formalized, the esport industry developed more slowly. The primary venues for large-scale video-game competition were esport tournaments. And, while these grew gradually in the late 1990s and 2000s, they never reached the kind of saturation point or durability of the Korean industry. To dismiss the international video-game competition industry during this period would ignore the global interest in esport preceding and following the Korean boom. The *Red Annihilation Quake* event held in 1997, for example, drew national attention in the United States (Davidson, 2018). Crucially, though, these events took place in temporary spaces such as hotels and convention centers. While the United States and Europe never developed cable channels for games or studios in the same way as Korea, the launch of what would become the *Twitch.tv* video livestreaming platform in 2011 introduced an alternative to this broadcast infrastructure outside Korea. Twitch aggregated video-game content on a single platform, circumventing the need for older kinds of broadcast infrastructure. This meant esports and esport-adjacent content, such as speedrunning or video playthroughs like *let's plays*, which had circulated online within their corresponding niches, now had something like a hub to collect and catalyze an interest in game spectatorship.

Twitch collected a range of spectator-game content, from livestreams and pen-and-paper roleplaying, to speedrunning and esports competition. It was an alternative to proprietary streaming platforms where esports used to circulate outside of Korea. *StarCraft II* events, for example, previously aired only on GOM TV's streaming interface, which was clunky and required a downloaded software package. In the absence of cable channels, livestreaming platforms created a centralized place where esport content could develop. This allowed esports to draw more consistent audiences globally, made tournaments more visible, and allowed audiences from adjacent niches to find content complementing their interests.

Interestingly, the growth of Twitch coincided with a global economic collapse, a wave of emergent streaming media platforms, and the launch of *LoL* in 2009 and *StarCraft II* in 2010. In some ways, it retraced the ecosystem that boosted Korean esports production, only on a more diffuse global scale. The global economic collapse of 2008 meant that when Twitch launched in 2011, its core demographic (predominantly young men) was looking at an unemployment rate nearly double what it had been three years earlier. It was also joining a wave of streaming platforms that were changing media consumption patterns.[11] Collectively, this produced a mixture of a broadened player base, new infrastructures that conditioned this base, and excitement around new competitive games, all within an economic climate making emergent media strategies more appealing. This was a boon for Twitch, for esports productions, and for the game publishers taking a managerial stake in competitions.

On the LCS Studio—A Monolith to Modern Esport Spectatorship

While esports had been growing steadily following buzz around *StarCraft II*, the launch of Twitch, an expanding number of international esports tournaments, and a general maturation of video gaming within the cultural register, the esports industry outside Korea was still very young and prone to fluctuation. In *Raising the Stakes: E-Sports and the Professionalization of Computer Gaming*, T. L. Taylor recounts having to keep track of tournaments on sticky notes posted to her office wall because new tournaments emerged and became defunct so quickly that keeping abreast of the current milieu became an ongoing task (2012). A number of prominent tournaments and leagues were being shuttered or relocated, with the Cyberathlete Professional League, World Series of Videogames, and Championship Gaming Series all folding between 2008 and 2009. In an article on esports in the aftermath of the recession, former professional gamer Marc-Andre Messier criticizes structural weaknesses in the industry for this decline, namely an ambiguity about the value of esports commodities (2015). By late 2012, though, when the *LCS* officially launched, esports was recovering strongly, more than quadrupling the number of tournaments hosted in 2008 (Popper, 2013).

By 2013, *LoL* was already wildly successful (Tassi, 2013).[12] But as a relatively young esport, *LoL*'s commercial potential as a spectator text was still being tapped. *The NA LCS* had just concluded what had been dubbed its "third season" with a record-setting championship held in LA's Staples Center (Pitcher, 2013; Sarkar, 2013). "Season 3" was something of a misnomer. The years 2011 and 2012 saw competitive esports play of *LoL* capped with "World Championships" but competitive play leading up to the championship was attached to larger tournaments and events like *PAX, MLG* and the *ESL*.[13] Season 3, in 2013, marked a departure from this format; for the first time, Riot Games took a managing role in organizing competitive play leading up to the finals, at least in the US and European markets. Korea had been running its own *LoL* competitive series since 2012 with weekly competitions held at the OnGameNet (OGN) e-stadium in Seoul, South Korea under the joint supervision of Riot and the KeSPA, a Korean state agency that regulates and oversees esport.[14]

The 2013 North American and European competitions sought to emulate the Korean format in the West. To do this, Riot leased two small sound stages—one in Cologne, Germany and one in Culver City, California—to host weekly competitions.

These two original studios aimed for a minimal kind of mediamaking designed for a rudimentary production of esports competitions. Larger esports competitions featured live audiences, but these early studios used in the production of the *LCS* did not accommodate a live studio audience (Kelly, 2011).

There was just enough space for the two competing teams, referees, cameras, and esports-casters.[15]

For the sake of competition in a virtual playing field, the material space of this competition did not need to be robust. It was visible only in brief moments at the beginning and end of competitive play as teams geared up for battle or reacted to the outcomes of the game. At a time when this production was more experimental than proven, it made more sense to develop it in an environment catering only to the obvious audience—the viewers tuning into the online stream.

And online, these streams were quite successful. Regular season play on Twitch for Season 3 consistently drew over 100,000 concurrent spectators, with many more tuning in via other platforms like YouTube and Azubu.tv. Though income from these streams did not cover the cost of this production, the success of *LCS* streams grew the profile of *LoL*, feeding fan participation in the game itself (Volk, 2016). The streams became a kind of synergistic marketing device to complement the world of the game, encouraging players to see themselves in relation to professional players.

Coincidentally, Riot reorganized the player ranking system within *LoL* in January 2013 to match the launch of *LCS* Season 3. The system eliminated a universal ladder that ranked all players together and implemented a tiered rating system with players moving between ranks from "bronze" at the lowest level to "challenger" at the highest.[16] In an official forum post about the change, Riot explained:

> We decided to move to the new league system for a few reasons. For starters, having a single ladder with all ranked players doesn't provide a lot of incentive for advancement. When you're ranked 290,000 and have 289,999 opponents left to pass on the way up, that process can seem meaningless and interminable. (Yegg, 2013)

This made competition more legible, but it also coincided with Riot's formal foray into esports. Its Challenger League was a feeder league for the *LCS*, meaning that top-ranked players could see themselves in direct relationship to professional teams. This also corresponded to a reworking of the timing of the game's ranked seasons to coincide roughly with the seasons of *LCS* competition.

In this initial year of esport production, there was a synergistic relationship developing between Riot's esport media and the world within *LoL*. While Riot had yet to implement a studio audience, it was already considering audiencing in other ways, including training its player base to see itself in relationship to its professional players. The reorganization of its ranking system better to reflect individual player relationships to professional play and the adoption of a season schedule in *LoL* that followed *LCS*'s timing acted as a kind of virtual audiencing that incentivized esport spectatorship and used esport to encourage game players. Its esport production placed *LoL* consistently near the top of Twitch's games directory, especially on days when *LCS* competition was running, building the game's profile among gamers who otherwise might not have been inclined to adopt *LoL*. The world of *LCS* esport also had an impact on the *metagame* in *LoL* as players emulated strategies used in professional play.[17] And the growth of a highly visible professional scene encouraged participation around *LoL*'s primary profit model, of purchasing in-game cosmetic items for real money—a microtransaction ingame economy (Altay, 2017).

The payoff for this preliminary year of esport production culminated with the World Championship for *LCS* Season 3. It drew thirty-two million online viewers across all of its streaming broadcasts in addition to 13,000 venue attendees (Sarkar, 2013). Where the *LCS* in its regular season drew consistent viewership, averaging in hundreds of thousands of concurrent Twitch viewers, the finals event demonstrated the vitality of *LoL* as an esport and validated Riot's move to manage its esport competition internally.

In December 2013, I visited a small sound stage on the MBS (Manhattan Beach Studios) Media Campus in Manhattan Beach, California. It was a collection of temporary studio spaces for rent. In a sound stage near the middle of the campus, avid video-gamer fans gathered for the 2013 *League of Legends Battle of the Atlantic*, an exhibition tournament pitting North American and European teams against one another. While the event was novel, it was also a landmark moment for the future of the US esports industry. The studio where this event was held served as Riot's semi-permanent studio for the subsequent season (Season 4) of competition in the *NA LCS*. The MBS studio was one of the first semipermanent venues for esports competition outside of Korea, and the first to hold regular spectator video-game competition for a studio audience. It was an upgrade from a small studio in nearby Culver City that had been home to the *LCS* for Season 3. Rather than a space for one-off tournaments, the MBS studio would serve as the centerpiece of North American competition over the course of the next year.

Following the success of Season 3's World Championship, Riot opted to build a semipermanent studio for its ongoing esports production. In their history of the Battle Arena, Riot's Bear Schmiedicker and Dave Stewart, lead producer for the *NA LCS*, explain, "[W]e knew how much the energy of a passionate crowd enhanced the experience for pros and fans watching at home. We created the opportunity for fans to gather and watch games live from the studio, bringing new energy to the weekly broadcasts" (2017). The MBS studio still temporarily reflected a need to identify and contextualize the practice of watching. *The Battle of the Atlantic* was one of the first live events staged in this new space. When viewed by the online audience, the broadcast made tactical use of the studio's affordances. In transitions leading up to play, a boom camera demonstrated the presence of the live audience (EpicSkillshot—LoL VOD Library, 2013). And the audience plays a large part in the look and energy of these competitions. They are given light-up boomsticks, they carry signs, and they wear clothes and costumes celebrating the game. At *The Battle of the Atlantic*, studio crew instructed members of the audience wearing *LoL* clothing and carrying signs to sit in the front-center section of the audience, with the intention of having the camera focus on them during transitions.

Offstage were two other parts of the production apparatus: a caster's booth where the game's commentators provided live, play-by-play analysis, and a second space used for commentary during transitions between games. While out of the audience's line of sight, the shout-casters were still present in the space of the audience, occupying a platform in the back of the room where they could watch the game and where the audience was visible to the online viewers. In moments where the broadcast switches to sportscaster commentary, the game screen and the auditorium remain focal. Online viewers see the stage and hear the audience behind the casters, with signs and players on stage frequently visible as the casters speak.

This studio, while temporary, was a robust media apparatus designed for serialized weekly production. The space changed the pace of production and the consistency of esports competition against an industry focused more on eventized competition than consistent and centralized mediamaking. *The Battle of the Atlantic* and an All Star Event in winter 2013–14 would serve as testing grounds for weekly *LCS* production in the MBS studio beginning in early 2014. The payoff was a practice of esports consumption connecting fandom more directly to the lifespan of the game. Dave Stewart says this model is geared towards building a fan base with a strong relationship to the game which will appreciate in value and return to *LoL* to play games of their own, buy cosmetic items, and grow the player base/fan base (2018).

Playing up Presence

In 2015, Riot launched the successor to its MBS studio, one directly across the street from its LA headquarters. This LCS Battle Arena, constructed in what had most recently been a Hulu television studio, would undergo a series of structural changes, but served as the permanent production site for *NA LCS* competition going forward.[18] While it continued to produce esports media in the same format as the MBS studio, its construction and implementation reflected an understanding of what audiencing meant for the *NA LCS*, especially pertaining to the experience of the studio audience. For viewers watching on a livestream, the translation of gameplay into spectator text is fairly seamless, where a television, phone, or computer screen corresponds to the natural

Figure 12.1 The most recent iteration of the *NA LCS* Studio; secondary screens annotate gameplay and emulate the look of the game world.

perspective of playing the game. By comparison, a studio is an unnatural space for video-game spectatorship. As Taylor suggests, the live studio audience is deployed for the benefit of the online viewer; its members serve as counterparts and points of reference for online spectators. But unlike sports where the action exists on the field of play and in a stadium, the esports live audience does not experience this, as gameplay also has to be represented to the crowd using screens and sports commentary. The studio begins to emulate the look of the game and annotate the experience of watching for the crowd. Where the development of the modern esports studio is broadly about hailing an online, livestreaming viewer, the development of these studio spaces has also demonstrated a more robust understanding of how an experience of presence and immediacy might be developed specifically for this studio audience.

What I would like to focus on here is audiencing in and through the space of the studio itself. Where the move towards a serialized, weekly format and studio-based production reflected a kind of audience-building practice aimed at a broadly conceived online market, the subsequent changes to the studio reflect a deepening understanding of the experience of presence in the studio itself, and how this presence contributes to the cultivation of synergistic fan practice. In keeping sight of the theme of speculation, these developments are not reducible to a singular goal or collective effect. Instead, they function rhizomatically, as minor adjustments that build out different parts of the spectatorial or spatial experience of the whole. What gets implemented are minor practices that seek to trade up toward a kind of collective threshold for audience engagement.

Riot's implementation of the LCS Battle Arena introduced a number of additions targeted at the studio audience, both to make it function like a sport arena and also to make the game-watching experience more immersive.[19] The launch of the Blizzard Arena in Los Angeles two years later also suggests a telos for this emergence, adopting and adapting many of the adjustments made to the LCS Studio. I see these developments as experiments with branching ways of reaching and engaging the studio audience.

I should like to focus on the refinements to the design of the studio space in particular. First, the development of an emphasis on player and team narratives, especially those rooted in the space of the studio itself, is significant because these narratives orient the experience of watching games and work towards a kind of locating, turning the studio into a destination or a physical node in the fandom. The development of esports stories and later of esports team franchises roots this production in legacies of competition, solidifies team positions, and ties this media to specific organizations and places. But these are changes also visible in the look of the studio, as they introduce a contemporary legacy or history for esports fandom on which to focus. And second, the development of visual adaptations augments the experience of studio video-game spectatorship. These adjustments to the studio affect the spectatorial experience of the studio audience, especially how the virtual space of the game is represented to this audience. These developments recast the virtuality of esports as the building blocks for specific sites, specific histories, and specific viewing practices.

When Riot built its permanent studio, the *LCS* was entering its fifth season and its third as a weekly, serialized esport based in permanent studios. As such, the *NA LCS* had developed a significant history from which to cite. This legacy of competition became a way to develop and ground fandom around specific teams, players, and regions. One of the most striking changes to the studio space was an emphasis on this history. In the move from the MBS Studio to the Battle Arena, the decor of the studio changed from game characters to images of past teams, past events (such as the World Championship), and star players. The design change shifted focus to the competition and the players as emergent celebrities and emphasized the corresponding teams and franchises associated with competition. In breaks between games, fans at the studio would line up in the parking lot to meet, take photographs, and shake hands with players. Dave Stewart explains that "people are interested in the stories," and narratives about players, teams, and rivalries make the fandom compelling (2018). As the lead producer for the *NA LCS* since Season 4 and a former producer and writer at Fox Sports, he speaks to the narratives sports media and emphasizes: player drama, legacies, and the sports stories of the game. This emphasis on narrative and character became part of the studio's design as well. It gestured retrospectively to the legacies of competition and concurrently to the players as the embodiment of this narrative. In subsequent seasons, the LCS studio would add championship banners commemorating major events the teams had won. Building both the legacies of competition and emphasizing team fandom over a general interest in esports, the studio served as a site around which esport spectatorship was organized, as well as an arena in which to develop and ground the durable kind of fan practice it was cultivating.

When Blizzard opened its studio in late 2017 to feature *Overwatch* competition, the *OWL* did not have the same legacy of competition. Nonetheless, the studio opened with regalia featuring its franchised teams and their star players. At the time of my first site visit, about four weeks into its inaugural season of competition, the studio was decorated with logos of the competing teams, and monitors around the lobby streamed video highlights of past matches. Other screens listed team rankings and match histories. Just as the *OWL* was launching, it was also generating its own highlights and history for the sake of fans visiting the studio, turning its emergent production into media with a legacy and weight. It was generating sports statistics and data to concertize and enable the kinds of knowledge diehard fans of a sports franchise would be able to recall. The effect of reciting its history through the studio space is to play down the nascent precariousness of this kind of media production. Though the *OWL* is new, this emphasis on statistics, sports data, and legacy suggests ways for fans to invest in the culture of esports. It reproduces the familiar look of sports reporting in the studio itself. In this case, the esports studio oscillates between a place where this legacy of competition is made and where the same history is cited as fodder for fan practice.

With both the *LCS* and the *OWL,* this practice of reciting esports stories through the studio space begins to encapsulate some of the virtuality of esports production. Just as the studio becomes a site to verify and represent the audience to itself, it also becomes a place to verify and produce the narrative of esports sportiness, as well as counter the immateriality of gameplay. For Blizzard's *OWL* in particular, the immediate shift from event to history compresses these stories so that events become immediate for esports reporting. The production of an esports history is a hope for the possibility of a longer history. The *OWL* is so new and untested that the citation of events as immediate history reflect a hope that this kind of legacy might be possible around the game. Or, to put it more modestly, the studio itself begins to do the work of producing fan knowledge and fan practice, equipping any visitor to the studio with the insider knowledge of a dedicated sport fan. These practices mimic sports reporting but situate it within the design of the studio space.

On a larger scale, this is also what takes place around both leagues' franchising systems. The hopeful or speculative use of permanent franchises is meant to signal the durability of esports as these developers are moving to enact a structural shift to make these productions more durable. The creation of permanent teams produces nodes of fan practice. In 2018, the *LCS* restructured its league system to give teams permanent spots rather than rotating the lowest-ranking teams out of the league at the end of a season. In the past, Riot had funded teams through stipends, and teams would make up the difference through branding, merchandising, and sponsorship. The new model reorients production so that all profits are pooled and split to support teams, broadcast production, and pro-player salaries. In a post on the change, Riot explained, "We see this as a positive step forward for the league for multiple reasons, but mainly because teams will no longer have to juggle uncertainty while planning for their future" (Riot Games, 2017). The rationale for the change is posited as a way of mitigating uncertainty for teams and encouraging long-term investment in their brands. While this is true, it also coincides with an attempt to make the *LCS* as a whole more dependable for Riot and more legible to fans. For Riot, the *LCS* served as a device to drive game participation rather than as a distinct revenue stream in its own right. The old model left the work of producing profit largely to teams who received support from Riot but had to find most of their support from advertising and sponsorship. The new model solidified team positions, allowing them to cultivate legacies that would generate returns in coordination with Riot's own production and established revenue streams, and thereby consolidating franchise investment. Specifically, Riot hopes that professional players will become the material loci where advertising and fandom collect in the service of the *LCS* as a whole.

When Blizzard launched the *OWL*, they adopted a similar model of franchising, with some crucial amendments. Notably, the *OWL* was an entirely new production, and any investment, from either Blizzard or its partners, was entirely speculative. Investors were anticipating the potential value of a partnership with Blizzard Entertainment and the potency of esports as a booming area of development. In a similar way to Riot's *LCS*, franchised teams were meant to be a point of identification for fans; only these franchises were tied to specific cities in a way that sought to make this fandom geographically particular. The teams in the *OWL* correspond to global cities with names such as "The New York Excelsior," "Seoul Dynasty," and "Los Angeles Gladiators." Where permanent franchises seek to give fan practice durable points of reference (and to underwrite esport production), these franchises also sought to create a kind of identity of place around these games. Mirroring traditional sports teams, many of which were the principal investors behind these esports franchises,[20] this enactment of team franchising is a kind of artificial locating. Though the capital for these teams in many cases comes from owners with geographic ties to the cities they cite, all of the teams participate in competition in Los Angeles, and their team rosters include players who do not necessarily share any relationship to the cities they represent. The London Spitfire, for instance, employed professional players only from Korea. Instead, this approach to affiliating franchises with cities aspires to a global appeal. Like an attempt to use the studio as a site simultaneously to generate and recite esport history, the geographic naming of franchises attempts to insert place into a media production that, on so many levels, exists in the absence or near-absence of location (Webster, 2018).[21]

Permanent franchises and franchises attached to notions of place reflect an effort to give nascent forms of fandom around esport something to latch on to. Because these teams are new, however, and because they share only a nominal connection with place or history, the studio itself becomes a physical site around which these things can be grounded and represented. Geographical identity presupposes a place that does not exist. The studio in this case stands in as a kind of universal site where team rivalries can be played out. It is the precondition for production practices seeking to create nodes of connection for fandom. The studio eases the oscillation from the virtual to some kind of located place and practice, enabling the marketing of place-based team identities.

While all of this is largely figurative, the studio also facilitates a much more immediate kind of oscillation between physical and simulated space. Where the studio audience is a point of identification for fans consuming a broadcast online, the studio itself also has to adapt the experience of watching for its audience. Esport is screen media, and unlike other forms of sport, it does not correspond directly to the physicality or specularity of an arena. Where sports fans watching other kinds of competition attend stadium games to see the action play out in person, the nature of esport competition makes this impossible. The studio provides a place for fans to gather and a site for the physical performance of play, but a screen always mediates game actions. This is true for the fans watching online, for the ones in the studio, and even for the players of the game who only see the impact of their play on their own screens. Naturally, then, the development of a more robust and more audience-focused spectator practice means a development of more engaging kinds of specularity in the studio space.

One of the major changes in Riot's permanent studio was the addition of more screens to supplement the experience of game watching. In terms of the translation of the virtual playing field into something meaningful for the audience, the proliferation of screens is one of the most palpable adaptations to these kinds of spaces. Each season and each new studio seemingly reflect a deeper understanding of how to make the studio space a more engaging place to watch video-game competition through the addition of screens to complement game action. In subsequent configurations of the LCS Arena, the main action continued to play out on a single central screen, but the experience of immersion and presence shifted to secondary screens.

The development of the studio space following Riot's initial MBS studio reflected a deepening understanding of how esports might be translated for a live, physically present audience. The number of screens increased, allowing spectators in the audience to see more information than they would have in the MBS studio or if they were watching online. For example, in 2016, the studio added a banner of screens along the wall behind each team to announce the achievement of in-game objectives. Familiar to players of the game, these screens announce "First Blood," "Baron," or "Victory," signaling crucial moments where a team has achieved one of these objectives in the game. In the studio, they also orient the audience, reminding them that a specific event is tied to the players on the stage, but these screens also serve a more ambient function. During regular gameplay, they display a 3-D image of the flora of the game world. Despite being just an animated image of a row of trees, it suggests the design of the studio is moving to make the connection between the computational space of gameplay and the physical space of competition more enveloping.

We can think of this development in terms of the history of sports-stadium screens. In an essay on large-screen video displays in sports arenas, Greg Seigel suggests the implementation of screens in sport stadiums beginning in the 1980s is a trickling of television aesthetics into the stadium (2002). In sports stadiums, screens encourage spectators to shift their attention from the field to the video display for replays, to see events in the stadium, and for close-ups on the action. For fans in an esports studio, the proliferation of screens moves the field of play into the studio, such as making player keyboard actions more visible and adopting the backdrops and flora of the game world. The use of this screen is less about making the stadium more television-like but rather more game-like. The ambient backgrounds and announcements of "Killing Sprees" and "Baron" function like moving the map and HUD[22] announcements of the game interface into the studio space.

On a much larger scale, this is what Blizzard Arena does with its own massive screen to produce a panorama of the game world. The screen runs from either side of the 450-seat, 50,000 square-foot auditorium housing the *OWL*, wrapping partially on to the walls at either side. Taking a page from Cinerama,[23] the screen wraps around the front of the studio in a gigantic "U" so that the stage, the players, and the first few rows of the audience sit within its span. It is more an LCD wall than a screen itself. At different points of play the screen is used in different ways. When competition is taking place, the game is displayed on a smaller screen in the center of this wall, and the edges serve as background. Sometimes this background shows a panorama of the game world but, at other points, it shows detailed information about the match taking place. During my visit, as the game announcers analyzed play from the perspective of New York Excelsior star player Pine, this screen projected a detailed in-game breakdown of his performance, accompanied by a gigantic rendering of his character's

Figure 12.2 Blizzard's *OWL* studio shows a panorama of the game world on its massive screen before competition resumes.

image. Another function of the screen is to offer studio fans a virtual tour of a map before competition begins. Like a Google Maps Street View of the playing field, a wide-angle camera flies through the map, allowing the audience to see the world the competitors will inhabit. And like the *LCS*'s ambient trees and rendition of HUD announcements, these supplementary screens envision ways to make the practice of watching in the studio more immersive and, crucially, more game-like. The panoramic screen of the *OWL* demonstrates this most directly, its technological capacity drawing the entire studio space directly into the game world.

This visual capability is not shared with the online audience. The broadcast feed going out over livestreams—for both the *LCS* and the *OWL*—shows only what studio audiences would see on the main screen. These other elaborations set the studio apart and recognize a need to make the studio site compelling for the audience present. Using screens to supplement this experience blurs distinctions between the world of the game and the world of the studio. This is not the sports stadium becoming more like television, but more like the world of the game, closing or mediating the distance between the virtual space of play and the physical space of spectatorship. And this is how we can start to see all these practices of locating and grounding begin to converge. The consolidation of audience, of franchises, of history, of place, and even of virtual worlds coalesces around the studio space. As esport franchises grow, studios serve as the sites where esports futures are imagined. Where Riot's first esports studio was a modest space designed to reach an online audience as efficiently as possible, the more robust studios that have followed from it are engaging with a more ambitious kind of audiencing. They are producing sites that can begin to grow a practice of esports spectatorship that builds upon itself, that seeks to spawn franchises and durable sites of connection, all of which cyclically reinforce gameplay through spectatorship and competitive play.

Conclusion

The creation of LA's first permanent esports venues reflects a growing confidence in esport as a media mode—not so much the development of an industry based on clear lines of profitability, but rather a speculative and emergent kind of media-making driven by margins, futures markets, and media synergy. While the legacies of esport and even esports studios are now over two decades old, their current formation elaborates on the conventions of the form and pioneers different ways of making this media profitable. The roots of the esport studio as a formal space in Korea provide the blueprint for Riot's LCS Studio and the Blizzard Arena, but the production rooted in these spaces strikes out in a new direction. This is not media made for cable television or for commercial profitability in the same way. Esport remains a turbulent and uncertain practice in the West, and what Riot and Blizzard capitalize on are their own ties to the games they make. The studio becomes a hopeful and still experimental space where the development and crystallization of tactics for hailing an audience reflect a growing understanding of how this production reaches its spectators, and furthermore, what producers hope spectators will invest in the consumption of esport media.

Neither the *LCS* nor Blizzard's *OWL* are internally profitable—at least for the producers of this media and at least not yet.[24] For the team owners and sponsors, profit models are a bit more complex but, for Riot and Blizzard, esport media-making involves hailing an audience that will support the ecosystem around the core game. This hailing occurs at several levels but it ultimately serves the games themselves. Esport is a nearly secondary kind of media, sustaining interest and activity around a primary media: the games at the heart of this competition. In this respect, studios themselves are also secondary spaces or translational spaces operating to make, ground, and contextualize this play. They provide sites of identification and participation through which the relationships and influences esport encourages can be developed. Riot's *LCS* has worked to foster interest and participation in *LoL*, to build its profile, and to extend its lifespan. Blizzard's *OWL* seeks to do the same for *Overwatch*. In this configuration, esport becomes a kind of transitional and fluid media that oscillates between the virtual and the material. The players, studio, and audience are live but the game action itself is made meaningful in the virtual space of a game that must then be translated back into the studio or stream.

This production creates a back and forth between virtuality and materiality. In the spirit of liquidity, it is both an oscillation from virtual worlds to material studios, as well as material media to virtual, speculative capital.

Notes

1. The inaugural season of the OWL, which ran from January to July 2018, featured twelve teams representing different cities, with many owned by existing sports franchise owners or established esport teams. They are divided into Pacific and Atlantic divisions, and Blizzard has indicated that, in the future, more teams might be added to the league (Blizzard Entertainment, 2018).
2. Riot's LCS remains the industry leader in esport, drawing viewership figures that remain the high watermark for competitive gaming and that generally double each year (Perez, 2018).
3. Microtransactions are optional ingame purchases that unlock content, and many free-to-play games are designed around this profit model.
4. "MOBA" refers to a Multiplayer Online Battle Arena, a genre that borrows formally from real-time strategy games. MOBA players control a single character and work as a team to complete objectives.
5. These franchises are designed to support the media ecologies of the video games at the heart of these competitions. Esports supports player bases and encourages participation in consumer practices around these games (such as the purchase of cosmetic items).
6. Nicholas Thiel Taylor deployed the term "audiencing" in his essay on esports event configurations to describe practices and infrastructures that house and hail an audience with the goal of developing an audience and instructing audience behaviors (Taylor, 2016).
7. An early and prominent example of an esports organization, MLG sold in 2016 to Activision/Blizzard Entertainment and thereafter shifted focus to Activision's Call of Duty series while also helping to promote Blizzard's OWL.
8. In 1961 researchers at MIT wrote *Spacewar!* which used a vector display to represent the game world and is often cited as the first video game.
9. Agon is a concept deployed by sociologist Roger Caillois to describe games of competition (1961).
10. It is worth noting that Korean cable television, while still driven by advertising, functions a bit differently from American cable television. Whereas American cable is a vestige of the American television industry of the 1970s, Korea's cable industry is more modern. It arrived in the early 1990s with the rapid expansion of broadband infrastructure and has catered to a much more culturally homogeneous audience. When I visited Korea for research, I remember getting to my AirBnB in Seoul and trying to find esports on television. A channel surf turned up stations dedicated to the board game Go, a style channel, and of course several dedicated exclusively to video games (at the time, these channels included SpoTV Games and OGN). These channels also owned production studios where esport competition was recorded.
11. Netflix launched its streaming service in 2007, and Hulu began streaming in 2010, meaning that a wider base of users was conditioned to accessing their media through streaming platforms.
12. In 2012 *LoL* boasted a player base of eighty million and an average of twelve million daily users. In 2014 these numbers would grow again, with *LoL* claiming twenty-seven million daily players (Tassi, 2013).

13. *Penny Arcade Expo* (*PAX*), *Major League Gaming* (*MLG*), and *European Star League* (*ESL*) are/were all major esport tournaments and general gaming conventions.
14. A year before Riot sought to build a regular live broadcast of *LoL* competition in North American and European markets, the game was already fully functioning with regular competition in Korea under a cooperative arrangement between Riot, KeSPA, OGN, and the cable network SPOTV Games.
15. "Esports-casters" refers to sportscasters, analysts, and announcers for esport.
16. Players would rank up or down within each rating, from "Bronze 5" to "Bronze 1," before moving to Silver 5, and so on. At the time, it very closely replicated the rating system in place for *StarCraft II* which also had ranks from bronze to diamond, and a "master tier" (instead of challenger).
17. "Metagaming" refers to strategies players develop above and around games, such as leveraging understandings of game algorithms to gain an advantage. Stephanie Boluk and Patrick LeMieux write extensively on the concept of metagaming, suggesting it as a way of seeing how games play-players, which I find to be a good encapsulation of esports marketing synergy (2017).
18. I think the connection between esports streaming media and Hulu, as a post-broadcast platform for television, is a fortunate coincidence that begins to express the trajectory that I am trying to sketch here, one that moves away from a static audience and toward new forms of distribution and consumption.
19. Among the changes that I address at length, the studio added a gift shop stocked with merchandise, some of it exclusive and only available in this space, but all of it focused on the characters and world of LoL and the professional players of the game—this would even come to include player jerseys and team memorabilia. The studio also added a permanent concessions stand, meaning that fans did not have to leave the venue during competitions, which often lasted upwards of five hours.
20. Both the *OWL* and the *LCS* franchises have deep ties to traditional sports teams in the NFL, NBA, NHL, and MLB. For the *OWL* some of these include the Miami Heat, the New York Mets, the New England Patriots, the Texas Rangers and Philadelphia Flyers. For the *LCS* this includes the Golden State Warriors, the Milwaukee Bucks, and the Cleveland Cavaliers.
21. At the time of writing it was imagined that eventually each team in the *OWL* would have its own stadium in the future but this, too, is obviously speculative so naming precedes locating (Webster, 2018).
22. "HUD" (Heads-Up Display) refers to generally non-diegetic elements of video games' visual semiotics that provide useful game information (for example, available skills as well as health and mana levels) but are not part of the game world.
23. Cinerama, a portmanteau of cinema and panorama, refers to an ultra-widescreen film projection technology that used a curved screen and multiple synchronized projectors to screen film in a 146-degree field of view.
24. Eventually, ticket sales, esports merchandising, and broadcast deals might offset these costs but mostly this kind of profitability is not the point. The value of esports is in its promotional capacity, supporting the games themselves.

References

Altay, Omer (2018), "Top Free to Play Games by Revenue, 2017—Superdata Research," *MMOS.com,* 30 January, <https://mmos.com/news/top-free-play-pc-games-revenue-2017-superdataresearch> (last accessed 20 April 2018).

Blizzard Entertainment (2018), "Welcome to Overwatch League," *Blizzard Entertainment*, <https://overwatchleague.com/en-us/about> (last accessed: 13 April 2018).

Boluk, Stephanie and Patrick LeMieux (2017), *Metagaming*, Minneapolis: University of Minnesota Press.

Burrows, Benjamin and Ramma, Paul (2015), "The eSports Trojan Horse: Twitch and Streaming Futures," *Journal of Virtual Worlds Research* 8 No. 2. <https://doi.org/10.4101/jvwr.v8i2.7176> (last accessed 15 April 2017).

Caillois, Roger (1961), *Man, Play, and Games*, Chicago: University of Illinois Press.

Davidson, John (2016), "How 'Quake' Changed Video Games Forever," *Rolling Stone*, 22 June, <https://www.rollingstone.com/culture/news/how-quake-changed-video-games-forever-20160622> (last accessed 15 April 2018).

EpicSkillshot—LoL VOD Library (2013), "Alliance vs Dignitas Game 1 | EU vs NA Battle of the Atlantic 2013 | ALL vs DIG G1 Bo3," YouTube, 14 December, <https://www.youtube.com/watch?v=bU7flfLeBL8> (last accessed 15 April 2018).

Garren, Patrick (2017), "A List of All the 2018 NA LCS Teams, Their Owners and Who's Backing Them," *The Esports Observer*, 21 November, <https://esportsobserver.com/na-lcs-2018-reported-team-owners/> (last accessed 13 April 2017).

Grayson, Nathan (2017), "Overwatch Still Isn't Ready To Be A Top Esport," *Kotaku*, 5 June <https://compete.kotaku.com/overwatch-still-isnt-ready-to-be-a-top-esport-1795833289> (last accessed 13 April 2017).

Hiltscher, Julia (2015), "A Short History of eSports," in Julia Hiltscher and Tobias Michael Scholz (eds), *Esports Yearbook 2013/14*, Norderstedt, Germany: Books on Demand GmbH. pp. 9–14.

Jin, Dal Yong (2010), *Korea's Online Gaming Empire*, Cambridge, MA: The MIT Press.

Kelly, Christina (2011), "Spectator Experiences at MLG Dallas vs. Korea," in Julia Hiltscher and Tobias Michael Scholz (eds), *Esports Yearbook 2010*, Norderstedt, Germany: Books on Demand GmbH, pp. 50–3.

Kishner, Irwin A. (2017), "Esports Leagues Set To Level Up With Permanent Franchises," *Forbes*, 3 October, <https://www.forbes.com/sites/kurtbadenhausen/2017/10/03/esports-leagues-grow-up-with-permanent-franchises/#2673723d21d6.> (last accessed 13 April 2017).

Messier, Marc-Andre (2011), "The Lessons eSports Should Learn From The Recession," in Julia Hiltscher and Tobias Michael Scholz (eds), *Esports Yearbook 10*, Norderstedt, Germany: Books on Demand GmbH. pp. 54–61.

Perez, Matt (2018), "Report: Esports To Grow Substantially And Near Billion-Dollar Revenues In 2018," *Forbes*, 21 February, <https://www.forbes.com/sites/mattperez/2018/02/21/report-esports-to-grow-substantially-and-near-a-billion-dollar-revenues-in-2018/#78b96ead2b01.> (last accessed 13 April 2018).

Pitcher, Jenna (2013), "League of Legends Finals a Sell Out at Staples Center, North American Regionals Conclude," *Polygon*, 2 September, <https://www.polygon.com/2013/9/2/4685046/league-of-legends-finals-a-sell-out-at-staples-center> (last accessed 15 April 2018).

Popper, Ben (2013), "Field of Streams: How Twitch Made Video Games a Spectator Sport," *The Verge*, 30 September, <https://www.theverge.com/2013/9/30/4719766/twitch-raises-20-million-esports-market-booming> (last accessed 15 April 2018).

Riot Games (2018), "Evolution of the LCS," *Riot Games*, <https://www.lolesports.com/en_US/articles/evolution-of-the-na-lcs.> (last accessed 15 April 2018).

Sarkar, Samit (2013), "League of Legends Season 3 Finals Drew 32M Total Viewers," *Polygon*, 19 November, <https://www.polygon.com/2013/11/19/5121688/league-of-legends-season-3-world-championship-finals-viewers.> (last accessed 16 April 2018).

Siegel, Greg (2002), "Double Vision: Large-Screen Video Display and Live Sports Spectacle," *Television and New Media*. 3.1, pp. 49–73.

Sinclair, Brendan (2016), "Friction Between Riot and League of Legends Team Owners," *Gamesindustry.biz*, 24 August, <http://www.gamesindustry.biz/articles/2016-08-24-friction-between-riot-and-league-of-legends-team-owners> (last accessed 15 April 2018).

Stewart, Dave (2018), personal interview, 4 February.

Stubbs, Mike (2017), "A Year On From Launch, Overwatch Is A Struggling eSport," *Eurogamer*, 28 June, <http://www.eurogamer.net/articles/2017-06-27-a-year-on-from-launch-overwatch-is-a-struggling-esport> (last accessed 13 April 2017).

Tassi, Paul (2014), "Riot's 'League of Legends' Reveals Astonishing 27 Million Daily Players, 67 Million Monthly," *Forbes*, 27 January, <https://www.forbes.com/sites/insertcoin/2014/01/27/riots-league-of-legends-reveals-astonishing-27-million-daily-players-67-million-monthly/#670ad5016d39> (last accessed 15 April 2018).

Taylor, Nicholas Thiel (2016), "Now You're Playing with Audience Power: The Work of Watching Games," *Critical Studies in Media Communication* 33, no. 4. pp. 293–307.

Taylor, T. L. (2012), *Raising the Stakes*, Cambridge, MA: The MIT Press.

Van Allen, Eric (2017), "Pro League Of Legends Team Says It's Losing More Than $1.1 Million A Year," *Kotaku Compete*, 2 September, <https://compete.kotaku.com/pro-league-of-legends-team-says-its-losing-more-than-1-1798727989> (last accessed 15 April 2018).

Volk, Pete (2016), "Riot: Esports Still Isn't Profitable, and We Don't Care," *Rift Herald*, 13 September, <https://www.riftherald.com/2016/9/13/12865772/lol-esports-profit-money-riot> (last accessed 13 April 2017).

Webster, Andrew (2018), "At the Overwatch League's Opening Night, I Witnessed the Strange, Thrilling Future of Esports," *The Verge*, 16 January, <https://www.theverge.com/2018/1/16/16897500/overwatch-league-opening-night-blizzard-arena> (last accessed 15 April 2018).

Yegg (2013), "New League System in Ranked," Official Post to the *North American League of Legends Forums*, 15 January, <http://forums.na.leagueoflegends.com/board/showthread.php?t=3004520> (last accessed 18 April 2018).

13. HOLLYWOOD'S VR VISION: NEW FRONTIER OR VIRTUALLY THE SAME THING?[1]

James Fleury

Set in 2045, the film *Ready Player One* (Steven Spielberg, 2018) features virtual reality (VR) users stepping into another film, *The Shining* (Stanley Kubrick, 1980). While this concept remains a fantasy, American media conglomerates have embraced VR—defined here as the use of a headset to enter into a simulated, immersive space—as one of the latest in a long line of emerging technologies for content production, distribution, and marketing. This chapter charts the history of Hollywood's VR marketing efforts from 2014 to 2018; after embracing other forms of immersion (for example, video games, "4-D" film screenings, and alternate reality games [ARGs]), studios during this time began experimenting with VR through licensing partnerships and the establishment of in-house teams. Though the VR market offers an immersive new medium, its commercial infancy has encouraged risk aversion. That is, studios gradually have engaged the technology with promotional, rather than original, content in service of existing intellectual property (IP) with an emphasis on media franchises. While the slow pace at which studios have embraced VR suggests a lack of confidence in the medium, their deployment of VR as a subservient marketing vehicle has left open the opportunity for other companies to dominate the VR market. Based on media-industry trade and popular press articles, statements from personnel involved in VR content, and analysis of the content itself, I argue that this promotional emphasis reflects the tendency for the American film and television industry to enter emerging formats through low-risk experiments in the form of promotional content before making a larger-scale push with original material.

I posit VR as a growing, yet neither unique nor exclusive, channel in the contemporary era of media convergence. Studios' VR marketing efforts, however, extend from historical precedents. This chapter begins with the demystification of the concept of VR as an entirely new phenomenon by contextualizing why, after earlier interest and subsequent disinterest in the 1990s, the film and television industries have once again turned to the medium. Their interest in the technology, in fact, represents a remediation—or recreation of one medium in another—of event marketing to capitalize on social media hype (Bolter and Grusin, 2000: 17). After providing this context, I examine three strategies Hollywood has experimented with in its use of VR as a tool for promoting existing IP and media franchises: experiential (that is, free content that allows users to inhabit diegetic spaces, usually with little control over the experience), consumer (that is, fee-based VR that typically employs video-game-like interactivity), and location-based (that is, premium, social experiences available only at specific venues and that, like experiential VR, cannot be replicated at home). Finally, this chapter considers the implications of studios' limited use of VR, particularly given that other companies appear poised to dominate the market on the one hand and that competing technologies, such as augmented reality (AR), offer the potential for even greater immersion on the other. I conclude that historical continuities mark Hollywood's take on VR, and that these continuities gain new traction in what Henry Jenkins has called convergence culture, in which digital technologies have encouraged cross-platform content, stakeholder cooperation, and audience engagement (2006: 2).

Setting the Virtual Stage

Though virtual reality has become a popular topic in contemporary American media industry discourse, it is not a new phenomenon. In the 1990s, VR had already wired the public's consciousness. Nintendo, for instance, introduced the Virtual Boy in 1995 as a VR gaming console but failed to capitalize on the technology's potential as a new frontier for entertainment capable of reengineering the media experience. As Steven Boyer has noted, the Virtual Boy struggled because of hardware limitations (such as a red-tinted monochromatic display), a lack of compelling content, and a tendency to cause headaches among users (2009: 30). As a result, the hype machine around the Virtual Boy swiftly overheated, crashing and burning the emerging VR business.

In Hollywood, however, VR remained the talk of the town. Dystopian, science-fiction films such as *The Lawnmower Man* (Brett Leonard, 1992), *Strange Days* (Kathryn Bigelow, 1995), and *The Matrix* (The Wachowskis, 1999) suggested ambivalence toward VR, showcasing its benefits (such as the ability to enter another world) as well as its dangers (such as becoming too "plugged in"). More recently, *Ready Player One* narratively shared this ambivalence even while its production and marketing reflected a renewed interest in the technology; for example, even though the film ends with a message about the importance of spending more time in the real world, director Steven Spielberg and the cast relied on the HTC Vive VR headset to plan out sequences set in the film's virtual world (Seymour, 2018). Television shows have similarly taken on VR themes and concepts, with the holodeck from *Star Trek: The Next Generation* (1987–94) representing a particularly notable example. For two decades and counting, VR has served as a narrative and aesthetic engine for American film and television.

As *Ready Player One* demonstrates, VR has evolved into one of Hollywood's new industrial engines. The technology has moved beyond the screen to become a consumer technology, one with a projected market value of $75 million by 2021 (Roettgers, 2017). The VR headset has turned from a fantasy into a reality, with major technology and media companies, such as Facebook, Google, HTC, Samsung, and Sony, all releasing their own hardware. For these players, VR represents a new media platform. As such, the technology offers a significant opportunity for both Silicon Valley and Hollywood. Essentially, VR headsets present an additional content distribution pipeline. As tech-media hybrids such as FAANG (that is, Facebook, Apple, Amazon, Netflix, and Google) continue to disrupt the traditional media supply chain, studios have harvested the power of VR to retain audiences while supporting their IP and franchise economy.

The turn to VR underscores shifts in the media landscape. Just as declining domestic box-office attendance in the early 1980s encouraged Hollywood to embrace the then nascent medium of video games for tie-in merchandise (Harmetz, 1982: A40), contemporary studios have adopted VR to compensate for falling theatrical revenues (Bond, 2017). The primary appeal of VR to studios remains its capacity to immerse users in cinematic and televisual environments, which has encouraged a promotional use of VR designed to drive theatrical and ancillary (such as home video) revenues. That is, VR allows studios to sell their films and television series as interactive experiences. While a film's trailer might encourage audiences to "experience it in IMAX," VR differs from IMAX and other large-screen formats by inviting users not just to watch a film but to inhabit its storyworld; furthermore, the immersive potential of VR headsets and multisensory hardware (such as controllers that provide haptic feedback) also distinguishes the format from other interactive entertainments such as video games, theme parks, and ARGs. VR also supports Hollywood's IP ecosystem through branded content available for home purchase and through installations (that is, location-specific "activations") housed at movie theaters and related venues (such as shopping malls). Studios have partnered with technology giants and startups to distribute VR content based on media IP and franchises; as such, studios have adopted VR as a strategy to engage audiences and thereby combat FAANG's continued remapping of Hollywood.

Though VR has the potential to reshape the media economy, Hollywood's tendency toward promotional content reflects a level of ambivalence in investing in original material for the format until it emerges as a viable market. At present, VR has presented major challenges for film and television stakeholders. Its viability remains an open question, as producers wait to see if VR will become a novelty or a utility following consumer rejection of once-hyped products such as laserdisc and 3-D televisions. Furthermore, VR software and hardware continue to undergo iteration at a rapid rate, already having fragmented the market into smartphone-integrated mobile technology and dedicated, PC- and gaming-console-powered headsets. Also, the format's immersive nature requires significant changes to content creation, presentation, and monetization. Specifically, VR not only provides a different and more expansive field of vision but it also allows for a multisensory experience. Content cannot simply be remediated from existing platforms into VR; instead, creators need to consider the technology's unique features and visual vocabulary (such as long takes instead of cuts). In addition to its lack of established standards, VR production remains expensive, with premium content costing as much as $1 million per minute (Donnelly and Pressberg, 2017). Finally, consumers have not yet embraced the technology on a wide scale which has inspired software and hardware companies to expand beyond the home market into location-based VR available at shopping malls, theme parks, and dedicated VR centers (Spangler, 2018).

Still, the potential of the VR space has encouraged Hollywood to establish a presence in the emerging marketplace and invest in VR companies to secure future partnerships. For instance, Jaunt VR's 2015 funding from The Walt Disney Company, Participant Media, and Creative Artists Agency demonstrates the early entrance of film and television stakeholders in the consumer VR market (Spangler, 2015). In addition to investment, some studios have launched internal innovation laboratories to experiment with VR technology and content, starting in 2014 with Twentieth Century-Fox's establishment of Fox Innovation Lab to create standalone VR experiments based on non-franchise films such as *Wild* (Jean-Marc Vallée, 2014) and *The Martian* (Ridley Scott, 2015) (Roettgers, 2016a). Because these short-form experiences proved commercially unsuccessful, Fox founded FoxNext in 2017 to produce commercial, rather than experimental, content, including location-based VR tie-ins to Fox franchises such as *Alien* (1979–present) (Roettgers, 2018b). Furthermore, Sony has created content for its PlayStation VR hardware (an add-on to the Sony PlayStation 4 console); while Sony's VR efforts primarily have focused on original gaming IP (such as 2018's *Animal Force*), it has also leveraged established media franchises (for example, 2017's *Ghostbusters VR:*

Now Hiring). Whether through investment or internal production, studios have remained involved in VR, albeit as an ancillary marketing platform.

Because of VR's challenges, studios have sought to avert risk by prioritizing promotions over riskier, original content. Their ambivalent embrace of VR for distributing promotional content reflects the film and television industry's assessment of the medium's viability. Historically, the industry has used marketing initiatives to capture and test new trends. For example, media conglomerates largely have regarded video games as derivative merchandise for their films and television series—a trend that began when Gulf+Western used its Sega subsidiary to release *Fonz* (Sega/Gremlin, 1976), an arcade tie-in to the Paramount Domestic Television series *Happy Days* (ABC, 1974–84). Only recently have studios started to treat video games as a distinct medium, such as the Warner Bros. Interactive Entertainment (WBIE) gaming division of Warner Bros. positioning games, such as *Batman: Arkham Asylum* (WBIE/Eidos, 2009) and *Middle-Earth: Shadow of Mordor* (WBIE, 2014), as associated with a general IP rather than with any specific film or television series.

That these standalone video games have continued to rely on existing brands suggests that VR will also support Hollywood's franchise and IP economy in the future. In their introduction to *Connected Viewing: Selling, Streaming, & Sharing Media in the Digital Era* (2014), Jennifer Holt and Kevin Sanson describe how Hollywood stakeholders view emerging media as "a necessary evil" and "promotional tool" in service of their existing film and television properties (2014: 6). As Holt and Sanson explain, producers open to new media opportunities "can be more flexible in adapting to the incredible pace of change taking place in digital media distribution right now" (2014: 6). Toward this flexibility, studios have adopted VR at an early stage. Doing so helps studios stand a better chance of avoiding divergence, or negative convergence, with an emerging media format. News Corporation's failed ownership of MySpace, which it bought in 2005 for $580 million but sold in 2011 for only $35 million, demonstrates the consequences of a traditional Hollywood company not adapting quickly enough to new media trends, as MySpace already had begun to cede social media market share to newer, more feature-rich platforms such as Facebook and Twitter (Stelter, 2011). To integrate VR into a studio system focused on the cultivation of franchises, Hollywood has taken a risk-averse, promotional approach.

Prehistory of Promotional VR Experiences

Studios have positioned VR as subservient to traditional media, with promotional content designed to create "buzz" and conversation in support of films and television series, especially those belonging to franchises and set within rich storyworlds. Prior to Hollywood's adoption of VR for tie-ins, brands had experimented in the virtual space through collaborations with digital studios, including Marriott and Framestore partnering on a 2014 campaign in which a VR-powered "teleporter" delivered users at select Marriott hotels to exotic travel destinations (Rubin, 2014). These marketing experiments demonstrated VR's innovative advertising potential and the production expertise of digital studios such as Framestore. After the Marriott "Virtual Travel" campaign, Framestore debuted at the 2015 Sundance Film Festival apparel company Merrell's "TrailScape" experience which *Advertising Age* noted as "the first-ever 'walk around' commercial virtual-reality experience" (Rodriguez, 2015). Through partnerships with digital studios, brands drew attention to the opportunity of reusing innovative VR content at multiple experiential events continually to drive consumer conversation around brands and products.

Framestore's early marketing examples influenced the explicit promotional direction in which Hollywood would take VR. According to Mike Woods, former head of the Virtual Reality Studio at Framestore, "Using VR for marketing is nirvana because you can put somebody in somebody else's world" (2015). This sense of immersion attracted content companies interested in experimenting with VR based on Hollywood IP; as Woods puts it, VR for studios provides "an easy win . . . because [there's] nothing the fanbase is going to love more than being in the thing they love" (2015). Hollywood had long sought to engage consumers through immersive formats, with VR promising a richer, multisensory experience.

Prior to expanding into VR content, studios had already immersed audiences in recreations of diegetic spaces through theme parks, video games, and physical installations. That is, experiential VR content, or virtual experiences tied to media brands, grew out of earlier forms of immersion. As Heather Lea Birdsall's chapter in this volume analyzes, theme parks and video games have a rich history of remediating cinematic and televisual environments into physical and virtual spaces. With video games, Jonathan Gray has argued that media tie-ins such as *The Thing* (Vivendi Universal Games, 2002) "allow audiences to set foot in their various storyworlds' diegetic spaces" (2010: 176). These sites, he specifies, can even extend beyond those seen in the original source text and thereby expand a given fictional universe. Gray points out that *The Simpsons: Hit & Run* (Vivendi Universal Games, 2003) provides the opportunity to see the spatial layout of the fictional Springfield setting from *The Simpsons* franchise (1987–present) (2010: 193); similarly, the video game *South Park: The Stick of Truth* (Ubisoft, 2014) required *South Park* franchise (1997–present) creators Trey Parker and Matt Stone to map out the titular town for the first time (Claiborn, 2014). VR iterates upon theme parks and video games by combining their immersive qualities with the explicit promotional imperative of event marketing campaigns.

In fact, promotional VR content grew out of earlier forms of event marketing. Events like red-carpet premieres historically have functioned as advanced marketing for Hollywood content owing to the press coverage they generate. More recently, studios embraced larger-scale events such as San Diego Comic-Con (SDCC) and South by Southwest (SXSW) as sites for showcasing their upcoming tentpole film and television releases. Their promotions at these events have grown increasingly elaborate to encourage social-media activity and consequently to drive attention to the advertised product. In 2009 and 2010, SDCC's first environmental installation appeared, with a recreation of Flynn's Arcade from the Disney film *Tron: Legacy* (Joseph Kosinski, 2010). By 2013, multiple walkthrough activations appeared, including the "*Ender's Game* Experience" for Lionsgate's *Ender's Game* (Gavin Hood, 2013), "Dracula's Crypt" for *Dracula* (NBC, 2013–14), and a "Godzilla Encounter" maze set in the world of Warner Bros.' *Godzilla* (Gareth Edwards, 2014) (Knowledge@Wharton, 2013). These examples of so-called "experiential marketing" share a similar promotional aim of theme-park attractions in their immersion of audiences in physical recreations of cinematic and televisual spaces. At the following year's convention, studios introduced VR installations that used multisensory technology to immerse users more deeply in simulated environments based on films and television series.

Experiential VR

By SDCC 2014, VR technology had developed enough to allow for a less ephemeral and more immersive form of experiential marketing. The box-office disappointment of *Tron: Legacy* and *Ender's Game,* coupled with low ratings for *Dracula,* probably further prompted this move from the real to the virtual. Moreover, content companies might have adopted VR because virtual tie-ins, compared to physical installations, more easily enable promotions to appear at multiple events prior to a consumer release. That year, Twentieth Century-Fox Home Entertainment showcased a virtual experience tied to the home-video release of *X-Men: Days of Future Past* (Bryan Singer, 2014), and Framestore's "Ascend the Wall" campaign transported users into the world of HBO's *Game of Thrones* (2011–19). In addition to advertising film and television properties, these tie-ins encouraged social-media conversations around the hardware, as they all used the Oculus Rift headset from Facebook's Oculus Rift VR subsidiary (Knowledge@Wharton, 2014). As such, promotional installations have become media events in their own right, with social-media sharing creating free marketing for the associated media properties and for the VR technology itself. Though studios have expanded from purely consumer VR to a location-based strategy, experiential VR continues to represent a means of attracting consumer attention. More specifically, Hollywood's use of VR marketing functions as a sales engine for hardware manufacturers.

Experiential VR tie-in activations emerged out of collaborations between studios and the same boutique digital agencies, such as Framestore, that previously had partnered with brands on VR-marketing campaigns. While studios provide access to their IP, the specialized digital studios remain responsible for the content creation. On the one hand, Hollywood's subsequent consumer and location-based strategies would move beyond the relatively passive immersion of experiential VR to integrate more gaming features (Wallenstein, 2015); on the other hand, early activations set precedents for these later iterations. In particular, early promotional VR relied on episodic, short-form content to acclimate users to the technology—which carries the risk of headaches and motion sickness—and to alleviate the learning curve for creators, given that VR obviates the need for filmmaking conventions like cuts and pans (Schwartzel, 2016; Shaw, 2015). Furthermore, shorter experiences have helped to control the high cost of production, while maintaining positive relations with exhibitors combating collapsing theatrical attendance and therefore loathe to compete with yet another distribution format (Donnelly and Pressberg, 2017).

"*Interstellar*: Oculus Experience," a tie-in to the Paramount Pictures science-fiction film *Interstellar* (Christopher Nolan, 2014), demonstrates one of the earliest examples of this industrial logic. Using what it calls "the world's first VR and Immersive Content Studio," Framestore created the project after having gained recognition for its visual effects contributions to films such as *Avatar* (James Cameron, 2009) and *Gravity* (Alfonso Cuarón, 2013) (Framestore VR Studio, 2014). Paramount's marketing department had hired the studio to create the "*Interstellar*: Oculus Experience" after the Marriott "Virtual Travel" campaign and "Ascend the Wall" had demonstrated Framestore's VR production expertise. Film and television companies quickly embraced a licensing model for VR content rather than take on the risk of in-house production, precisely because of digital studios like Framestore's experience in VR promotions. This licensing approach averted risk and enabled Hollywood to observe, and learn from, the experiments of boutique producers in an emerging medium before eventually entering into the VR market themselves.

"*Interstellar*: Oculus Experience" captures the experimental phase of Hollywood's use of VR for promotional purposes. The project's brevity and viscerality underscore the conceptual inspiration that early tie-in VR experiences took from physical, tie-in activations and theme-park attractions. Showcased as an installation at events such as the 2014 versions of SXSW and SDCC, the three-minute "*Interstellar*: Oculus Experience" allows users to float around in zero gravity aboard the *Endurance* spacecraft from the film but ends just as the ship prepares to enter a wormhole (Horn, 2014). A review from *USA Today* noted that the Oculus VR technology and the installation's moving chair created a zero gravity sensation "so real that headset-wearing participants often reach[ed] for floating objects that [were] not truly present"

(Alexander, 2014). This contextual dependence suggests that the project had been designed for a limited period of use and not as a longer-term product. The "*Interstellar*: Oculus Experience," in other words, had been destined for ephemerality, given that the experience could take place only as part of a larger installation incapable of replication in any consumer release.

The project's temporary nature reflects its basis as a promotional vehicle in service of a more privileged media product. As such, unlike some tie-in video games (such as *The Thing*) and other more story-based forms of diegetic immersion, "*Interstellar*: Oculus Experience" and similar experiential VR qualify as more purely ancillary marketing pieces and not as examples of transmedia storytelling. According to Henry Jenkins, "A transmedia story unfolds across media platforms, with each new text making a distinctive and valuable contribution to the whole" (2006: 97–8). Jenkins has also acknowledged the rarity of narratives that fit this definition, arguing that transmedia texts tend to focus less on narrative than on the process of "world-building" (2007). Because it merely extends the experience, or world, of *Interstellar* and not the film's narrative, "*Interstellar*: Oculus Experience" suggests how, early on, studios viewed VR as a marketing tool instead of as a unique medium.

The level of collaboration between Framestore and the team behind the film underscores the project's marketing imperative. Whereas Framestore's Mike Woods took part in interviews promoting "*Interstellar*: Oculus Experience," Paramount limited the participation of *Interstellar* director Christopher Nolan to a prepared statement that emphasized the content's fidelity to the film: "I was impressed with how the experience is able to put you in the cockpit of *Interstellar*'s spaceship. It was just like being back on set" (Alexander, 2015). Additionally, Framestore primarily communicated not with Nolan or other members of the film production but, rather, with Paramount's marketing department. In fact, the studio marketing group remained involved throughout the project's development; as Woods explains, "It's their assets. It's their story, and it's their characters. So they need to safeguard that, one, it's completely authentic, and, two, that you're not giving away too much about the secrets in the film" (2015). Like most relationships between Hollywood IP holders and licensors, the studio sought to protect its property. This protectionist stance and limited level of access to talent recall the production of tie-in video games whose explicit promotional purpose tends to result in poorly received products (Elkington, 2009; Fleury and Mamber, 2019). Owing to possible desires to explore the potential of VR more fully, however, Paramount, according to Woods, encouraged Framestore to be "completely free" with the technology and trusted the company's expertise with a medium unfamiliar at the time to Hollywood (2014).

In addition to shedding light on to how Hollywood studios initially interfaced with external VR vendors, the "*Interstellar*: Oculus Experience" also demonstrates one method of distributing promotional VR experiences. In contrast to the ninety-second "Ascend the Wall"—which appeared at events in the spring and summer of 2014—the "*Interstellar*: Oculus Experience" toured the United States in late fall 2014 prior to the film's November opening. This roadshow strategy, recalling Classical Hollywood's distribution for special-event films such as *Gone with the Wind* (Victor Fleming, 1939) (Neale, 2003: 51), allowed the project to promote the film closer to its release date and delivered the experience to audiences across the country who might not have had the chance to visit SXSW or SDCC.

After debuting at these events, "*Interstellar*: Oculus Experience" appeared in the lobbies of select movie theaters. These spaces have continued as a centerpiece in the distribution strategy of VR tie-in installations, given that they can join posters and banners as "paratexts" that lead consumers into the experience of watching a particular film with one key difference: instead of inviting audiences to enter a movie theater, they invite audiences to enter a movie world, albeit in a highly controlled and managed manner (Gray, 2010: 6). In this way, VR activations in cinema lobbies serve a similar role as video-game arcade units have in theater lobbies and other, often adjacent venues (such as arcades and bowling alleys) since the 1980s. Because tie-in installations merely integrate VR into the existing business model of the theatrical feature-film release, they provided little indication of how original VR content would perform with a wide audience. To grow the VR market beyond ephemeral, promotional content, film and television companies expanded into consumer and location-based strategies.

Consumer VR

Even though Hollywood studios continue to use a licensing and roadshow strategy for introducing experiential VR tie-ins, alternative production and distribution methods emerged in 2015 as VR technology entered the mainstream. That is, studios and vendors still partner for promotional VR projects, even as some studios have taken a more direct role in producing their own content, and even as commercial hardware has allowed for direct-to-consumer distribution. In 2015, for example, licensed partnerships—including Felix & Paul Studios's "*Jurassic World*: Apatosaurus" for Universal, and Framestore's "Battle for Avengers Tower" on behalf of Marvel—iterated upon the "*Interstellar*: Oculus Experience" with longer play durations and the ability to transport users into the familiar storyworlds of media franchises such as *Jurassic Park* (1993–present) and the Marvel Cinematic Universe (2008–present). While licensing reduces the risk of experimenting with new technology, some studios opened their own innovation labs to have greater oversight of the deployment of their IP in an emerging market.

Vacillating between licensed and in-house production enabled Hollywood content companies to experiment with distribution and monetization strategies. Free VR content initially continued the marketing imperative of experiential VR but in consumer hardware and not only event-specific installations. In the lead-up to Disney and Lucasfilm's *Star Wars: Episode VII—The Force Awakens* (J. J. Abrams, 2015), for instance, the *Star Wars* mobile app presented the "Jakku Spy" VR experience. A free, short-form promotion in which users take on the role of a Resistance spy alongside the BB-8 astromech droid on the planet Jakku, "Jakku Spy" debuted in December 2015 on Google Cardboard, an inexpensive attachment that converts smartphones into VR headsets. As such, both the "Jakku Spy" software and the Google Cardboard hardware worked to acclimate audiences to a new way of experiencing media content. To ease consumers into the technology, "Jakku Spy" immerses users in the *Star Wars* universe, a storyworld more familiar to mainstream audiences than those of the non-franchise films, such as *Interstellar* and *Wild*, that had inspired earlier VR experiences; furthermore, unlike the guided paths of older tie-ins, "Jakku Spy" introduces a minor degree of control, allowing users to turn their heads to direct movement within the virtual space. Also, whereas "*Interstellar*: Oculus Experience" stemmed from a collaboration between a studio marketing department and a licensed vendor, Lucasfilm itself produced "Jakku Spy" through its ILMxLab studio in partnership with Google and Verizon (Olanoff, 2015).

In addition to Lucasfilm, other film and television companies also began creating consumer-based VR tie-ins. This content grew out of studio accelerator programs (such as Disney Accelerator) and research and development units (for example, Fox Innovation Lab) that initially had focused on mobile and social applications prior to experimenting with VR. While some studio-funded startups can provide a return on investment (Disney, for example, has partnered with Sphero to release *Star Wars* merchandise, like a smartphone-controlled BB-88), their intangible value lies in delivering innovation and fresh thinking to studios' typically risk-averse divisions (Panzarino, 2017).

Twentieth Century-Fox continues to show particular interest in experimenting with both free and monetized VR tie-ins. First, its Fox Innovation Lab partnered with Felix & Paul Studios for "*Wild*—The Experience," a companion to the Fox drama *Wild*. Later made available to consumers as a free download, it debuted at the 2015 Consumer Electronics Show to promote the recently released film. Like "Jakku Spy," "*Wild*—The Experience" integrates some interactivity with a head-tracking interface and a gameplay mechanic of contextual triggers. Specifically, the 360-degree video begins with users following protagonist Cheryl (Reese Witherspoon) along a nature trail until an audio cue prompts users to turn their heads to see the ghost of her mother (Laura Dern) (Volpe, 2015). "*Wild*—The Experience," then, functioned as an experimental companion piece to its cinematic source and as a proof of concept for the Fox Innovation Lab; by contrast, "Jakku Spy" marketed *The Force Awakens* as well as Google Cardboard, and "*Interstellar*: Oculus Experience," as its title indicates, explicitly promoted both the *Interstellar* film and the Oculus Rift hardware.

Subsequent VR tie-ins from Fox Innovation Lab have followed a similar experimental strategy while also embracing a commercial logic. In an interview promoting "*Wild*—The Experience," Fox Innovation Labs producer David Greenbaum emphasized that the studio perceived its early VR tie-ins as experiments within an emerging medium:

> Everything, every project, every discussion, every meeting is always a learning experience. And the rules aren't really established. The narrative tools and rules and technologies are evolving very, very quickly. So, I think our hope is if we just keep working on these projects, we'll get better at them; we'll learn how to make them; and there'll be more of them. (Volpe, 2015)

Fox, of course, entered the VR space for commerce as much as for art; as Greenbaum's statement suggests, Fox stands a better chance than its studio rivals of introducing innovative, perhaps even original, content in a new market by learning how to create within VR at an early stage.

For its next consumer VR tie-in, Fox Innovation Lab iterated upon "*Wild*—The Experience" with a longer play time and a monetized, rather than free, distribution model. In November 2016, Twentieth Century-Fox Home Entertainment released "*The Martian* VR Experience," nearly a year after the Fox film *The Martian* had appeared on home video. While press reports of the project's debut at the January 2016 Consumer Electronic Show drew attention to the film's recent home-video release, its belated commercial rollout characterized "*The Martian* VR Experience" as a standalone, commercial product that stood in contrast to free, promotional ancillary VR tie-ins supporting upcoming films. In fact, Fox Innovation Lab head Ted Schilowitz referred to "*The Martian* VR Experience" as "a significant entertainment product" independent from the film (Terdiman, 2016). To underscore this independence, Fox submitted it to the Sundance Film Festival's New Frontier Program and, in media coverage, touted that the project had received the Cannes Silver Lion in Digital Craft and the Association of Independent Commercial Producers Next VR Award (Ewalt, 2016).

That Fox released "*The Martian* VR Experience" so long after the film appeared in theaters and on video suggests Fox's faith in VR as a dedicated platform. Instead of merely promoting an upcoming film, the project allowed the studio to try out the potential for a monetized model of consumer VR content. Launching at a cost of $19.99 for HTC Vive and PlayStation VR, "*The Martian* VR Experience" presented a twenty-minute experience that eclipsed the shorter length of more experimental tie-ins such as "'*Wild*—The Experience." Executive produced by Ridley Scott and directed by Robert Stromberg of The Virtual Reality Company, "*The Martian* VR Experience" places users in the role of marooned astronaut Mark Watney while attempting to escape from Mars. Unlike relatively passive VR tie-ins before it that had focused on worldbuilding, "*The Martian* VR Experience" not only adapts its source film but also introduces more interactive elements, such as controlling a crane and driving a rover. To encourage VR as a viable consumer product, the project integrates the cinematic perspective of experiential VR with simple gameplay mechanics. Like other monetized VR experiences, such as *Marvel Powers United VR* (Oculus Studios, 2018), it incorporates interactivity to drive audience engagement, given the tendency for users to return to games more than other VR content (Donnelly and Pressberg, 2017). Response to "*The Martian* VR Experience," however, proved mixed. The review from media-industry trade *Variety*, on the one hand, noted the project's distinction from earlier VR tie-ins: "Instead of following the studios' playbook to date on using VR for just a short promotional clip, this experience is a more fully fleshed out rendering akin to a franchised video game." The same review, on the other hand, also called into question the combination of cinematic and gaming elements in "*The Martian* VR Experience" (Roettgers, 2016b). While this review obscures the broader technological forces at work in the VR market, this critique of "*The Martian*: VR Experience" as a film–video game hybrid reflects concerns regarding the technology's integration with established media formats. "*The Martian* VR Experience" proved to be a commercial and critical disappointment, with possible factors including its late release date, simplistic gameplay, and adaptation of a non-franchise film. By April 2018, less than eighteen months after its consumer debut, it had dropped from $19.99 to only $4.99 and carried an average user rating of two-out-of-five stars across multiple VR storefronts (Roettgers, 2018b).

In response to this disappointment, Fox in 2017 established the FoxNext division to produce content of a less experimental nature and with ties to the studio's media franchises. As a sign of Fox's more moderate enthusiasm for the medium, VR represents only one of the formats for which FoxNext develops content. Specifically, the unit oversees consumer VR as well as mobile-gaming apps and theme-park attractions (Roettgers, 2018b). Acknowledging the challenges of the consumer VR market, FoxNext president Salil Mehta has referred to "*The Martian* VR Experience" as "a learning experience" (Roettgers, 2018b). Still, more recent content continues to provide the unit with additional learning experiences. Whereas "*Wild*—The Experience" and "*The Martian* VR Experience" shared the character-centered approach of their cinematic sources, FoxNext's 2018 VR shooting game *Crisis on the Planet of Apes* tonally deviated from the film trilogy that began with *Rise of the Planet of the Apes* (Rupert Wyatt, 2011). As *IGN* criticized in its review:

> The rebooted Planet of the Apes movies are a trilogy of thoughtful, morally complex blockbusters that pose serious questions about what it means to be human. By contrast, *Crisis On the Planet of the Apes* . . . is an often infuriating VR game that plays like a bad arcade rail shooter from the early '90s. (Marsh, 2018)

With a two-and-a-half-hour playtime and an original story set between two films in the trilogy, *Crisis on the Planet of Apes* returned to the easily digestible textual emphasis of transmedia storytelling by stylistically and narratively iterating upon Fox's earlier consumer VR experiences. As both it and "*The Martian*: VR Experience" suggest, however, promotional tie-ins may represent a transitional strategy as Hollywood moves toward embracing VR as a core media platform. Their extended narratives, monetization, and belated release dates point to Fox and other studios eventually embracing VR as a dedicated platform instead of as an adjunct to film and television content.

Location-based VR

As of 2018, VR as a commercial medium has yet to capture mainstream consumer adoption. Because hardware sales have disappointed (only eight million headsets sold worldwide in 2017) and because few frequent users exist (in 2017, only 2.9 percent of Americans used VR even once a month), studios continue to rely on VR for the purposes of marketing (Roettgers, 2018b; Spangler, 2018). Whether consumer VR has failed to find a wider audience due to its high cost, prohibition of multitasking, or potential for causing headaches and motion sickness, location-based VR has emerged as an alternative distribution outlet that combines virtual and physical forms of immersion.

Studios have also moved into location-based VR amid signs of waning interest in experiential tie-ins. Branded VR content continues to appear as roadshow installations but recent editions of SDCC have suggested a cooling market for VR. While attendees at the 2017 convention waited up to three-and-a-half hours to engage with a five-minute VR experience promoting the second season of *Stranger Things* (Netflix, 2016–present), virtual activations had to compete for audience attention against the reemergence of physical installations—such as an escape room based on *The Expanse* (Syfy/Amazon, 2015–present)—that provided attendees a greater sense of exclusivity due to their limited capacity and inability for consumer replication (Rubin, 2017; Knowledge@Wharton, 2017). The following year's SDCC found VR merging with other marketing trends. To promote the debut of its *Tom Clancy's Jack Ryan* television series (2018–present), Amazon presented a hybrid VR/escape room activation, or what the studio called "the world's first hyperreality spy experience" (Knowledge@Wharton, 2018). Even if Hollywood continues to move away from event-based VR promotions, studios increasingly have embraced location-based VR, housed within movie theater lobbies, theme parks, shopping malls, and dedicated VR centers.

Compared to the disappointing revenues of consumer VR, location-based VR represents a new frontier for the medium. High admission costs (for example, tickets for "*Star Wars*: Secrets of the Empire," range from $30 to $35) and optimistic projections for location-based VR—such as a 2018 Greenlight Ventures report estimating growth from a $1 billion annual business into one worth $12 billion by 2023—encouraged studios to expand into location-based products (Porges, 2018; Giardina, 2018). With location-based VR, media franchises—such as the first-person shooter "John Wick Chronicles" based on the Keanu Reeves films—continue to outnumber other Hollywood licenses, potentially as a risk-averse strategy in light of the VR market's previous failure to reach projections (for example, the $2.7 billion global revenues for 2016 fell far short of Digi-Capital's earlier estimate of $3.8 billion) (Donnelly and Pressberg, 2017). In addition to its potential revenues, location-based VR has expanded, in part due to interest from shopping malls seeking to recapture consumers who have migrated to online retail. Shopping-center corporation Westfield Group, for example, has invested in Dreamscape Immersive, a location-based VR company whose flagship exhibition center resides in Los Angeles's Westfield Century City mall (Barnes, 2017). For exhibitors, these experiences help drive consumer traffic to the movie theaters housed within shopping malls.

Location-based VR's potential for growth has similarly attracted attention from numerous film and television companies. Aside from Westfield, investors in Dreamscape Immersive include Steven Spielberg, 21st Century Fox, MGM, Warner Bros., and AMC Entertainment, all of whom have been drawn to its creation of a "virtual-reality multiplex" that allows up to six people to explore a VR environment simultaneously through a virtual avatar visible to the other users (Barnes, 2017). Entertainment attraction company The Void has also designed tie-in walkthrough VR experiences, including a collaboration with Sony on a *Ghostbusters* attraction in Times Square, and a partnership with Lucasfilm's ILMxLab to present "*Star Wars*: Secrets of the Empire" at shopping malls and Disney theme parks (Roettgers, 2018a).

In 2018, FoxNext introduced its first location-based VR experience, "*Alien*: Descent." Set in the storyworld of Fox's long-running *Alien* franchise, "*Alien*: Descent" particularly demonstrates distinctions between location-based and earlier tie-in VR formats. Whereas the ephemerality of experiential VR encourages experiences that promote upcoming releases, location-based VR products can function as standalone products capable of generating consumer attention, and income, even after a film's theatrical window has closed. "*Alien*: Descent," in fact, debuted at a Southern California mall eight months following the home-video release of the franchise's latest film, *Alien: Covenant* (Ridley Scott, 2017). As a further distinction from earlier formats, location-based experiences synthesize the activation setup of experiential VR with the gameplay and monetization of consumer VR. In a sign of how the technology had advanced since "*Interstellar*: Oculus Experience," "*Alien*: Descent" runs for fifteen minutes and requires a headset, as well as motion sensing sleeves and gun controllers that provide haptic feedback. Because the project involves free-roaming movement and physical stimuli tied to moments in the diegesis (for example, a moving platform simulates the sensation of a descending elevator), "*Alien*: Descent" depends on an installation context. As FoxNext president Salil Mehta told *Variety*, "It's not something you can experience at home" (Roettgers, 2018a). Hollywood studios as well as VR companies hope that consumers who had previously rejected consumer VR will seek out the more sophisticated and immersive experiences that the contextual affordances of location-based VR make possible.[2]

While studios at present have embraced location-based VR for derivative

Figure 13.1 Outside FoxNext's "*Alien*: Descent" at The Outlets at Orange in Southern California.

tie-ins, the format could inspire original Hollywood VR content. Dreamscape Immersive already seeks to create its own material but has partnered with studios to create VR experiences based on existing media franchises as a means to scale (Barnes, 2017). Dreamscape Immersive could take inspiration from IMAX's VR Centre venture. IMAX had created its own VR experiences, such as the first-person shooter "Skyfront" and the thrill-ride simulator "Richie's Plank Experience," and also worked with studios on franchise tie-ins such as "*Justice League*: An IMAX VR Exclusive," which opened in conjunction with the Warner Bros. film *Justice League* (Zack Snyder, 2017). Similar to "Alien: Descent," content at IMAX VR Centres adopted a gaming approach; game companies such as Skydance Media and Ubisoft even contributed experiences, with "Archangel" and "*Star Trek*: Bridge Crew," respectively.

IMAX entered into the VR market partially as a replacement for 3-D films whose box-office surcharge had provided a profitable revenue stream until consumer disinterest set in (Lang, 2018). Under a pilot program that launched in 2017, IMAX experimented with variable pricing models based on content length and concept at seven VR Centres. By March 2018, however, all locations except for the standalone center in Los Angeles had failed to generate steady profits; then in December 2018, IMAX announced the closure of all of its VR centers (Longwell, 2018; Roettgers, 2018c). Partly to blame was a lack of blockbuster VR content, aside from the *Justice League* tie-in—whose IMAX VR engagement expired after only two weeks before becoming available for consumer headsets. This exemplifies how, unlike "*Alien*: Descent" and other location-based VR experiences, IMAX VR content is not exclusively location-based and is instead replicable, to an extent, in the home (Roettgers, 2018c). Ironically, IMAX CEO Richard Gelfond had pitched Hollywood studios on the idea that his company's centers would provide "a jump-start" to the VR market, "a complex ecosystem" which he described as having "been in a holding pattern, (and) slow to go mainstream" (Faughnder, 2017).

Despite studios turning to the format to reignite consumer interest, location-based VR from Dreamscape Immersive might fall victim to the same fate as earlier location-based ventures like DisneyQuest. A subsidiary of Walt Disney Parks and Resorts that launched in 1998 as an "indoor interactive theme park," DisneyQuest closed its final location in 2017 after management turnover and a failure to replace its aging VR attractions at the right time (Williams, 2015). The DreamWorks SKG and Universal Pictures collaboration GameWorks serves as another cautionary tale to location-based VR stakeholders. Founded in 2001 as an immersive arcade center featuring VR games, GameWorks filed for bankruptcy in both 2004 and 2010 before closing most of its locations. After several ownership changes, GameWorks sold in 2017 to software startup Oomba, which seeks to transform the existing centers into arenas dedicated to esports (that is, competitive video gaming)

(Grubb, 2017). Hollywood may discover that location-based VR sites, such as DisneyQuest and GameWorks before them, represent only a temporary solution for engaging audiences until other emerging technologies, such as AR, provide a greater sense of immersion.

Conclusion

The challenges IMAX experienced with its VR Centers suggest that virtual reality ultimately might not grow into a sustainable distribution outlet capable of protecting Hollywood's studios against Silicon Valley upstarts such as Amazon and Netflix. As further signs of VR's decline, in November 2017 HTC and Google canceled their Daydream headset (Roettgers, 2017) and, in February 2018, Viacom closed its Viacom Next VR Studios (Roettgers, 2018d). Also, despite their innovation in the VR space, the future of the Fox Innovation Lab and FoxNext remains in question following Disney's acquisition of their parent studio.

It remains to be seen if Hollywood will embrace VR as a platform for original storytelling and not just derivative promotions. Hailing from VFX companies such as Framestore, the earliest VR content focused more on technological, rather than narrative, experimentation. More recently, storytellers have introduced original content, even if much of it—like "Carne y Arena" (Alejandro G. Iñárritu, 2017), a seven-minute immigration documentary—resembles installation art more than mainstream products (Schwartzel, 2015). At the studios, innovation labs and accelerator programs might one day move beyond marketing tie-ins and into original VR experiences. In 2015, Fox Innovation Lab Head of Production Ted Gagliano professed his hope that, while early VR may have been "a marketing tool," studios like Fox would one day move into original, standalone VR content, with serialized short-form experiences comprising a longer-length story (Volpe, 2015).

The rapid pace of technological change, however, could pose a risk to studios that remain dedicated to VR.[3] This change has created an unstable market, which has only further encouraged studios to limit their involvement to low-risk promotions of their film and television properties, particularly those tied to media franchises. As an example of how quickly the technology can change, Mike Woods explained that the "*Interstellar*: Oculus Experience" represented "a considerable step up" from "Ascend the Wall," which Framestore had completed only a couple of months earlier. According to Woods, the time between these projects saw the release of a new game engine and a new Oculus Rift development kit that, respectively, allowed for significant advancements in visual resolution and positional tracking "so that 'Interstellar[: Oculus Experience]' looks and feels more immersive than 'Game of Thrones[: Ascend the Wall]'" (Alexander, 2014). Not long after, "*The*

Martian: VR Experience" showcased its own technological innovation, with a feature called "volumetric VR" that allows users to walk around an entire film scene (Terdiman, 2016).

Even with all these developments, AR may overtake VR as a preferred consumer technology, given that its reliance on virtual overlays does not isolate users from the real world or prohibit multitasking (Robertson, 2015). As such, AR could prove even more disruptive than VR. The startup Magic Leap, for example, presents the opportunity for exhibition to occur anywhere, and therefore will challenge content companies wanting to exhibit their content only in specific contexts, such as the controlled spaces of a television screen, movie theater, or VR headset display. Digi-Capital has even projected that, by 2022, the annual AR market may reach upwards of $90 billion, compared to $10 billion for VR (Spangler, 2018). Such projections underscore the likelihood of film and television companies embracing AR over VR to provide consumers with immersive experiences.

VR will showcase its potential for creative storytelling and production once it moves beyond the boundaries of the current cinematic and gamic vision. Ongoing experimentation, however, suggests a pursuit for both original creativity and sustainable economics. VR's future in entertainment lies at the intersection of both. In this sense, Hollywood's vision is focused on long-term development, with the potential one day to transform the multisensory exploration of cinematic spaces featured in *Ready Player One* from a fantasy into a (virtual) reality. Some of what we see in VR may be virtually the same thing, but, as the technology evolves, Hollywood's iterative approach will probably scale the next media frontier.

Notes

1. This chapter began as a presentation that Matthias Stork and I gave at the 2016 Society for Cinema and Media Studies Conference in Montréal, Canada. I am indebted to Matthias for the initial collaboration and to the panel's other members and attendees for their feedback.
2. The rise of location-based VR as an alternative to consumer VR reflects a simultaneous rise of VR in game centers (for example, American "barcades" such as Dave & Busters as well as Japanese arcades) to attract console and mobile gamers.
3. Similarly, due to its own rapid technological changes, the video-game industry operates according to a logic of predetermined technological obsolescence. Specifically, hardware manufacturers (for example, Microsoft, Nintendo, Sony) have established a product cycle of releasing new consoles approximately every five years. More recently, however, the quickening sophistication of mobile devices has inspired a more iterative strategy of introducing incremental upgrades to existing hardware (Kohler, 2016).

References

Alexander, Bryan (2014), "3D "'nterstellar' in Oculus Rift Virtual Reality Exhibit," *USA Today*, 6 October, <https://www.usatoday.com/story/life/movies/2014/10/06/interstellar-oculus-rift/16806167/> (last accessed 28 June 2018).

Barnes, Brooks (2017), "Coming Soon to AMC Theaters: Virtual Reality Experiences," *The New York Times*, 26 September, <https://www.nytimes.com/2017/09/26/business/media/amc-theaters-virtual-reality.html> (last accessed 28 June 2018).

Birdsall, Heather Lea (2019), "The Happiest Plays on Earth: Theme Park Franchising in Disneyland Video Games," in James Fleury, Bryan Hartzheim, and Stephen Mamber (eds), *The Franchise Era: Managing Media in the Digital Economy*, Edinburgh: Edinburgh University Press, pp. 77–104.

Bolter, Jay and Richard Grusin (2000), *Remediation: Understanding New Media*, Cambridge, MA: The MIT Press.

Bond, Paul (2017), "Virtual Reality Set to Become a $5 Billion Industry by 2021," *The Hollywood Reporter*, 7 June, <https://www.hollywoodreporter.com/news/virtual-reality-set-become-a-5-billion-industry-by-2021-1010193> (last accessed 28 June 2018).

Boyer, Steven (2009), "A Virtual Failure: Evaluating the Success of Nintendo's Virtual Boy," *The Velvet Light Trap*, 64, pp. 23–33.

Claiborn, Samuel (2014), "Map of South Park Revealed," *IGN*, 7 March, <http://www.ign.com/articles/2014/03/08/map-of-south-park-revealed> (last accessed 28 June 2018).

Donnelly, Matt and Matt Pressberg (2017), "Why Hollywood Studios Are Slow to Embrace Virtual Reality," *TheWrap*, 23 July, <https://www.thewrap.com/hollywood-studios-virtual-reality-slow-embrace/> (last accessed 28 June 2018).

Elkington, Trevor (2009), "Too Many Cooks: Media Convergence and Self-Defeating Adaptations," in Bernard Perron and Mark J. P. Wolf (eds), *The Video Game Theory Reader 2*, New York: Routledge, pp. 213–35.

Ewalt, David M. (2016), "Fox Releases 'The Martian' VR Experience," *Forbes*, 14 November, <https://www.forbes.com/sites/davidewalt/2016/11/14/fox-releases-the-martian-vr-experience/#71f86aaa3f94> (last accessed 28 June 2018).

Faughnder, Ryan (2017), "Virtual Reality Industry 'In Need of a Jump-Start,' Imax CEO Says at New VR Center," *Los Angeles Times*, 15 February, <http://www.latimes.com/business/hollywood/la-fi-ct-imax-vr-20170214-story.html> (last accessed 28 June 2018).

Fleury, James and Stephen Mamber (2019), "The (Im)Perfect Organism: Dissecting the *Alien* Media Franchise," in James Fleury, Bryan Hartzheim, and Stephen Mamber (eds), *The Franchise Era: Managing Media in the Digital Economy*, Edinburgh: Edinburgh University Press, pp. 31–51.

Framestore VR Studio (2014), "Introducing the World's First VR Studio," *Framestore VR Studio*, 4 June, <http://framestorevr.com/news2/introducing-the-worlds-first-vr-and-immersive-content-studio> (last accessed 28 June 2018).

Giardina, Carolyn (2018), "Location-Based Virtual Reality Is Growing: How Big Is the Potential Market?," *The Hollywood Reporter*, 19 May, <https://www.hollywoodreporter.com/behind-screen/location-based-virtual-reality-is-growing-how-big-is-potential-market-1113113> (last accessed 28 June 2018).

Gray, Jonathan (2010), *Show Sold Separately: Promos, Spoilers, and Other Media Paratexts*, New York: New York University Press.

Grubb, Jeff (2017), "The GameWorks Arcade Chain Is About to Get an Esports Makeover," *VentureBeat*, 31 August, <https://venturebeat.com/2017/08/31/the-

gameworks-arcade-chain-is-about-to-get-an-esports-makeover/> (last accessed 28 June 2018).

Harmetz, Aljean (1982), "Hollywood's Video Gamble," *The New York Times*, 28 March, p. A40, ProQuest.

Holt, Jennifer and Kevin Sanson (2014), "Introduction: Mapping Connections," in Jennifer Holt and Kevin Sanson (eds), *Connected Viewing: Selling, Streaming, & Sharing Media in the Digital Era*, New York: Routledge, pp. 1–15.

Horn, Leslie (2014), "Oculus Rift Took Me Inside Interstellar and I Wanted to Stay There," *Gizmodo*, 6 October, <https://gizmodo.com/oculus-rift-took-me-inside-interstellar-and-i-wanted-to-1643053176> (last accessed 28 June 2018).

Jenkins, Henry (2006), *Convergence Culture: Where Old and New Media Collide*, New York: New York University Press.

Jenkins, Henry (2007), "Transmedia Storytelling 101," *Confessions of an Aca-Fan: The Official Weblog of Henry Jenkins*, 21 March, <http://henryjenkins.org/blog/2007/03/transmedia_storytelling_101.html> (last accessed 28 June 2018).

Kohler, Chris (2016), "Soon You'll Buy Consoles the Way You Upgrade iPhones," *Wired*, 26 May, <https://www.wired.com/2016/05/future-of-game-consoles/> (last accessed 28 June 2018).

Knowledge@Wharton (2013), "Marketing at Comic-Con: Experience the 'Experiences,'" *Knowledge@Wharton*, 29 July, <http://knowledge.wharton.upenn.edu/article/comic-con-marketing-experience-the-experiences/> (last accessed 28 June 2018).

Knowledge@Wharton (2014), "Marketing at Comic-Con: Virtual Reality Gets Real," *Knowledge@Wharton*, 1 August, <http://knowledge.wharton.upenn.edu/article/marketing-comic-con-virtual-reality-gets-real/> (last accessed 28 June 2018).

Knowledge@Wharton (2017), "Marketing at Comic-Con: Virtual Reality Gets Real (Again)," *Knowledge@Wharton*, 3 August, <http://knowledge.wharton.upenn.edu/article/marketing-comic-con-gets-real/> (last accessed 28 June 2018).

Knowledge@Wharton (2018), "Marketing at Comic-Con: More Is More," *Knowledge@Wharton*, 8 August, <http://knowledge.wharton.upenn.edu/article/marketing-comic-con/> (last accessed 15 August 2018).

Lang, Brent (2018), "3D Box Office Hits Lowest Levels in Eight Years," *Variety*, 4 April, <https://variety.com/2018/film/news/3d-box-office-1202743603/> (last accessed 28 June 2018).

Longwell, Todd (2018), "Imax Explores New Tech Horizons Beyond Big Screen," *Variety*, 8 March, <https://variety.com/2018/film/features/imax-explores-new-tech-horizons-beyond-big-screen-vr-streaming-1202721123/> (last accessed 28 June 2018).

Marsh, Calum (2018), "Crisis on the Planet of the Apes Review," *IGN*, 11 April, <http://www.ign.com/articles/2018/04/11/crisis-on-the-planet-of-the-apes-vr-review> (last accessed 28 June 2018).

Neale, Steve (2003), "Hollywood Blockbusters: Historical Dimensions," in Julian Stringer (ed.), *Movie Blockbusters*, New York: Routledge, pp. 47–60.

Olanoff, Drew (2015), "Star Wars Dips Into Virtual Reality with 'Jakku Spy,' Created In Parallel with The Force Awakens," *Techcrunch*, 2 December, <https://techcrunch.com/2015/12/02/star-wars-gets-its-first-virtual-reality-experience-with-jakku-spy/> (last accessed 28 June 2018).

Panzarino, Matthew (2017), "Disney Puts a New Spin on its Accelerator," *Techcrunch*, 11 July, <https://techcrunch.com/2017/07/11/disney-puts-a-new-spin-on-the-accelerator/> (last accessed 28 June 2018).

Porges, John (2018), "Inside The Void's Fully Immersive 'Star Wars' VR Experience: A Review," *Forbes*, 4 April, <https://www.forbes.com/sites/sethporges/2018/04/04/

inside-the-the-voids-fully-immersive-star-wars-vr-experience-a-review/#1863c9a9 6199> (last accessed 28 June 2018).

Robertson, Adi (2015), "Virtual Reality Is for Watching Things. HoloLens Is for Making Them," *The Verge*, 5 May, <https://www.theverge.com/2015/5/5/8546957/augmented-reality-vs-virtual-reality-hololens-oculus-rift-vive> (last accessed 28 June 2018).

Rodriguez, Ashley (2015), "Apparel Brand Merrell Uses Virtual Reality to Refresh Brand," *Advertising Age*, 6 February, <http://adage.com/article/cmo-strategy/apparel-brand-merrell-virtual-reality-refresh-brand/297014/> (last accessed 28 June 2018).

Roettgers, Janko (2016a), "Fox Innovation Lab Explores Virtual and Augmented Reality with Two-Day Summit," *Variety*, 2 February, <https://variety.com/2016/digital/news/fox-innovation-lab-virtual-reality-augmented-reality-summit-1201694350/> (last accessed 28 June 2018).

Roettgers, Janko (2016b), "Review: The Martian VR Experience," *Variety*, 11 January, <https://variety.com/2016/digital/reviews/vr-review-the-martian-vr-experience-1201674354/> (last accessed 28 June 2018).

Roettgers, Janko (2017), "HTC Cancels Google Daydream Standalone VR Headset," *Variety*, 14 November, <https://variety.com/2017/digital/news/htc-google-standalone-daydream-vr-1202614626/> (last accessed 28 June 2018).

Roettgers, Janko (2018a), "'Alien: Descent' Location-Based VR Experience Coming to Southern California Mall," *Variety*, 18 April, <https://variety.com/2018/digital/news/alien-descent-vr-foxnext-1202756431/> (last accessed 28 June 2018).

Roettgers, Janko (2018b), "Fox's FoxNext Unit Makes Strides in VR, Mobile Games," *Variety*, 3 May, <https://variety.com/2018/biz/news/fox-vr-mobile-games-1202795693/> (last accessed 28 June 2018).

Roettgers, Jank (2018c), "Imax Is Shutting Down its VR Business, Closing Remaining Three VR Centers in Q1," *Variety*, 13 December, <https://variety.com/2018/digital/news/imax-vr-is-dead-1203089351/> (last accessed 14 December 2018).

Roettgers, Janko (2018d), "Viacom Shuts Down Viacom Next VR Studio," *Variety*, 7 February, <https://variety.com/2018/digital/news/viacom-next-vr-shutting-down-1202690420/> (last accessed 28 June 2018).

Rubin, Peter (2014), "The Future of Travel Has Arrived: Virtual-Reality Beach Vacations," *Wired*, 18 September, <https://www.wired.com/2014/09/marriott-vr-teleporter/> (last accessed 28 June 2018).

Rubin, Peter (2017), "Hollywood VR Fantasy Was Everywhere at Comic-Con—With a Catch," *Wired*, 26 July, <https://www.wired.com/story/comic-con-vr-fantasy/> (last accessed 28 June 2018).

Schwartzel, Erich (2015), "Virtual-Reality Projects Get Hollywood Treatment," *The Wall Street Journal*, 15 April, <https://www.wsj.com/articles/hollywood-plays-to-virtual-reality-1429125999> (last accessed 28 June 2018).

Seymour, Mike (2018), "Ready Player One: Inside the Oasis," *Fxguide*, 16 April, <https://www.fxguide.com/featured/ready-player-one-inside-the-oasis/> (last accessed 6 July 2018).

Shaw, Lucas (2015), "Virtual Reality Inches Closer to Homes as Studios Embrace," *Bloomberg*, 6 January, <http://www.bloomberg.com/news/articles/2015-01-07/virtual-reality-inches-closer-to-homes-as-studios-embrace> (last accessed 28 June 2018).

Spangler, Todd (2015), "Virtual-Reality Startup Jaunt Raises $65 Million from Disney, Others," *Variety*, 21 September, <https://www.bloomberg.com/news/articles/2015-01-07/virtual-reality-inches-closer-to-homes-as-studios-embrace> (last accessed 28 June 2018).

Spangler, Todd (2018), "Could 'Ready Player One' Awaken Sleepy VR Market?," *Variety*, 12 March, <https://variety.com/2018/digital/news/ready-player-one-virtual-reality-vr-market-growth-catalyst-1202724167/> (last accessed 28 June 2018).

Stelter, Brian (2011), "News Corporation Sells MySpace for $35 Million," *The New York Times*, 29 June, <https://mediadecoder.blogs.nytimes.com/2011/06/29/news-corp-sells-myspace-to-specific-media-for-35-million/> (last accessed 28 June 2018).

Terdiman, Daniel (2016), "Fox's The Martian as Hollywood's Most Ambitious VR Experience," *Fast Company*, 5 January, <https://www.fastcompany.com/3054924/how-foxs-the-martian-vr-experience-became-hollywoods-most-ambitious-vr-experiment> (last accessed 28 June 2018).

Volpe, Joseph (2015), "Fox's Journey into Virtual Reality Begins with 'Wild,'" *Engadget*, 9 January, <https://www.engadget.com/2015/01/09/fox-searchlight-virtual-reality-wild/> (last accessed 28 June 2018).

Wallenstein, Andrew (2015), "Hollywood Tries Its Hand At Virtual Reality Content," *Variety*, 6 January, <https://variety.com/2015/digital/features/hollywood-tries-its-hand-at-virtual-reality-content-1201392775/> (last accessed 28 June 2018).

Williams, Kevin (2015), "End of an Era: Disney's First VR Attraction Set to Close," *Road to VR*, 10 July, <https://www.roadtovr.com/end-of-an-era-disneyquest-first-vr-attraction-set-to-close/> (last accessed 28 June 2018).

Woods, Mike (2015), telephone interview, 3 March.

INDEX